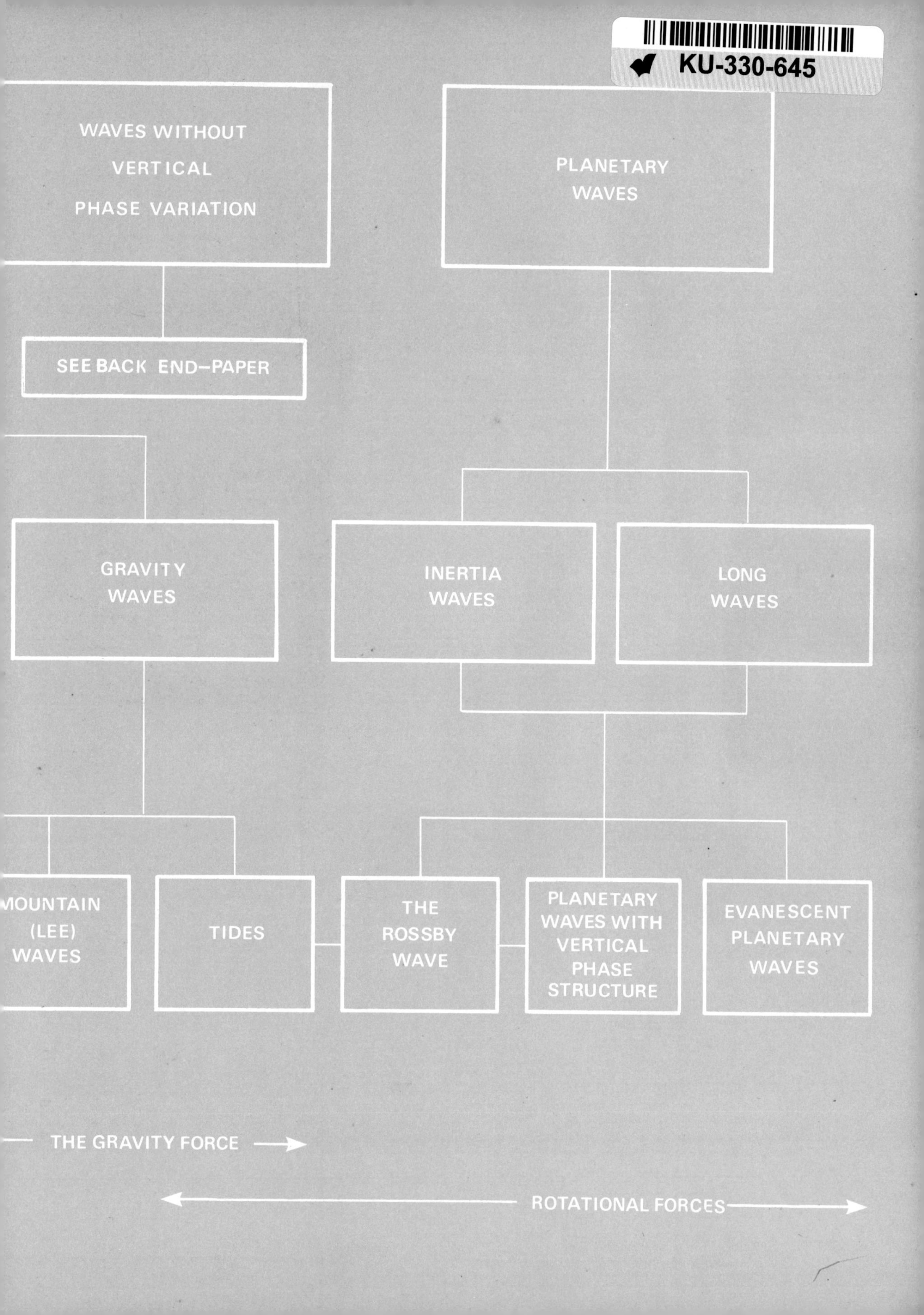

WAVES WITHOUT VERTICAL PHASE VARIATION
PLANETARY WAVES
SEE BACK END–PAPER
GRAVITY WAVES
INERTIA WAVES
LONG WAVES
MOUNTAIN (LEE) WAVES
TIDES
THE ROSSBY WAVE
PLANETARY WAVES WITH VERTICAL PHASE STRUCTURE
EVANESCENT PLANETARY WAVES
THE GRAVITY FORCE
ROTATIONAL FORCES

PATET OMNIBUS VERITAS

ATMOSPHERIC WAVES

****
For this book, the author was
awarded an
ADAM HILGER PRIZE
****

Billowed banner cloud on Mt. Everest

TOM BEER Ph.D
PHYSICS DEPARTMENT
UNIVERSITY OF GHANA
LEGON, GHANA

Atmospheric Waves

ADAM HILGER
LONDON

First published June 1974

ISBN 0 85274 238 X

Published in Great Britain by
ADAM HILGER LTD
Rank Precision Industries Ltd
29 King Street, London, WC2E 8JH
Set on Monophoto Filmsetter and printed by
J. W. Arrowsmith Ltd., Bristol, England

Contents

I have looked at clouds from both sides now,
From up and down, and still somehow
It's cloud's illusions that I recall
I really don't know clouds
At all.

Preface

This book is intended for both the tyro and the experienced research worker in atmospheric physics. At the same time I fervently hope that advanced undergraduate and graduate students as well as workers from the fields of telecommunications, oceanography and space science will all find something in here that is both useful and rewarding.

With such diversity of readership in mind the book has invariably turned into a compromise between the scientific text-book, which is a compendium of well known and accepted facts and between the advanced research monograph, which aims at a complete coverage of the field. I hope that the balance that has been achieved is acceptable. I have tried to give a good guide to the literature for the acolyte and at the same time to provide a useful handbook for the specialist.

In order to make this work as up to date as possible, a number of items have been included in this volume that have not as yet been published in the open scientific literature. One example is the theory of equatorial spread F which ascribes it to a spatial resonance mechanism (the pousse-café effect). Should this or some of the other new items included herein later be shown to be incorrect then I hope that the opprobrium of my colleagues will be mitigated by their respect for my good intentions.

Finally, I would like you to contemplate the whole basis of our natural philosophy. All the work of each individual scientist is built on a superstructure constructed by others. If the whole foundation is wrong then all the work is for nought. G. K. Chesterton wrote:

> ... All the terms used in the science books, 'law,' 'necessity', 'order', 'tendency', and so on, are really unintellectual because they assume an inner synthesis which we do not possess. The only words that ever satisfied me as describing Nature are the terms used in the fairy books, 'charm', 'spell', 'enchantment'. They

express the arbitrariness of the fact and its mystery. A tree grows fruit because it is a *magic* tree. Water runs downhill because it is bewitched. The sun shines because it is bewitched.

His words are thought-provoking. Perhaps more wisdom is to be gained by examining the broad foundations of our discipline than by continuing in this ever narrower spiral of specialisation.

Tom Beer

Legon
October 1, 1972

Acknowledgements

We wish to thank the following Journals, Publishers, Organizations and individuals for their permission to reproduce previously published material or for providing illustrations:

Annales de Géophysique (Centre National de la Récherche Scientifique, Paris)
Canadian Journal of Physics (National Research Council of Canada)
H. H. Coutts, Sittingbourne Laboratories, Kent
Dr. M. J. Curry, Atmospheric Environment Service, Downsview, Canada
Geophysical Journal of the Royal Astronomical Society (Blackwell Scientific Publications, Oxford)
N. Goodman, Adam Hilger Ltd.
Institute for Defence Analyses, Arlington
Journal of Atmospheric and Terrestrial Physics (Pergamon Press Ltd., Oxford)
Journal of Geophysical Research (American Geophysical Union)
McGraw-Hill Book Company, New York
M. G. Pearson, University of Edinburgh
Radio Science (American Geophysical Union)
D. Reidel Publishing Company, Dordrecht, The Netherlands
Reviews of Geophysics and Space Physics (American Geophysical Union)
Professor R. S. Scorer, Imperial College, University of London
The Quarterly Journal of the Royal Meteorological Society
The Royal Society, London
US Government Printing Office, Washington D.C.
Weather (Royal Meteorological Society)
World Meteorological Organization, Geneva
Wing Commander J. R. C. Young, Riyadh, Saudi Arabia.
Dr. J. Witcombe, University College of North Wales, Bangor

References to the sources of individual figures are given in the captions. The lines which precede the Preface are taken from the song 'Both Sides Now' by Joni Mitchell.

List of Symbols

With one exception, the officially agreed SI units are used herein. The exception lies in my use of the kilogram-mole instead of the official gram-mole. It strikes me as being more logical.

Some confusion may arise from my choice of horizontal coordinates. Due to typographical exigencies the set (e, n, z) is used for cartesian coordinates pointing eastward, northward and upward. Though this usage is highly unusual it does have the advantage of keeping the right hand coordinate system in alphabetical order. The disadvantage occurs when transferring into the spherical coordinate system (r, θ, ϕ). Since the colatitude increases southward we have the velocity components $U_n = -U_\theta$.

A_0	Wave amplitude
$\mathbf{B}$	Earth's magnetic field
c	Speed of sound
C_p	Specific heat at constant pressure
C_v	Specific heat at constant volume
D	Polarization term for wind divergence
$\mathscr{D}$	Dispersion coefficient
$\underline{\underline{\mathbf{D}}}$	Diffusion tensor
e	Partial pressure of water vapour
$\lvert e \rvert$	Absolute value of electronic charge
$\mathscr{E}$	Energy
$\mathbf{E}$	Electric field
f	$\omega/2\Omega_E$
f	Wave frequency
$\mathscr{f}$	Coriolis parameter
F	Energy spectrum as a function of wavenumber
F	Froude number
$\mathscr{F}$	Scorer parameter
$\mathbf{F}$	External forces per unit volume
g	Acceleration due to gravity
G	Amplitude of $-\dfrac{Dp}{Dt}/(\gamma p_0)$
h	Equivalent depth
H	Scale height
i	Square root of -1
I	Dip angle
I	Impedance
J	External heating
$\mathbf{J}$	Current density
k	Real wavenumber
$\mathscr{k}$	Boltzmann's constant
K	Complex wavenumber
$\mathscr{K}$	Thermal conductivity

L	Characteristic length
$\mathscr{L}$	Lagrangian density
m	Particle mass
M	Molecular weight
n_x, n_z	Refractive index
N	Electron density
N	Neutral particle concentration
$\mathscr{N}$	Polarization relation for electron density perturbations
N_a	Avogadro's number
N_l^m	Neumann form of the Associated Legendre polynomial
p	Pressure
P	Polarization term for pressure perturbations
P_l^m	Fully normalized Associated Legendre polynomial
$\mathscr{P}_l^m$	Schmidt form of the Associated Legendre polynomial
q_j	Generalized coordinate
q	Vertical component of transformed wavenumber
$\hat{Q}$	Stoke's derivative operator
Q_l^m	Associated Legendre polynomial of the second kind
r	Humidity mixing ratio
R	Universal gas constant
R	Polarization term for density perturbations
$\mathscr{R}$	Rate of energy dissipation
R_E	Radius of the earth
$\mathrm{R_e}$	Reynold's number
$\mathrm{R_g}$	Geophysical Reynold's number
$\mathrm{R_i}$	Richardson number
S	Entropy
S_l	Spherical or tesseral harmonic
t	Time
t	Cosine of the colatitude
T	Temperature
$\mathscr{T}$	Eddy lifetime
$\mathbf{U}$	Wind velocity
$\mathbf{U}_g$	Geostrophic wind velocity
$\mathscr{U}_{1z}$	Transformed vertical perturbation velocity
$\mathbf{v}$	Charged particle velocity in a reference frame moving with the neutral wind
$\mathbf{V}$	Charged particle velocity
V_p	Phase velocity
V_g	Group velocity
V_R	Rossby wave phase velocity
W	Work done by gas
W	Height-dependent amplitude of vertical perturbation wind
x	Scale height units
X, Y, Z	Polarization terms
α	Angle between transverse coordinate and east
α	Recombination coefficient
α	Temperature gradient
α^*	Adiabatic temperature gradient
β	Coriolis parameter gradient
β	Attachment-like coefficient
β^1	Transverse coriolis parameter gradient
γ	Ratio of specific heats
$\varGamma$	Imaginary part of complex vertical wavenumber
Γ	Circulation
$\boldsymbol{\Delta}$	Three-dimensional Laplacian
ε	Incremental volume change
ζ	Vorticity
ζ_a	Absolute vorticity
η	Kinematic viscosity
θ	Colatitude
θ	Potential temperature
$\varTheta$	Hough function

λ	Wavelength
λ	Elastic potential energy
Λ	Attenuation coefficient
μ	Coefficient of viscosity
$\underline{\underline{\boldsymbol{\mu}}}$	Mobility tensor
ν	Collision frequency
ξ	Disturbance
ρ	Density
σ	Conductivity
$\underline{\underline{\boldsymbol{\sigma}}}$	Stress tensor
Σ	Integrated layer conductivities
τ	Dissipation time scale
ϕ	Velocity potential
$\boldsymbol{\Phi}$	Tidal potential
χ	Wind divergence
χ	Geopotential altitude
ψ	Stream function
Ψ	Stream function amplitude
ω	Angular frequency
Ω	Doppler shifted frequency
$\boldsymbol{\Omega}$	Angular velocity
ω_a	Acoustic cutoff frequency
ω_{an}	Non-isothermal acoustic cutoff
ω_B	Vaisala–Brunt frequency
ω_g	Isothermal Vaisala–Brunt frequency
ω_H	Gyrofrequency
Ω_E	Earth's angular velocity

Subscripts and Coordinates

(x, y, z)	Arbitrary horizontal components with z directed vertically upward
(e, n, z)	Eastward, northward, upward
(me, mn, z)	Magnetic eastward, magnetic northward, upward
(r, θ, ϕ)	Radial, colatitude, longitude
(me, $\parallel$, $\perp z$)	Magnetic eastward, parallel to magnetic field, perpendicular to magnetic field
(0, 1, 2)	Longitudinal, Pedersen, Hall conductivities and mobilities

Subscripts

0, 1	Zero order and first order perturbation
r, s	Rotating and fixed reference frames
g	Value of quantity at $z = 0$
e, i, n	Electron, ion, neutral collision frequency

Superscripts

$+, -$	Ion, electron

1

Introduction

1.1 Types of Waves

We are all familiar with the concept of a wave. For example, a sugar cube dropped into a cup of tea will generate waves that travel radially outwards; a breeze blowing over a river will produce waves that will move in the direction of the wind on the surface of the river, even though the current may be flowing in some other direction; and every day at the seashore there is a continuous panorama of incoming waves that become unstable and eventually break against the shore. These are all examples of wave motion in a fluid and they have two properties in common that they share with all other types of waves: firstly, energy is being propagated from one point to another; secondly, the disturbance travels through the medium without giving the medium as a whole any permanent displacement. The wave may be considered as a perturbation on the steady slowly changing background. The mathematical analysis of wave motion is one of the greatest triumphs of mankind's powers of synthesis. Wave motion can be reduced to the solution of a differential equation, yet there still remain a myriad of different waves that can exist, all of which are solutions of the appropriate differential equations. This book deals with the types of waves that are important in a particular compressible, rotating, spherical fluid that is permeated by density gradients and temperature gradients—the earth's atmosphere.

The atmosphere is capable of sustaining a large number of wave phenomena. In this book, over twenty types of atmospheric waves are listed, and yet no attempt has been made to include all the possible types exhaustively. However, atmospheric waves can be broken up into three classes: waves that propagate horizontally and are composed of vertical displacements (vertical transverse waves), waves that propagate horizontally with horizontal displacements perpendicular to the propagation direction (horizontal transverse waves), and waves whose displacements are in the same direction as the propagation (longitudinal). These waves are depicted in Fig. 1.1. Atmospheric wave motions may be thought of as being a combination of these three types of waves.

All these waves can exist in the form of very small perturbations on a steady state of the atmosphere, so that the waves will then satisfy

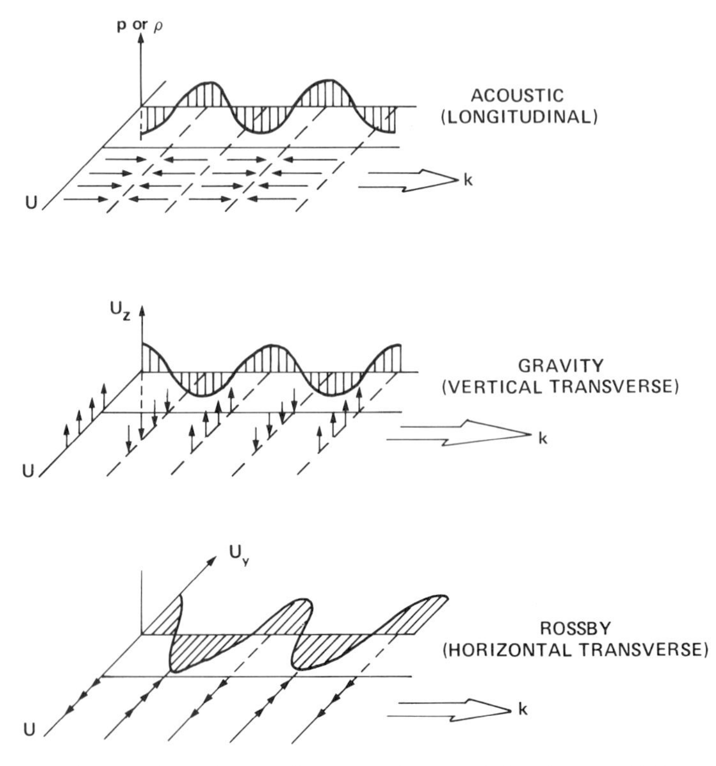

1.1 Three principal types of atmospheric waves (after Georges, 1967)

linear equations. The meaning of this is that small disturbances of different amplitude, wavelength or frequency can then be superimposed without interacting; one wave, for example, can go straight through another without noticing it. In these circumstances any small disturbance, however complex, can be analysed into regular sinusoidal components called Fourier components, each with its own frequency and wavelength. However, as the wave progresses it may become attenuated or may actually grow in amplitude. It may then eventually grow to an amplitude for which non-linear effects become important, including interactions of one wave with another and with itself.

The best known atmospheric waves are the sound waves which play such a vital role in our methods of communicating. Sound, or acoustic waves, are longitudinal waves formed by a balance between a fluid's resistance to changes in its volume—called the fluid's compressibility—and inertia, which is the resistance to a change of velocity and is expressed by Newton's second law of motion.

In a homogeneous stationary fluid in the absence of an external force (gravitational, magnetic, etc.), acoustic waves are the only type of waves that can exist. Acoustic waves include sound waves whose frequency spectrum covers the response range of human hearing. They also include those waves whose frequency lies above the threshold of human hearing, called ultrasonic waves, and those that lie below the lower limit of hearing, termed infrasonic.

The law of propagation of sound in a homogeneous stationary fluid may be written in terms of the fluid density ρ and the pressure p as

$$\frac{\partial^2 \rho}{\partial t^2} = c^2 \boldsymbol{\Delta} \rho \tag{1.1.1}$$

where $\boldsymbol{\Delta} = \partial^2/\partial x^2 + \partial^2/\partial y^2 + \partial^2/\partial z^2$ is the three-dimensional Laplacian.

$$c^2 = \frac{dp}{d\rho} \tag{1.1.2}$$

defines the speed of sound, c. Newton tried to find the speed of sound by a method that was equivalent to evaluating equation (1.1.2) at a constant temperature and obtained an answer 20 per cent too low, but it later became clear that the thermodynamic quantity that remains constant in these changes is the measure of molecular disorder known as entropy. The compressions and rarefactions occur adiabatically so that energy is not available to increase the molecular disorder.

The earth's atmosphere is a fluid that is continually being acted upon by the force due to the earth's gravity. There is therefore a distinct decrease in density with altitude, which is generally referred to as a density stratification, though in fact it is more likely to be a smooth variation in density than a layered one. This density gradient also endows the atmosphere with a stability that is completely lacking in a homogeneous fluid. When the force of the earth's gravity and the magnitude of the stabilizing restoring force introduced by the atmospheric density gradient become comparable with compressibility forces, the resulting waves are sometimes called acoustic gravity waves. These waves are no longer purely longitudinal (except when they propagate vertically) because gravity has produced a component of the air particle motion that is transverse to the propagation direction. The Froude number, F, is a measure of the significance of the force of gravity. It is defined by

$$\mathrm{F} = \frac{|\mathbf{U}|}{\sqrt{(gL)}} \tag{1.1.3}$$

where $\mathbf{U}$ is a characteristic velocity and L is a characteristic length. In the case of atmospheric gravity waves, the wind speed may be chosen as $|\mathbf{U}|$ while L could be taken as the height required for the density to drop to one half of its value. Hydrostatics operates in the range of small values of F, i.e. $\mathrm{F} \ll 1$, when gravity is the force of dominant importance with respect to the compressibility. Conversely, $\mathrm{F} \gg 1$ represents the case when gravity forces are dwarfed by pressure and inertia; most aerodynamic and hydrodynamic flows are in this latter category. In our cases for acoustic waves $\mathrm{F} > 1$, for acoustic gravity waves $\mathrm{F} \sim 1$, and for gravity waves $\mathrm{F} < 1$.

Gravity waves occur when the dominant motion in the atmosphere is due to the stable restoring force of the density gradients. There has long been meteorological interest in these waves (Haurwitz, 1951). The incidence of clear air reflections (radar angels)* during radar probings of the lowest 15 km of the atmosphere and the deformation of rockets' smoke trails in the troposphere have been used as evidence for the existence of these atmospheric waves at low altitudes (Naito, 1966). Clear air turbulence, which is a particularly troublesome intermittent phenomenon experienced by aircraft, can also be caused by gravity waves. In the presence of large wind shears it is possible for the gravity waves in these shear layers to become unstable (Bretherton, 1969a). The amplified gravity waves eventually break under the action of this instability, called the Kelvin–Helmholtz

* While developing radar during the second world war spurious reflections were often observed and nicknamed "angels". Later it was discovered that most "angels" are caused by the movements of birds, though gravity waves are believed to account for some of them. At sufficiently short wavelengths, the water content of birds will reflect electromagnetic waves.

instability, and form patches of turbulence. At lower heights similar unstable waves become visible in the form of billow clouds (called *Wogenwolken* by Helmholtz).

There also exist long waves that are influenced by the curvature of the earth and its rotation (the Coriolis effect). The variation of the strength of the Coriolis effect with latitude acts as an external force field which results in horizontally transverse waves with wavelengths thousands of kilometres long. These waves are known as Rossby waves or planetary waves; they provide a meteorologically useful theory for the description of the pressure distribution associated with moving wave-like high-pressure and low-pressure systems.

Rossby waves have a phase velocity that is always directed towards the west and is often directed in the opposite direction to the background wind. Thus the planetary wave concept is primarily of meteorological interest in regions of prevailing eastward blowing wind (i.e. westerlies)*.

Shearing waves can occur in the atmosphere. They are also very common in streams where there is an interaction between the water waves and the initial sheared flow of the water. The sheared flow arises because the water at the bottom and at the edges of a stream moves substantially slower than the water at the centre. In the atmosphere, shearing waves are important at the interface between two adjacent masses of air whose physical properties are different. This interface is called a front and it is along the frontal zones that unstable sheared Rossby waves convert their potential energy into the kinetic energy associated with large travelling storms known as cyclones. Though this will be counteracted to a certain extent by the stability of the inertia waves (Fig. 1.2) that are a consequence of the earth's curvature, the shear waves dominate for wavelengths in the interval between 500 km and 3000 km and unstable cyclone waves form (Pettersen, 1940; Eliassen and Kleinschmidt, 1957).

Atmospheric shear flows are always unstable (Fig. 1.2) and any regular wave is rolled up by non-linear effects into discrete vortices. A certain class of vortices, known as Taylor vortices, can be produced in the laboratory in the flow between two concentric cylinders rotating at different speeds—i.e. sheared rotary flow. Internal gravity waves and surface gravity waves, which are of

* In describing winds, the aeronomic and oceanographic convention of naming the direction towards which the wind is blowing will be used (e.g. eastward). This is in contrast to the meteorological convention of describing the wind by the direction from which it blows (e.g. westerly).

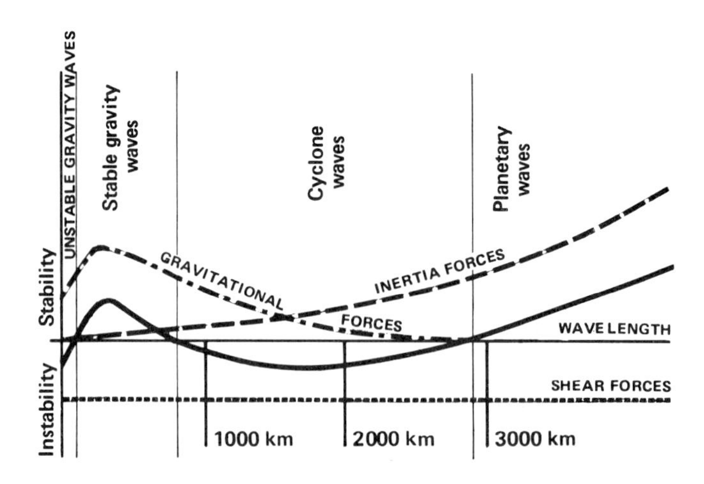

1.2 Qualitative diagram of the resultant forces (heavy curve) for meteorologically relevant waves (after Pettersen, 1940)

much smaller wavelength than the Rossby waves, can develop unstable vortices as well. The typical situation of gravity waves on a sheared flow occurs in the lee of a mountain. The lee wave can grow in amplitude to the stage where it becomes a discrete vortex, the so-called rotor. This is depicted in Fig. 1.3, which also depicts various possible cloud formations. One of these is the cap cloud which sits

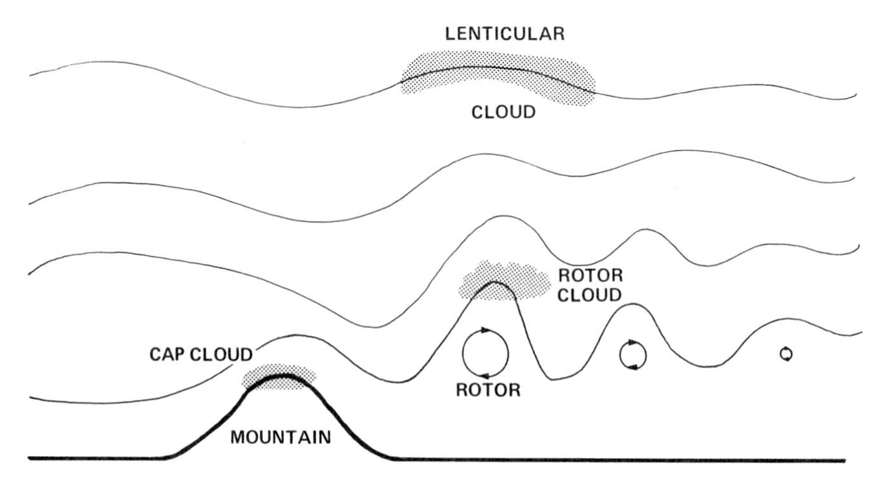

1.3 The idealized airflow over mountains (after Alaka, 1960)

on the mountain top. At times it is possible for the cap cloud (or cap clouds) to extend well into the windward side of the mountain where the rising air has become saturated. In this case, there appears to be a steady bank of clouds on the windward side of the mountain which abruptly ceases at the mountain itself. This phenomena is known as a foehn wall, so called because the Romans called the warm dry wind descending in the lee of the European alps the foehn. Though the term foehn has become generic it is often known by a variety of local names such as the Chinook of the Rockies, the Koschava and Ljuka of Yugoslavia, the Germich of the south-western part of the Caspian sea, the Afganet and Ibe of Central Asia, the Kachchan of Ceylon, the Berg wind of South Africa, the Santa Ana of the California Sierra Nevada mountains and the nor'-wester of New Zealand's Southern Alps.

Atmospheric motions whose periods are submultiples of the solar or lunar day are called atmospheric tidal oscillations. The sea tides can be gauged by measuring the changing height of the water surface. The air tides cannot be measured in this way because the atmosphere has no such clearly defined boundaries. The alternative is to use a barometer at the bottom of the aerial ocean. The vertical accelerations of the air are so small that the barometer effectively measures the weight of the overlying air.

Table 1.1 Principal Atmospheric Waves

Wave	*Period*	*Importance*
Acoustic	<270 sec	Speech
Gravity	270 sec—3 hrs	Ionosphere
Atmospheric tides	$24/m$ hrs ($m = 1, 2, \ldots$)	Geomagnetic variations
Rossby	>12 hrs	Meteorology

Therefore an above-average barometer reading implies a high tide of air whereas the low tide produces a low barometric reading. By analogy with the sea tides one would expect a rise and fall of the barometer of lunar semi-diurnal period. In the case of atmospheric tides however, the lunar atmospheric tides, which are gravitational in origin, are extremely small compared with the atmospheric tides whose periods are fractions of the solar day. The atmosphere is primarily driven by the heating effects of the sun rather than by the gravitational effects of either the sun or moon, and the equatorial barometers show this solar variation (Fig. 1.4) as being a semi-diurnal one. This is not what would be expected. As the heating effect of the sun has a 24-hour periodicity one would expect the solar diurnal tide to dominate. This problem will be dealt with in Chapter 5.

Atmospheric tidal oscillations and atmospheric gravity waves both have amplitudes that grow upwards as $\rho^{-\frac{1}{2}}$, where ρ is the neutral gas density. This increase in amplitude is a direct consequence of the

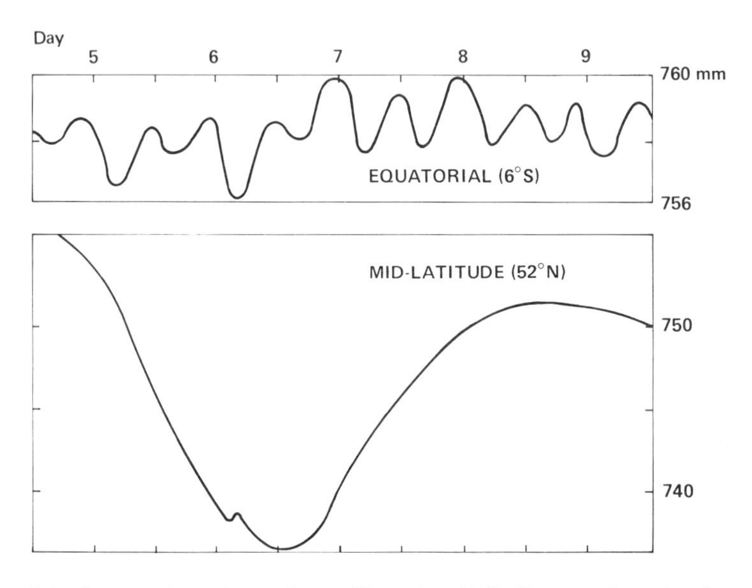

1.4 Barometric variation during November 1919. Equatorial results show tidal effects, whereas planetary waves dominate at mid-latitudes (after Chapman and Lindzen, 1970)

conservation of energy. If the amplitude of the wave is represented by A_0 then the kinetic energy is given by $\rho A_0^2/2$ which remains constant. In the real atmosphere ρ decreases almost exponentially with height so that the amplitude also has an exponential variation. The upward amplification of the tides and gravity waves indicates that there will be very important components of the wind velocity in the upper atmosphere provided that these waves can actually propagate to these heights. There is evidence, which will be discussed later, to indicate that both these types of waves actually do exist at ionospheric heights (60–500 km altitude).

The solar semidiurnal component of the atmospheric tide manifests itself on ground-based magnetograms, which measure the small-scale variation of the earth's magnetic field. It is believed that this is caused by the dynamo effect in the ionosphere. The tidal winds set in motion the charged particles that comprise the ionosphere, giving an electric field and a current, which then affects the background magnetic field so as to produce the small-scale variations in the magnetic field. These variations are around 4×10^{-4} of the background field.

Though this book deals only with waves in the atmosphere of the earth, much of the theory is sufficiently general to apply to any gravitationally gradated fluid. For example, gravity waves and acoustic gravity waves exist in an ocean whose temperature and salinity vary with depth and the theory of underwater acoustics is very similar to the theory of atmospheric acoustics. The waves that appear on the surface of the ocean also possess a counterpart in the theory of atmospheric waves. In the atmospheric case they form at the boundary between two media and have an energy that decreases exponentially away from the boundary. Atmospheric surface waves turn out to be a special case of evanescent wave. Evanescence in this context implies that waves within the fluid have no phase variation in the vertical direction.

Waves propagating in the earth's atmosphere would be expected to be both anisotropic and dispersive. Anisotropy—which means that the wave properties are not the same in all directions—is a consequence of the preferred direction of many of the geophysically important forces. For example, gravity is an anisotropic force, and so is the virtual force, the Coriolis force, produced by the effect of the earth's rotation.

Dispersion occurs when the wave frequency is not independent of the wavelength. This is often written in terms of the angular velocity ω to show that it is a function of the wavenumber k,

$$\omega = \omega(k),$$

where k is $2\pi/(\text{wavelength})$. Dispersion can arise from two distinct effects. It can be structural in origin and depend only on the properties of the medium or it can be geometric in origin and arise from interference effects due to reflections at the boundaries of the medium. The dispersion of atmospheric waves is structural in origin and is connected with the internal resonant frequencies that the atmosphere possesses. These are the Vaisala Brunt resonant frequency (§ 1.4) for gravity waves, the acoustic resonance for acoustic waves and a gyroscopic resonance due to precessional motions in a rotating fluid that influences the planetary waves. One consequence of dispersion is that the energy flow is in a different direction to the phase propagation of the wave.

In the non-dispersive case for plane waves, the variations of some quantity ξ can be expressed in a fourier expansion of the function $\exp[i(\mathbf{k}\cdot\mathbf{r})]$, so that each harmonic has the form, in one dimension, of

$$\begin{aligned}\xi &= A_0 \exp[i(\omega t - kx)] \\ &= A_0 \cos(\omega t - kx) + iA_0 \sin(\omega t - kx)\end{aligned}$$

where A_0 is a constant amplitude, ω a constant angular frequency (often just called the frequency), k is a constant wavenumber, and t and x are time and space variables respectively. In this particular case, the amplitude is constant on surfaces defined by

$$\omega t - kx = \text{constant}$$

and these surfaces move with the phase speed ω/k in the x direction.

However it has already been mentioned that in many physical situations the phase velocity is not a constant, but is determined by some dispersion relation which gives the variation of ω as a function of the wavenumber. Consider the case of a disturbance which produces waves of a variety of wavelengths and amplitudes. Waves whose periods and wavelengths almost coincide will produce wave packets. If, for example, we consider only the sine terms, then the total disturbance is

$$\begin{aligned}\xi &= a \sin[\omega t - kx] + a_1 \sin[(\omega + \delta\omega_1)t - (k + \delta k_1)x - \varepsilon_1] + \dots \\ &= A \sin(\omega t - kx) + B \cos(\omega t - kx) \\ &= C \sin(\omega t - kx - \varepsilon)\end{aligned}$$

where

$$\begin{aligned}A &= a + a_1 \cos(t\,\delta\omega_1 - x\,\delta k_1 - \varepsilon_1) \\ &\quad + a_2 \cos(t\,\delta\omega_2 - x\,\delta k_2 - \varepsilon_2) + \dots \\ B &= a_1 \sin(t\,\delta\omega_1 - x\,\delta k_1 - \varepsilon_1) \\ &\quad + a_2 \sin(t\,\delta\omega_2 - x\,\delta k_2 - \varepsilon_2) + \dots\end{aligned}$$

$$C = A^2 + B^2 \qquad \tan \varepsilon = B/A.$$

Now

$$t\,\delta\omega_1 - x\,\delta k_1 - \varepsilon_1 = \delta k_1\left(t\frac{\delta\omega_1}{\partial k_1} - x\right) - \varepsilon_1 = \delta k_1(V_g t - x) - \varepsilon_1$$

$$t\,\delta\omega_2 - x\,\delta k_2 - \varepsilon_2 = \delta k_2(V_g t - x) - \varepsilon_2$$

so that A, B, and therefore C and ε are functions of $(V_g t - x)$. Therefore the wave packet moves with a velocity V_g (Fig. 1.5)

$$V_g = \frac{\partial \omega}{\partial k}.$$

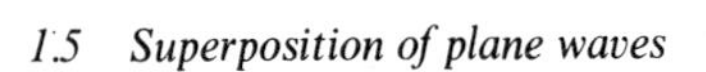
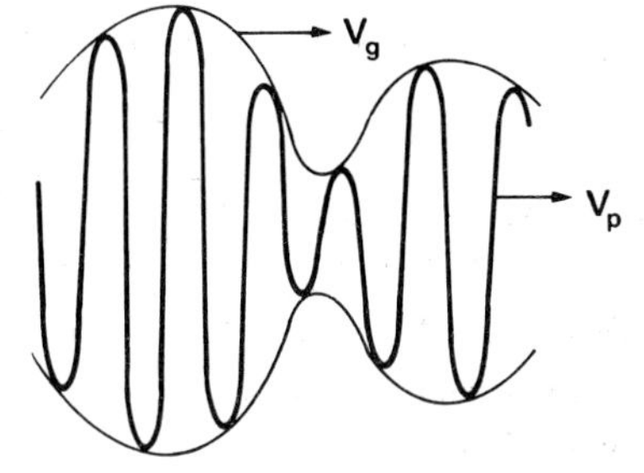

1.5 Superposition of plane waves

For waves in three dimensions, the characteristic speed of the wave-packet's motion in each direction is given by the group velocity

$$V_g = \left(\frac{\partial \omega}{\partial k_x}, \frac{\partial \omega}{\partial k_y}, \frac{\partial \omega}{\partial k_z}\right).$$

Under most conditions, the group velocity represents the direction and speed of energy propagation and is not necessarily equal to the phase velocity. In a highly dispersive medium, however, the concept of the group velocity becomes meaningless, because there are not enough waves of nearly equal phase velocities to allow wave packets to form. In a highly dispersive medium, each component travels individually at its own phase velocity. For three-dimensional waves, the true phase velocity is the phase velocity in the direction of the wave normal. If we denote the magnitude of the true phase velocity by V_p and if $V_{px} = \omega/k_x$, $V_{py} = \omega/k_y$, and $V_{pz} = \omega/k_z$ are the phase velocities in the x, y, and z directions, then

$$\frac{1}{V_p^2} = \frac{1}{V_{px}^2} + \frac{1}{V_{py}^2} + \frac{1}{V_{pz}^2}.$$

We shall return to this point in § 2.9.

1.2 The Earth's Atmosphere

Introduction

The absorption of the incoming and re-radiated solar radiation is responsible for practically everything that goes on in the earth's atmosphere. Absorption is important in well-defined and quite distinct regions: at levels above 100 km, where solar radiation of very short wavelength is absorbed by molecular oxygen; at levels between 35 and 70 km, where solar radiation in the near ultra-violet is absorbed by ozone; at ground level, where the solar radiation in the visible is absorbed by land and sea surfaces; and in the lower 15 km of the atmosphere, where the re-radiated infra-red part of the spectrum is absorbed by water vapour and carbon dioxide.

This absorption of radiation determines the temperature of the atmosphere as a function of altitude (Fig. 1.6). The lowest part of the atmosphere is the troposphere, in which the temperature decreases with altitude. The troposphere terminates at the tropopause, where a local minimum temperature is reached, usually at heights in the range 7–17 km, depending on the latitude and the season. The tropopause is a natural lid on the lower atmosphere; above it lies the upper atmosphere. The temperature then rises through the stratosphere until the stratopause is reached at 45–55 km, and then declines through the mesosphere to a minimum at the mesopause near 80–85 km. Above this height is the thermosphere. In the thermosphere, the neutral gas temperature increases to an altitude of 200 km and then remains constant to heights exceeding 1000 km. The isothermal behaviour of the upper thermosphere arises because its thermal conductivity is so high that most of the energy absorbed by the gas is removed downwards (Willmore, 1970).

The limiting thermospheric temperature is determined by the incoming ultra-violet radiation, so that there are seasonal variations in heat input and variations over the solar cycle as well as the diurnal

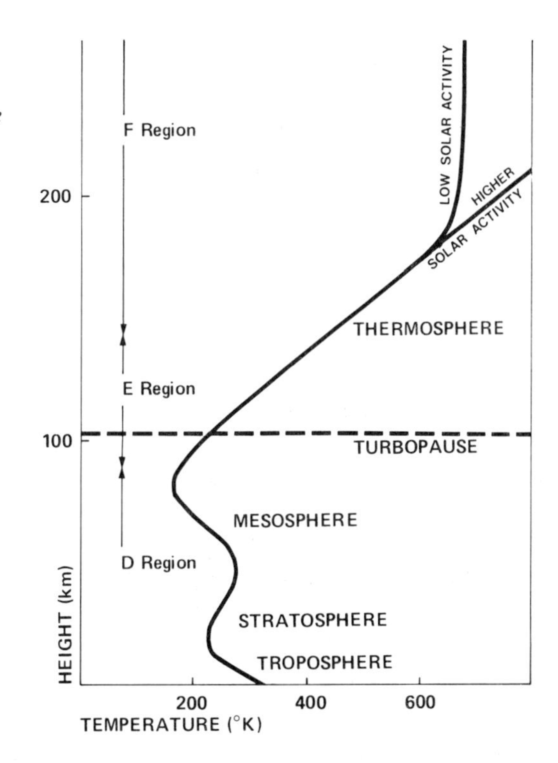

1.6 Atmospheric temperature profile

and latitude differences in temperature. During periods of very high solar activity, the sun emits larger amounts of ultra-violet radiation and the thermospheric temperature reaches 2200°K, but at the minima of solar activity, no part of the atmosphere exceeds 750°K (CIRA, 1965).

Above 600 km, the gas density has diminished to such low values that the air no longer acts as a fluid; at these heights, particles can attain sufficient velocity to escape the earth's gravitational field. In consequence, a transition from atomic oxygen ions to helium ions takes place at heights of the order of 1000 km and from helium to hydrogen ions at heights of the order of 3000 km. Therefore, because of the inapplicability of fluid dynamics above 600 km, this height will be considered to mark the top of the earth's atmosphere. The region above 600 km is variously called the metasphere, the protonosphere, the exosphere, or the geocorona.

The incoming solar radiation is also able to ionize the neutral particles, so that a region of free electrons, ions, and neutral particles exists in the upper atmosphere. This region is known as the ionosphere. It consists of three major regions. The D region at heights of 60 to 90 km, the E region from 90 to 150 km, and the F region at higher altitudes up to the top of the atmosphere. In addition to the regular layers of the ionosphere, there are several of a transient or irregular nature, as well as a large number of different types of ionospheric irregularities. Though in the D and E regions, the temperatures of all the atmospheric constituents are equal, this is no longer true of the F region, in which the electron temperature exceeds the ion temperature which in turn exceeds the neutral gas temperature.

Fig. 1.6 also shows the height of the turbopause—105 km. This is the boundary between the turbulent regions of the atmosphere and the regions in which laminar flow always exists. There is a transition region in the height range 85 km to 105 km in which it is also possible for there to be laminar flow, though below 85 km the atmosphere is always turbulent. Therefore, below 85 km there will be a convective mixing of the atmospheric constituents that will keep the mean molecular weight constant. Above this height, the separation of oxygen and nitrogen will produce a steady decline in the value of the mean molecular weight.

Thermodynamic Equations

The properties of a fluid like the atmosphere may be described in terms of any two of the four variables, ρ, p, T, and S, which are respectively its density, pressure, temperature, and entropy. Eckart has shown that the mathematics gains greatly in conciseness by the use of pressure and entropy. Nevertheless, I shall use the variables pressure and density, because the greater familiarity that most people have with them leads to a better intuitive concept of the physical processes in the atmosphere.

The atmosphere may be considered as an ideal gas whose molecules are all identical, interacting only by collisions and moving in straight lines between the collisions. The specific heats of an ideal gas remain constant.

Then the pressure and density are related to the temperature by the equation of state

$$p = \frac{\rho R T}{M} = \frac{\rho k T}{m}. \qquad (1.2.1)$$

The universal gas constant R may be defined in terms of Boltzmann's constant k,

$$k = 1{\cdot}38 \times 10^{-23}\ \text{joule/°K}$$

and the Avogadro or Loschmidt number N_a,

$$N_a = 6{\cdot}02 \times 10^{26} \text{ molecules/kmol-°K}$$

so that

$$R = kN_a = 8{\cdot}31 \times 10^3 \text{ joule/kmol-°K}.$$

m is the mean mass of a molecule, in kilograms, whereas M is the mean molecular weight, expressed in kilograms/kmol. The two are related: $M = N_a m$.

Throughout the lowest 80 km of the atmosphere, the mean molecular weight remains a constant 28·966 kg/kmol, whilst above 80 km it starts to decrease with height. This variation in M results from the dissociation of O_2 and the diffusive separation of O_2 and N_2 from atomic oxygen up to 400 km. Above 400 km altitude, the atomic oxygen (O) predominates. The values of M at these heights that are generally used are obtained from published standard atmospheres (e.g. CIRA, 1965, 1972). Lindzen (1970) used the following easily integrable and differentiable analytical formula for M at the altitude z (in km)

$$M = 28{\cdot}9 - 6{\cdot}45\left(1 + \tanh\frac{z-300}{100}\right). \tag{1.2.2}$$

If no external heat sources or sinks contribute to the energy balance of a parcel of atmospheric gas, then the atmospheric processes are adiabatic so that

$$d\mathscr{E} + dW = 0 \tag{1.2.3}$$

from the first law of thermodynamics. In an adiabatic process the entropy remains unchanged, $d\mathscr{E}$ is the increase in internal energy of the gas, and dW is the work done by the gas in changing its volume. So,

$$C_v\, dT - (p/\rho^2)\, d\rho = 0. \tag{1.2.4}$$

C_v is the specific heat per unit mass at a constant volume. By using (1.2.1) and (1.2.4) it may be shown that

$$p\rho^{-\gamma} = \text{constant} \tag{1.2.5a}$$

where

$$\gamma = C_p/C_v \quad \text{and} \quad R = M(C_p - C_v).$$

In the region of the atmosphere where the molecular oxygen and nitrogen is giving way to the atomic oxygen, γ will increase from its value of 1·4 for a diatomic gas to a value of 1·67 for a monatomic gas. In a similar manner to equation (1.2.2), an analytic form for γ may be used

$$\gamma = 1{\cdot}4 + 0{\cdot}135\left(1 + \tanh\frac{z-300}{100}\right) \tag{1.2.6}$$

where z is again measured in kilometres. By applying the general principles of thermodynamics it may be shown that, in terms of the entropy, S,

$$p\rho^{-\gamma} = (\gamma-1)A \exp\left[\frac{(\gamma-1)S}{R}\right] \tag{1.2.5b}$$

where A is another constant. Equations (1.2.5a and b) are together known as the equation of the isentropes.

Provided that equation (1.2.3) is true, then the speed of sound may be defined in terms of the time rate of change of a fluid element as

$$\frac{Dp}{Dt} = c^2\frac{D\rho}{Dt} \tag{1.2.7}$$

where D/Dt represents differentiation following the motion. If the process is not adiabatic, such as in the damping of atmospheric waves by viscosity or heat conduction, then equation (1.2.3) has to be modified by including a term that accounts for the net accession or loss of heat. With the exception of the above processes, which will not be dealt with in this chapter, the thermodynamic processes for short-period atmospheric waves are all nearly adiabatic, because the major atmospheric heating, which is due to the sun, has a time constant of the order of one day. For atmospheric tides, which may be thought of as long-period gravity waves, equation (1.2.7) needs to be modified, though Chapman (1932) showed that tides near the ground were approximately adiabatic.

The Barometric Equation and the Scale Height

The atmosphere, to great heights, is governed by a hydrostatic balance between the vertical pressure-gradient force and the force of gravity. This is expressed by the hydrostatic equation

$$\partial p/\partial z = -\rho g \tag{1.2.8}$$

where z is the altitude and g the acceleration due to gravity. The minus sign arises because the acceleration due to gravity is directed downwards whilst altitude is measured upwards. The density may be eliminated from the hydrostatic equation by using the equation of state to yield the pressure at an altitude z as

$$p = p_g \exp\left(-\int_0^z \left[\frac{Mg}{RT}\right] dz\right) \tag{1.2.9}$$

where p_g is the pressure at the altitude of the lower limit of integration, in this case at $z = 0$. Equation (1.2.9) is known as the barometric equation.

It is useful to define a quantity

$$H = \frac{RT}{Mg} = \frac{kT}{mg} \tag{1.2.10}$$

which is known as the scale height of the neutral atmosphere. In the case of an isothermal atmosphere, both pressure and density are given by Halley's Law, namely

$$\frac{p}{p_g} = \frac{\rho}{\rho_g} = \exp\left(-\frac{z}{H}\right). \tag{1.2.11}$$

Because T and M do not remain constant (and g also varies slightly) the scale height will vary at each height in the atmosphere. The scale height at any particular altitude, z, represents the height that a column of air of uniform density ρ_g would have if it produced the uniform pressure p_g at $z = 0$, provided that the column had the same temperature as exists at the height z, and also provided that variations in g can be ignored. This can be simply shown by equating $p_g = \rho_g gh = \rho_g RT/M$ for an incompressible atmosphere to give $h = H$.

The scale height may be rewritten in terms of the ratio of specific heats, γ, and the speed of sound, c, where

$$c^2 = \frac{\gamma kT}{m} = \frac{\gamma RT}{M} = \frac{\gamma p}{\rho} \tag{1.2.12}$$

so that the scale height

$$H = \frac{c^2}{\gamma g} \tag{1.2.13}$$

or

$$c^2 = \gamma g H.$$

Geopotential Altitude

The acceleration due to gravity, g, varies with both latitude and altitude. The sea level value is about 0·5 per cent higher at the poles than at the equator. Its variation in the vertical direction is given by

$$g(z) = \frac{g(0)}{[1+(z/R_E)]^2} \tag{1.2.14}$$

provided that the effects of the rotational forces may be neglected. $g(0)$ is the value at sea level and R_E is the radius of the earth. For example, at 100 km height the value of g is about 97 per cent of its value at sea level.

A vertical coordinate, called the geopotential altitude is often introduced to absorb the variability of g. It is defined by

$$g_0 \chi = \int_0^z g\,dz$$

where χ is the geopotential altitude and g_0 is chosen to be 9·8 in order to make the numerical value of geopotential similar to the numerical

Table 1.2 Values of g and χ in geopotential kilometres

z (km)	χ (geopotential km)	g m s^{-2}
0	0	9·80
50	49·61	9·65
100	98·45	9·50
150	146·54	9·36
200	193·90	9·22
250	240·54	9·08
300	286·48	8·94
350	331·73	8·81
400	376·32	8·68
450	420·25	8·55
500	463·54	8·43

value of the geometric altitude to which it corresponds. As g_0 is taken to be dimensionless, the dimensions of χ are energy per unit mass, so that χ represents the potential energy that would be gained by a unit mass lifted from the earth's surface to the height z.

It is also possible to assign the value 10 to g_0. In this case, χ represents the geodynamic altitude.

In this book, vertical variations of g will not be explicitly considered. This means that the altitude should be understood to be the geopotential altitude, so that $g_0\,d\chi$ replaces $g\,dz$ in the hydrostatic equation.

Atmosphere with Constant Temperature Gradient

If one assumes that the temperature T at an altitude z is given by

$$T = T_g + \alpha z \qquad (1.2.15)$$

where α is a constant then equation (1.2.9) yields a simple solution

$$p = p_g \left(\frac{T}{T_g}\right)^{-Mg/R\alpha}. \qquad (1.2.16)$$

In this case $p/p_g \neq \rho/\rho_g$ and in fact

$$\rho = \rho_g \left(\frac{T}{T_g}\right)^{-1-(Mg/R\alpha)}. \qquad (1.2.17)$$

In the case of an adiabatic, or isentropic atmosphere,

$$p\rho^{-\gamma} = \text{constant} \qquad (1.2.5)$$

so that

$$T\rho^{1-\gamma} = \text{constant}$$

and

$$Tp^{(1-\gamma)/\gamma} = \text{constant}. \qquad (1.2.18)$$

Differentiating logarithmically with respect to z and using the hydrostatic equation and the equation of state

$$\frac{\partial T}{\partial z} = \frac{\gamma - 1}{\gamma}\frac{T}{p}\frac{\partial p}{\partial z} = \frac{-(\gamma-1)Mg}{\gamma R} \approx -9{\cdot}8°\text{K/km} \qquad (1.2.19)$$

which is a constant in as far as g, γ, and R are constants.

Therefore the temperature must fall linearly as the height increases. This shows that the temperature gradient in an adiabatic atmosphere is a constant, which will be denoted by

$$\alpha^* = \frac{-(\gamma-1)Mg}{\gamma R} = \frac{-Mg}{C_p}. \tag{1.2.20}$$

The quantity $-(\partial T/\partial z)$ is called the lapse rate, so that $Mg/C_p \approx 9{\cdot}8°K/km$ is the adiabatic lapse rate for air that does not contain water vapour. It is often abbreviated to DALR (dry adiabatic lapse rate). Now direct use may be made of the results in equation (1.2.16) to give

$$p = p_g\left(1 - \frac{(\gamma-1)Mg}{\gamma R T_g}z\right)^{\gamma/(\gamma-1)} \tag{1.2.21}$$

and from (1.2.17)

$$\rho = \rho_g\left(1 - \frac{(\gamma-1)}{\gamma}\frac{Mg}{RT_g}z\right)^{1/(\gamma-1)}. \tag{1.2.22}$$

ρ becomes zero at 27·5 km which may be considered as being the height of the top of an adiabatic atmosphere.

Adiabatic equilibrium tends to be set up in atmospheres that possess convective motion (e.g. turbulence) because convection in gases proceeds more quickly than the conduction necessary to establish isothermal equilibrium. As the earth's atmosphere is known to be turbulent below 100 km, these regions should exhibit adiabatic equilibrium. To a certain extent, this is true of the troposphere, though it is not in perfect adiabatic equilibrium. It has a lapse rate of 5° K/km and the tropopause lies between 8 km and 20 km and is not equal to 27·5 km.

However, only in the troposphere is the temperature distribution controlled by convection. In the stratosphere it is controlled by radiation. This is because the atmosphere allows the visible and infra-red radiation emitted by the sun to penetrate with very little absorption.

Therefore the incoming radiation does not actually heat the atmosphere (except in very high regions) but heats the ground, which then re-radiates the energy in the far infra-red part of the spectrum.

The minor constituents of the atmosphere (water vapour, carbon dioxide, ozone) have strong absorption for the infra-red, so that the outgoing radiation from the ground is strongly absorbed by these gases. Therefore, if there were no convective motions, the temperature distribution would fall very rapidly near the ground where the water vapour and carbon dioxide predominate. However it will be shown in § 1.4 that such a rapid fall of temperature would make the atmosphere unstable. This instability sets up convective motions which bring the temperature distribution to a stable value, which is slightly above the adiabatic lapse rate. In adiabatic equilibrium, an element of gas, when it is transferred from one place to another, does not lose or gain any heat by conduction and takes up the requisite pressure and temperature in its new position. For an atmosphere in adiabatic equilibrium, the gases are mixed sufficiently well to exist in approximately the same proportions at all heights.

1.3 Atmospheric Wave Observations

Microbarographic Observations

Atmospheric waves are characterized by variations in the wind speed, the atmospheric density, and the atmospheric pressure. Therefore, besides the direct visual evidence for atmospheric waves that is afforded by the mountain wave clouds, it is possible to determine the existence of the waves by measuring the small changes in the atmospheric pressure produced during the passage of a wave. This is done by using a barometer that is capable of measuring pressure changes of the order of a few microbars*. It is interesting to realize that, though ten microbars is a small pressure change in the long-period meteorological systems, it corresponds to a sound pressure level of 60 decibels, which would be a loud sound at 1000 Hertz.

The spectrum of periods measured by microbarographs ranges from 10^{-1} to 10^{5} seconds. The high-frequency, generally non-periodic noise is due to turbulence, whereas the low-frequency portion of the pressure spectrum is characterized principally by the semi-diurnal and diurnal tides. Within the narrow band of periods between 5 and 10 minutes, uniform sinusoidal oscillations in the pressure variation frequently appear. These are the result of tropospheric gravity waves. Convection and the influence of nearby weather systems produce rather more irregular fluctuations with periodicities of 15 minutes to one hour. The vertical fluctuations in the temperature

* 1 bar = 10^{6} dyne/cm^{2} = 10^{5} N/m^{2} = 750 mm of mercury = 10^{5} Pa.

can also form waveguides within which the atmospheric wave system can become ducted and propagate its energy over very large distances.

A typical microbarograph record is shown in Fig. 1.7. The spectrum of waves from 0348 to 0452 hours was sorted into its frequency components and the results displayed as in Fig. 1.8. The microbarograph itself filtered out waves with periods less than two

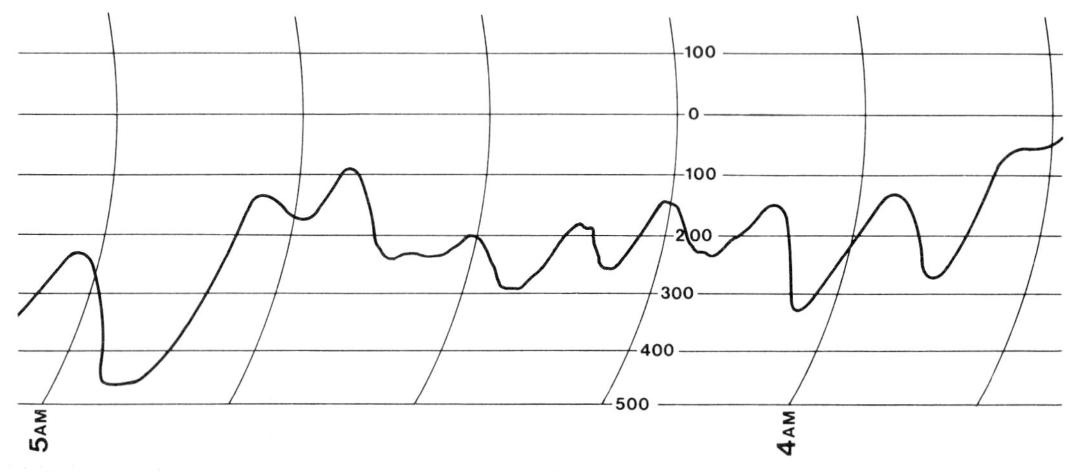

1.7 Typical microbarograph record

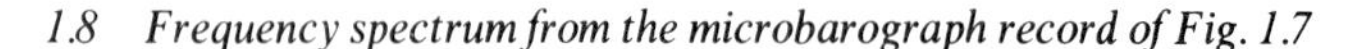

1.8 Frequency spectrum from the microbarograph record of Fig. 1.7

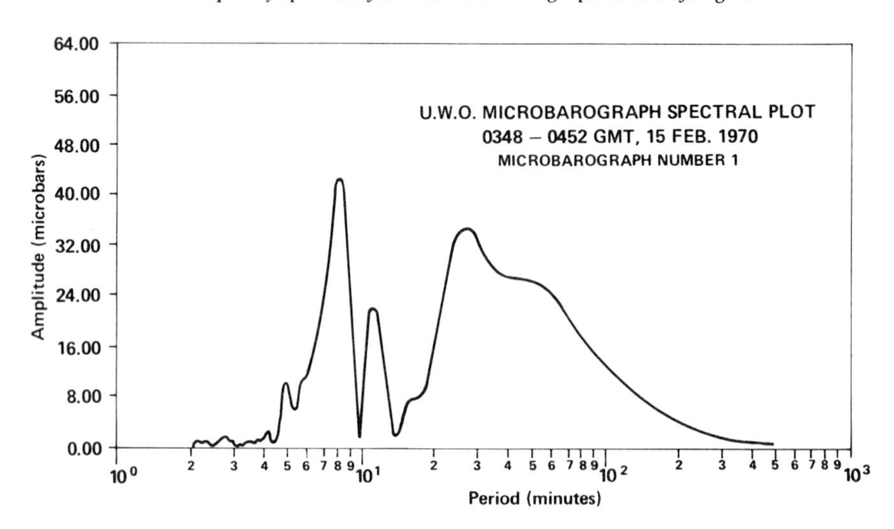

minutes or longer than two hours and any such waves are absent from the spectral plot. The band of tropospheric gravity waves is clearly visible.

It appears that the worldwide barometric observations of the atmospheric waves launched by the eruption of Krakatoa in 1883 stimulated Lamb (1945, p. 541) to investigate the general problem of propagation in an atmosphere with gravitational forces. He derived the hydrodynamical equations of motion that have formed the basis for more modern work. Microbarographic observations of pressure waves were also made after the great Siberian meteor of 1908. More recently there has been considerable work done on the microbarographic observations of the atmospheric waves generated by nuclear detonations.

Tolstoy and Lau (1970) point out that, under certain favourable conditions, it is possible for ionospheric disturbances to be measured by ground-based microbarographs, and they cite the work of Herron and Montes (1970) as an instance. Herron and Montes obtained sinusoidal microbarographic oscillations simultaneously with sinusoidal variations in their Doppler ionosonde readings. A Doppler ionosonde measures the frequency shift of a fixed-frequency signal as a function of time and gives a measure of the speed of the ionospheric irregularities detected at a particular operating frequency.

It also appears likely that during a solar eclipse, microbarographs would be able to record the atmospheric waves generated (Chimonas, 1970b). As the lunar shadow moves along the path of the eclipse, it produces an effective cooling spot that acts as a continuous source of gravity waves.

Because the moon's shadow will be moving at supersonic speeds the gravity waves will be in the form of a bow wave on either side of the lunar shadow. The gravity waves that are formed by the eclipse will also produce significant interactions with the ionization at ionospheric heights. The gravity waves will be focused because of the curved path of a solar eclipse and so observations conducted near these points of focusing should reveal marked atmospheric wave effects. Davis and Da Rosa (1970) found ionospheric effects which they claim were due to gravity waves initiated by the North American solar eclipse of 1970 March 7, and further analysis (Chimonas and Hines, 1971) confirms their interpretation. More recent experiments by Schoedel *et al.* (1973) at the focusing point of the 1973 Central African eclipse failed to detect any oscillations that could be ascribed to atmospheric waves. The reason for this discrepancy between the American and African results is a mystery, though it may be a result of differing ozone concentrations during the two eclipses.

Gravity Waves in the Ionosphere

In the troposphere, there appears to be a continuous generation of Rossby waves at mid-latitudes. There is also an intermittent occurrence of gravity waves that is related to tropospheric temperature inversions (Gossard and Munk, 1954). However, there is convincing evidence (Hines, 1960), based on studies of the upper atmospheric wind system, that there is a continuous occurrence of gravity waves at ionospheric heights. Planetary waves on the other hand, which are ubiquitous at meteorological heights in mid-latitudes, are liable to occur rather intermittently in the upper atmosphere. The existence of planetary waves in the upper atmosphere is discussed in § 4.3.

Observations of the dynamics of the upper atmosphere are conducted by visually observing the drift motions of noctilucent clouds (which are dealt with later in this section), long-lasting meteor trails, and the luminescent chemical trails laid by rockets. These techniques are unable to provide a picture of the temporal variations of the wind systems, so ground-based techniques have had to be devised.

In the height range 80–110 km meteors produce ionized trails. Except in the rare cases when the trails are aligned within less than a degree along the magnetic field lines, these ionized trails drift with the neutral air. It is then possible to map out the path of the drifting trail by continuously measuring the range of the meteor trail with radar pulses. Similarly it is possible to track the drifts of naturally occurring ionospheric irregularities. These ionospheric drifts are only likely to be related to the neutral air winds at heights below 100 km, and are usually measured only by day at these heights, owing to the lack of ionization at night.

These techniques indicate that the upper atmospheric motion can be divided into four categories, which may be considered as being produced by (i) a prevailing wind, (ii) a tidal wind, (iii) an irregular wind, and (iv) turbulence, which ceases rather abruptly above 105 km.

The prevailing wind is predominantly zonal (east–west) and varies throughout the course of a year in a pattern that tends to repeat in both hemispheres according to season. Strong eastward winds in winter give way to more moderate westward winds in summer, at 60–80 km altitude (Groves, 1969b). These prevailing winds may be expected to be geostrophic, i.e. to be in equilibrium under the combined effects of a pressure gradient, the Coriolis force, and gravity. In the thermosphere, the prevailing winds are primarily due to the daily temperature variations and therefore may be considered as constituting a thermally driven diurnal tide.

The tidal variations in the lower thermospheric winds are revealed most clearly by the radar observations of drifting meteor trails (Greenhow and Neufeld, 1956), which also show an irregular scattering of speeds about the smooth tidal oscillations.

This irregular component often makes the largest contribution to the total wind and, like the tidal component, tends to increase in amplitude as the height increases. Prior to 1960, these irregular variations were ascribed to turbulence, though once the existence of the turbopause was firmly established it became clear that some other mechanism must be responsible.

C. O. Hines (1960) showed that the irregular fluctuations in the thermospheric winds could be explained in terms of internal atmospheric gravity waves. Fig. 1.9 illustrates the result obtained by photographing a long-lasting meteor trail. This photograph clearly

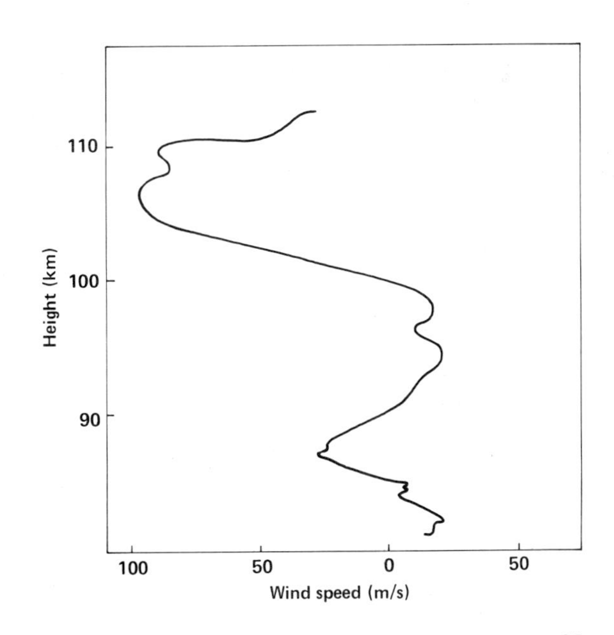

1.9 Wind velocity at meteor heights

showed that the irregular winds are predominantly horizontal and have amplitudes that increase with height. Hines (1960) then normalized the wind profile of Fig. 1.9 by removing the general wind shear and by diminishing the residual by a factor proportional to the square root of the atmospheric density. The result of this procedure is shown in Fig. 1.10.

Even though tidal winds have an upward amplitude-growth proportional to the square root of the atmospheric density, the waves depicted in Fig. 1.10 are not tidal in nature. Their wind velocity is

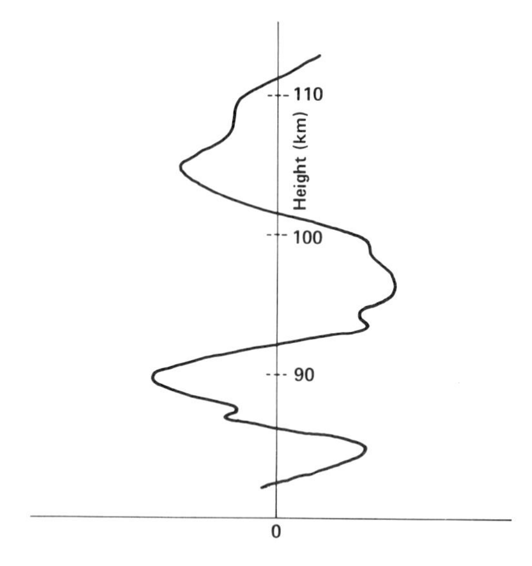

1.10 Normalized winds deduced from Fig. 1.6 by removing the shear and reducing the residual by $\rho^{\frac{1}{2}}$

too large. Measurements of the geomagnetic field variations produced in the ionosphere by tidal winds indicate that the tidal wind in the lower thermosphere is only of the order of 20–30 m/s, whereas the irregular winds that are found at these altitudes can have speeds of around 80–100 m/s. All of the observed properties of these irregular winds—horizontal motions, increase of amplitude with height, and increase of the dominant scale size of the vertical wavelength with height—are explicable in terms of gravity waves. Greenhow and Neufeld (1959) also show that the horizontal scale size of E-region irregular winds exceeds the dominant vertical scale size by a factor of 20 or more, and so must lie in the hundreds of kilometres. Once again, this is consistent with gravity-wave theory.

Noctilucent Clouds

Noctilucent clouds are a rare type of cloud that strongly resemble cirrus clouds but occur only on summer nights in a band of geographical latitudes between 45° and 75°. As their name implies, they are clearly visible during the night. Extensive photographic coverage has revealed that they occur at heights from 70 km to 90 km. They are seen only when the sun is about 6° to 16° below the horizon and only in the part of the sky where they are directly illuminated by sunlight, so that it appears that they are extremely tenuous and are actually made visible by the light that they scatter. The most frequent observations have been between 55° N and 60° N during June and July and there is strong evidence that the lowest temperature of the year in the mesopause occurs just at the time, the place, and the level of the atmosphere that is associated with the occurrence of noctilucent clouds.

Detailed analysis of the noctilucent cloud motion from the results of many observers shows a wave structure superimposed on the field of the cloud motion. The preferred direction of the movement of the cloud is towards the southwest, and the average velocity is 45 m/s, though speeds of up to 200 m/s have been reported. This motion is presumably caused directly by the background prevailing winds.

The wave nature of the cloud is visible in large bands, or striations, with a wavelength of the order of 50 km and amplitudes up to 4 km. These waves propagate in the direction opposite to the cloud system with velocities of the order of 10 to 20 m/s, which implies wave periods of the order of several minutes. These observations strongly indicate that the recurrent wave structure is caused by the wind fluctuations of an atmospheric gravity wave.

Direct measurements have been made on the atmospheric temperature at noctilucent cloud heights. When clouds were present, the temperature was 130° K, whereas it was over 140° K in their absence. This indicates that the noctilucent clouds may be formed when the temperature at the mesopause drops below the local frost-point temperature. This explanation relies on there being about two parts per million of water vapour at mesospheric heights. As the solar ultra-violet radiation destroys water vapour at these heights, there must be either a supply of water vapour from below or some source of water vapour at these heights. The existence of mesospheric water vapour has been strongly argued in the past, and recent results of rocket-borne mass spectrometers (Johannessen *et al.*, 1972) show that water cluster ions of the form $H^+(H_2O)_n$, where $n = 5$ and 6, certainly exist at the polar summer mesopause. These would be expected to be in equilibrium with water vapour and so their presence provides strong support for the existence of the water vapour. Frith (1968) claims that a combination of both water vapour transport and water vapour production is quite plausible.

There remains, however, a problem with the mesospheric temperature. Paradoxically, the temperature at the mesopause is known to be 100°K *lower* in summer than in winter, which explains why the noctilucent clouds are only observed in summer. However, the winter temperature is far higher than can readily be explained and this seasonal difference may have a significant effect on the atmospheric waves that can reach ionospheric heights from below. At present, no one knows what causes this mesospheric warming in winter. Kellog (1961) has suggested chemical heating through the recombination of oxygen, whilst Hines (1965b) and Gossard (1962) have suggested dynamical heating by atmospheric gravity waves from the troposphere.

The presence of clouds complicates the equations describing atmospheric dynamics. In order to allow for the water vapour content of the atmosphere, one defines the humidity mixing ratio, r, as the ratio of the mass of water vapour in a given sample to the mass of dry air with which it has been mixed to form the sample. Thus, if the density of water vapour is ρ_v in a sample of dry air of density ρ_d, then $r = \rho_v/\rho_d$. The mean molecular weight of the moist air is thus

$$\overline{M}_m = \frac{\rho_v + \rho_d}{\dfrac{\rho_v}{M_v} + \dfrac{\rho_d}{M_d}} = \frac{r+1}{\dfrac{r}{M_v} + \dfrac{1}{M_d}} \tag{1.3.1}$$

where M_d and M_v are the molecular weights of the dry air and water vapour respectively.

The virtual temperature T_v is defined by

$$\frac{T_v}{M_d} = \frac{T}{\overline{M}_m} \tag{1.3.2}$$

so that

$$T_v = T\frac{\left(\frac{M_d}{M_v}r+1\right)}{r+1}.$$

The virtual temperature is the temperature at which a fictitious neighbouring volume of dry air would have the same density, and therefore the same buoyancy, as the moist air. If virtual temperature is substituted for the actual temperature in a relationship derived for dry air, then that relationship becomes applicable to moist air.

Cloud formation depends on the partial pressure of the water vapour, e. If this equals the saturated vapour pressure, then it is possible for clouds to condense. Clouds are normally unable to form in the mesosphere and thermosphere because the atmospheric pressure, p, there falls below the saturated vapour pressure, so that the air can never become saturated. Under certain conditions, however, noctilucent clouds can form at the mesopausic temperature minimum.

The humidity mixing ratio is related to the pressures by

$$r = \frac{M_v e}{M_d(p-e)}. \tag{1.3.3}$$

For most practical meteorological purposes $M_v/M_d = 18/29$ and $e \ll p$, so that $r \approx 5e/8p$.

1.4 Compressible Fluid Dynamics

This section reviews some of the elements of geophysical fluid dynamics that will be required for the exposition in the subsequent chapters. Perhaps the most basic of these is the choice of the coordinate system used in the calculations. The motion of a fluid may be investigated by two different methods. In the Lagrangian method, we fix our attention upon an element of the fluid and follow its motion throughout its history, whereas, in the Eulerian method, attention is fixed upon a particular point in space and observations are made as to the fluid characteristics at this fixed point. Both approaches will be used in this book. The two formulations are connected through the equation

$$\frac{D}{Dt} = \frac{\partial}{\partial t} + \mathbf{U} \cdot \nabla$$

U is the wind velocity and the operator D/Dt is the Stokes derivative, which is used to give the Lagrangian formulation of the fluid motion. The Stokes derivative is useful because the basic equations that will be used retain their familiar mechanical form when expressed in terms of the Stokes derivative. On the other hand, the above-mentioned equality is valid only in a Cartesian coordinate system and can be applied only to scalars. A more general form is given in the next chapter.

For all the calculations that will be of interest to us when we insist on making observations at one location, it is necessary to use the Eulerian formulation. The operator

$$\mathbf{U} \cdot \nabla = U_x \frac{\partial}{\partial x} + U_y \frac{\partial}{\partial y} + U_z \frac{\partial}{\partial z}$$

can then be blamed for all the complexities that it introduces by making the basic equations non-linear. Non-linear equations are difficult to work with, so that a good deal of our ingenuity is going to be used in finding methods of making the equations linear.

There are then two ways to justify the linearizations. Mathematically, one can examine the magnitudes of the non-linear terms in the governing equations and decide that they are small and can be eliminated. This approach is normally acceptable, but can sometimes lead to errors when what starts out as a small term insists on growing into a large non-negligible term through resonance or some other amplification mechanism. The other approach to justify the initial linearizations is to compare the results obtained from the theory with experimental observations. If they agree, then one can relax and feel content with the linearization. This approach has much to recommend it.

Vaisala–Brunt Frequency

Static equilibrium can occur only when forces are properly balanced, but the converse does not hold. A balance of forces in no way guarantees that the equilibrium configuration does in fact occur. In the atmosphere, there will always be small extraneous forces to disturb a strict equilibrium condition. The manner in which a system responds to small disturbing influences is characterized by its stability. In the atmosphere, this stability is controlled by the temperature distribution.

Because the temperature in the troposphere decreases with height, meteorologists use the lapse rate, which is defined as $-\partial T/\partial z$. This conveniently gives a positive value for an upward temperature decrease. Another meteorological parameter is the potential temperature θ. This is the temperature that a parcel of air would attain if brought adiabatically to a reference pressure p_g so that one obtains Poisson's equation

$$Tp^{(1-\gamma)/\gamma} = \theta p_g^{(1-\gamma)/\gamma} \tag{1.4.1}$$

In practice p_g is generally chosen as being 1000 mb. It can then readily be seen that an atmosphere in adiabatic equilibrium has a constant potential temperature throughout.

Fig. 1.11(*a*) represents an atmosphere whose temperature gradient exceeds the adiabatic temperature gradient, so that $\alpha > \alpha^*$. Suppose that a packet of air at the cross-over point undergoes an adiabatic displacement upwards. The air packet, which follows the adiabatic

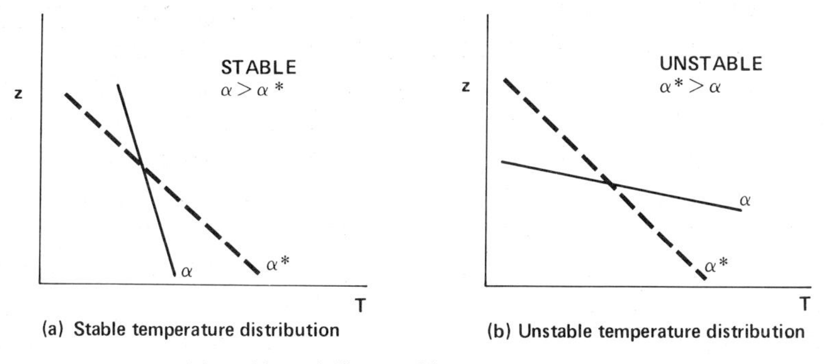

1.11 (a) Stable and (b) unstable temperature distributions

temperature gradient, is at a lower temperature than its surroundings and sinks back towards its initial position. Similarly, a downward displacement results in a higher packet temperature than the surroundings and the air packet rises. Such a situation is stable.

However the situation shown in Fig. 1.11(*b*) is unstable. When the packet is displaced in either direction then it will keep moving. In this case, convection will restore stability, though there are notable exceptions. One of them is in the immediate vicinity of the earth's surface, where the boundary inhibits vertical motions. Tornadoes represent the violent consequence of short-lived configurations that are statically unstable.

If a particle is displaced vertically in a stable atmosphere, then a restoring force will act. If the frequency of the resultant oscillation is large, then the effects due to gravitation are negligible and the wave propagates as an ordinary sound wave with the compressibility of the medium providing the restoring force. At lower frequencies, the gravitational restoring forces are comparable to those due to compressibility. Consider a small parcel of fluid at a height z. The parcel is displaced by an infinitesimal amount Δz. The density of the parcel is initially ρ and, after being displaced, the density is $\rho+\Delta\rho$, where $\Delta\rho$ is given by $\Delta\rho = (\Delta\rho/\Delta p)\Delta p$.

If the displacement takes place adiabatically, then the speed of sound is given by

$$c^2 = \frac{\Delta p}{\Delta \rho}. \tag{1.4.2}$$

In its initial state the density of the parcel, ρ, equals the density ρ_0 of the external atmosphere. After displacement, the density of the parcel in general no longer equals that of the environment. Since

$$\Delta\rho_0 = \left(\frac{d\rho_0}{dz}\right)\Delta z$$

the parcel is acted on by a buoyancy force that must balance the inertial force

$$\begin{aligned} \rho_0 \frac{d^2(\Delta z)}{dt^2} &= g(\Delta\rho_0 - \Delta\rho) \\ &= g\left(\frac{d\rho_0}{dz} + \frac{\rho_0 g}{c^2}\right)\Delta z \end{aligned} \tag{1.4.3}$$

provided that the hydrostatic equilibrium is not disturbed by the displacement. Provided

$$\frac{g}{\rho_0}\left(\frac{d\rho_0}{dz} + \frac{\rho_0 g}{c^2}\right) < 0 \tag{1.4.4}$$

the parcel will oscillate about its initial mean position with a frequency ω_B where

$$\omega_B^2 = -g\left[\frac{d(\ln \rho)}{dz}+\frac{g}{c^2}\right]. \tag{1.4.5}$$

The quantity ω_B is variously called the Vaisala frequency, the Brunt frequency, the Vaisala–Brunt frequency or the Brunt–Vaisala frequency. For a perfect gas, the Vaisala–Brunt frequency takes the form

$$\omega_B^2 = \frac{g(\alpha-\alpha^*)}{T}. \tag{1.4.6}$$

The symbol ω_g will be used for the Vaisala–Brunt frequency in an isothermal atmosphere where, recalling that $\alpha^* < 0$

$$\omega_g^2 = \frac{-g\alpha^*}{T} = \frac{(\gamma-1)g^2}{c^2} \tag{1.4.7}$$

so that

$$\omega_B^2 = \omega_g^2 + \frac{g}{c^2}\frac{\partial c^2}{\partial z}. \tag{1.4.8}$$

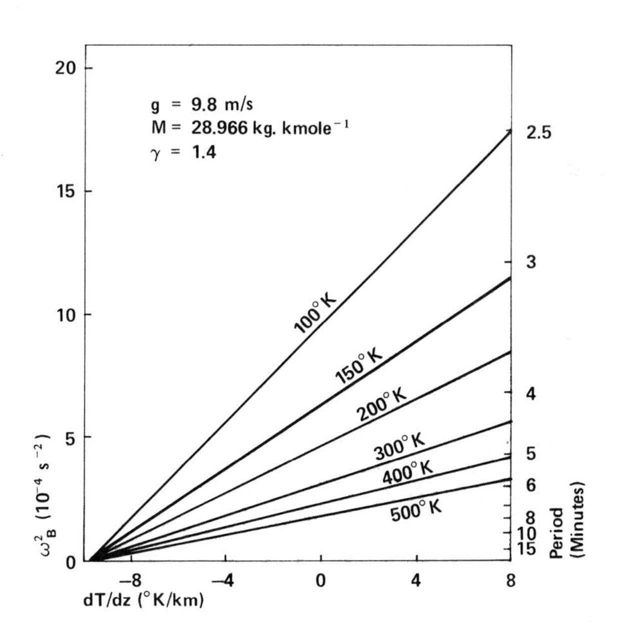

1.12 Vaisala–Brunt frequency and period plotted for various temperatures and temperature gradients

The Vaisala–Brunt frequency takes quite a simple form when expressed in terms of the potential temperaturc, θ,

$$\omega_B^2 = g\frac{\partial(\ln\theta)}{\partial z}. \tag{1.4.9}$$

The Equations of Motion

The momentum equation, in its simplest form, equates the rate of change of momentum with the external forces per unit volume*, **F**:

$$\rho\frac{D\mathbf{U}}{Dt} = \mathbf{F}$$

where **U** is the fluid velocity. This assumes that the phenomena being examined take place on a scale much larger than the molecular, so that fluid characteristics may be used. The choice of **F** then allows one to make this equation virtually as complicated as one desires.

It is however possible to simplify the problem by a technique known as scaling. Scaling consists of the evaluation of various dimensionless parameters, like the Froude number that we have met in equation (1.1.3), in order to determine which forces should be included in **F** for any given situation. For example, atmospheric tides and planetary wave motions have very large horizontal wavelengths. The Rossby number for these waves, defined in § 4.3, then becomes very small, which indicates that the Coriolis force term that arises because of the earth's rotation is important and needs to be retained in the equations of motion. For acoustic waves and gravity waves of periods less than three or four hours, the Rossby number substantially exceeds unity and the Coriolis force terms may be neglected in **F**.

By restricting consideration to acoustic and gravity waves, the external forces per unit volume become

$$\mathbf{F} = -\nabla p + \rho\mathbf{g} \tag{1.4.10}$$

when additive and dissipative forces, such as solar heating, viscosity, etc., are neglected.

* The equation of motion is not $D(\rho\mathbf{U})/Dt = \mathbf{F}$, as one might initially expect. If we recognize that the equation of motion is the equation of continuity for the momentum then we see that it is $D(\rho\mathbf{U})/Dt + \rho\mathbf{U}\nabla\,.\,\mathbf{U} = \mathbf{F}$, but then by using the equation of continuity of mass the left-hand side can be reduced to $\rho D\mathbf{U}/Dt$.

The addition of an equation representing the conservation of mass

$$\frac{D\rho}{Dt}+\rho\nabla \,.\, \mathbf{U} = 0 \qquad (1.4.11)$$

and one representing conservation of energy

$$\frac{Dp}{Dt} = c^2\frac{D\rho}{Dt} = -c^2\rho\nabla \,.\, \mathbf{U} \qquad (1.4.12)$$

gives a set of equations that can be solved by making a certain number of further assumptions. This is done in the next chapter.

Equation (1.4.12) can be derived from (1.2.4), the first law of thermodynamics, and it is possible to express the equation in terms of T, the temperature as the dependent variable, so that

$$\frac{DT}{Dt}+(\gamma-1)T\nabla \,.\, \mathbf{U} = 0.$$

Any external heating (or cooling) of J kcal kg^{-1} s^{-1} transforms the energy equation to

$$\frac{DT}{Dt}+(\gamma-1)T\nabla \,.\, \mathbf{U} = \frac{J}{C_v}.$$

If one includes the effects of viscosity, thermal conduction, radiation, cooling, and ion drag, then both the energy equation and the momentum equation become considerably more complicated. These effects are discussed in subsequent chapters.

Coriolis Force

The surface of the earth is not fixed in space but rotates about an axis with an angular velocity $\mathbf{\Omega}_E$. The equations of motion given in the previous section are applicable to coordinates fixed in space and may be used for waves of short period and wavelength where the rotational effects will be negligible. In all other cases, the apparent force on a body in motion that arises due to the rotation of the observer on the earth's surface must be included. This force is known as the Coriolis force. The Coriolis force acts everywhere at right angles to the motion of the object and is proportional to the speed. Though the Coriolis force can change the direction of the motion it cannot alter the speed.

The relation between a vector measured in an inertial frame fixed in space, represented by the subscript s, and the vector measured in a rotating frame, represented by the subscript r, is (Goldstein, 1950)

$$\left(\frac{d}{dt}\right)_s = \left(\frac{d}{dt}\right)_r + \mathbf{\Omega}_E \times . \qquad (1.4.13)$$

Thus

$$\mathbf{U}_s = \mathbf{U}_r + \mathbf{\Omega}_E \times \mathbf{r} \qquad (1.4.14)$$

where $\mathbf{r}$ is the radius vector from the centre of the earth to the body. The acceleration $\mathbf{a}_s$ is given by

$$\begin{aligned}\mathbf{a}_s &= \left(\frac{d\mathbf{U}_s}{dt}\right)_r + \mathbf{\Omega}_E \times \mathbf{U}_s \\ &= \mathbf{a}_r + 2(\mathbf{\Omega}_E \times \mathbf{U}_r) + \mathbf{\Omega}_E \times (\mathbf{\Omega}_E \times \mathbf{r})\end{aligned} \qquad (1.4.15)$$

where

$2(\mathbf{\Omega}_E \times \mathbf{U}_r)$ is the Coriolis term

$\mathbf{\Omega}_E \times (\mathbf{\Omega}_E \times \mathbf{r})$ is the centrifugal term.

To an observer in the rotating system it therefore appears as if the particle is moving under the influence of an effective force $\mathbf{F}_{\mathrm{eff}}$:

$$\mathbf{F}_{\mathrm{eff}} = m\mathbf{a}_r = \mathbf{F} - 2m(\mathbf{\Omega}_E \times \mathbf{U}_r) - m\mathbf{\Omega}_E \times (\mathbf{\Omega}_E \times \mathbf{r}) \qquad (1.4.16)$$

where $\mathbf{F} = m\mathbf{a}_s$ and in future the subscript r on $\mathbf{U}_r$ will be dropped.

The earth rotates anti-clockwise about the North Pole with an angular velocity

$$\mathbf{\Omega}_E = 7{\cdot}29 \times 10^{-5}\ \mathrm{s}^{-1}$$

and it has a radius

$$R_E = 6{\cdot}37 \times 10^6\ \mathrm{m}$$

so that the maximum value of the centrifugal force

$$\mathbf{\Omega}_E^2 R_E = 3{\cdot}38 \times 10^{-2}\ \mathrm{m/s^2}.$$

Therefore the centrifugal force $\mathbf{\Omega}_E^2 R_E \ll g$ and will be neglected in all the calculations.

In the northern hemisphere, $\mathbf{\Omega}_E$ points out of the ground and the Coriolis force deflects a moving body to the right of its direction of travel. The Coriolis deflection reverses direction in the southern hemisphere where it deflects the moving body to the left. The importance of the Coriolis force does not arise from its strength, because it is quite weak. For a velocity of 10^3 m/s the maximum Coriolis force is only about $0{\cdot}015g$. However, in meteorological applications the time scale of motions is quite long, so that the small deflective force acts over a sufficiently long period to produce marked effects. The anti-clockwise motion of the wind around northern hemisphere cyclones is a consequence of the Coriolis force.

It is necessary to neglect the horizontal component of the Coriolis vector $\mathbf{\Omega}_E$ (i.e. to neglect the vertical component of $2\mathbf{\Omega}_E \times \mathbf{U}$). This neglect is often referred to as the 'traditional approximation'. If this is not done, then the equations of motion when solved would indicate that an airflow that starts out by being horizontal would be purely vertical six hours later. This is not what actually happens because the airflow follows the spherical motion of the earth's surface. The problems associated with the neglect of the vertical Coriolis force, which is intricately related to the vertical wind's structure, are discussed in Eckart's book (1960, p. 96). In the next section, we shall justify the neglect of the vertical wind velocity in the large-scale background motion. Phillips (1966) has justified the traditional approximation for general hydrodynamic motions in a shallow atmosphere and this is quite adequate for our purposes. We shall invoke the Coriolis force only for large-scale waves, but to a wave with a very large wavelength the atmosphere appears shallow. Whether the traditional approximation would suffice for a deep atmosphere has not yet been determined.

The vertical component of the Coriolis vector will be a function of the latitude because the rotation of a point on the surface of a sphere about a vertical axis drawn through the point is a function of latitude. It is $\Omega_E \cos\theta$, where θ is the co-latitude defined by θ (in degrees) = 90° − latitude (in degrees). The reason for the introduction of the co-latitude is that later in this book we shall have occasion to employ spherical coordinates of which θ is one component. The vertical component of the Coriolis vector is proportional to the projection on the polar axis of a line drawn from the centre of the earth to the specified point on the earth's surface. The magnitude of the Coriolis force on a mass m is therefore

$$\begin{aligned} |2m(\mathbf{\Omega}_E \times \mathbf{U})| &= 2m\Omega_E|\mathbf{U}|\cos\theta \\ &= mf|\mathbf{U}| \end{aligned} \tag{1.4.17}$$

where

$$f = 2\,\Omega_E \cos\theta \qquad (1.4.18)$$

is called the Coriolis parameter. It is positive in the northern hemisphere and negative in the southern hemisphere.

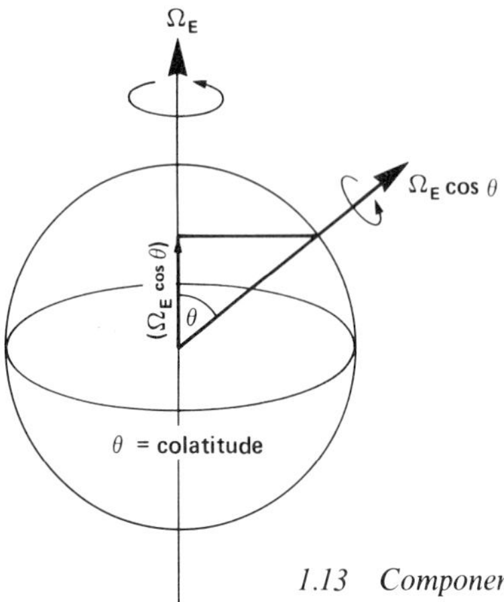

1.13 Component of rotation of a point on the earth's surface about the polar axis

Thermal Wind Equation

Before proceeding to analyse the wave behaviour that is inherent in the equations of motion, it may be useful first to discuss some of the steady-state solutions to the equation of motion:

$$\frac{D\mathbf{U}}{Dt} = -\frac{1}{\rho}\nabla p - 2\boldsymbol{\Omega}_E \times \mathbf{U} - \boldsymbol{\Omega}_E \times (\boldsymbol{\Omega}_E \times \mathbf{r}) - \mathbf{g}. \qquad (1.4.19)$$

Let us assume

(1) A flat earth described by the Cartesian coordinates e, directed to the east, n to the north, and z vertically upwards.
(2) The traditional approximation.
(3) That the centrifugal force is negligible. This is legitimate in large-scale air motions, provided that they are not strongly curved (e.g. hurricanes).
(4) Hydrostatic balance in the vertical direction so that

$$\frac{\partial p}{\partial z} = -\rho g. \qquad (1.4.20)$$

Now equation (1.4.20) taken in conjunction with assumptions (1), (2) and (3) and the equation of motion shows that $dU_z/dt = 0$. The vertical component of the wind velocity is then a constant that may be determined by applying boundary conditions. At the ground there can be no large-scale vertical wind, so that $U_z = 0$ throughout the atmosphere. There will, however, be small-scale departures from hydrostatic equilibrium that will set up vertical motions in an attempt to restore the equilibrium. These motions will be dealt with later.

The equation of motion may now be used to find the components of the wind that exists when pressure gradient forces are balanced by Coriolis forces. This wind, called the geostrophic wind, is $\mathbf{U}_g$ where

$$U_{ge} = -\frac{1}{f\rho}\frac{\partial p}{\partial n}; \qquad U_{gn} = \frac{1}{f\rho}\frac{\partial p}{\partial e}; \qquad U_{gz} = 0 \qquad (1.4.21)$$

at mid-latitudes; f is defined in equation (1.4.18), $f > 0$ in the northern hemisphere and $f < 0$ in the southern hemisphere. At the equator $f = 0$ and there can be no geostrophic wind there, so that the air must flow directly across the isobars (lines joining points of equal pressure).

The geostrophic equations describe the motion of a parcel of air that is initially moving directly from a high pressure area to a low pressure area. The motion is sufficiently slow, however, for the Coriolis force to deflect the air into a motion that is parallel to the isobars. Just as temporary departures from hydrostatic equilibrium can lead to oscillations about a mean state with periods of minutes or more, departures from geostrophic equilibrium lead to oscillations with periods of several hours or more. These oscillations are planetary waves.

The geostrophic wind is too idealized to be of significant use in predictive meteorology. On the other hand, the departure of the actual wind from the idealized wind provides a measure of the role of the other unconsidered forces. If the motion of the wind is along a circular arc, then the centrifugal forces should be included. The resulting wind is called the gradient wind.

Now consider what happens if the pressure variations from which the geostrophic wind arises are due to temperature variations through the equation of state

$$p = \rho\frac{RT}{M}, \qquad (1.4.22)$$

air being assumed an ideal gas. Then the components of the geostrophic wind velocity may be written in the vector form

$$\mathbf{U}_g = \frac{RT}{M f}\left(-\frac{\partial(\ln p)}{\partial n}, \frac{\partial(\ln p)}{\partial e}, 0\right). \qquad (1.4.23)$$

If the quantity $\mathbf{U}_g/T$ is differentiated with respect to z and the hydrostatic balance equation applied, then we obtain the thermal wind equation

$$\frac{\partial}{\partial z}(\mathbf{U}_g/T) = \frac{-g}{f T^2}\left(\frac{\partial T}{\partial n}, -\frac{\partial T}{\partial e}, 0\right) \qquad (1.4.24)$$

which relates the *vertical* shears in the geostrophic wind to the *horizontal* gradients in temperature.

As an example, consider the earth at ground level to be hot at the equator and cold at the poles. Then in the northern hemisphere, $\partial T/\partial y < 0$, so that $\partial(U_{ge}/T)/\partial z > 0$, so that up to the tropopause the winds increase in strength towards the east. Even a westward wind at the ground switches at some height into an eastward wind. In the southern hemisphere $\partial T/\partial y > 0$ but f, the Coriolis parameter, also changes sign so that in this case as well the winds increase in strength towards the east as one ascends.

At the stratosphere the situation will be different. The summer pole is continually heated at a higher temperature than the stratosphere at the equator, which in turn is hotter than the totally unilluminated winter pole. Thus the thermal winds in the summer stratosphere blow in the opposite direction to the tropospheric winds.

The wind pattern and temperature distribution up to the mesosphere is fairly well known, and except for the lowest 1000 metres of the atmosphere the zonal (east–west) winds appear to be adequately described by the thermal wind equation (Murgatroyd, 1957), though there still remain problems associated with the warm winter mesopause. It is at about 65 km that the summer and winter temperatures are roughly equal. Above this height in the mesosphere the winter temperatures are higher, whereas below this height in the stratosphere the summer temperatures are higher. There are also meridional (north–south) components of prevailing wind, leading to a net latitudinal flow of air which must presumably be balanced by return flows at other heights. At higher heights, the acceleration and viscosity terms as well as the interaction with the ionization become important and render the thermal wind equation in the form of equation (1.4.24) inappropriate.

Within the bottom 1 km of the atmosphere, the frictional drag along the earth's surface plays a dominant role. One of the effects of friction upon the geostrophic wind is to cause the air to move with a component across the isobars, from high to low pressure. Another effect of friction is to reduce the speed of the wind in comparison with the geostrophic wind. As we ascend from the earth's surface, the frictional forces weaken and, by an altitude of 1 km, the wind is very nearly geostrophic and may be analysed by the thermal wind equation. A theory for the variation due to friction was developed by Ekman (1905), who found that the variation of the wind with height could be represented by an equiangular spiral, now known as an Ekman spiral. The lowest kilometre of the atmosphere is known as the Ekman boundary layer.

1 Background Reading

Lighthill, M. J. (1965) *Waves in fluids*, Imperial College of Science and Technology, London.
Hines, C. O. (1972) Gravity waves in the atmosphere, *Nature*, **239**, 73.

2

Theory of Atmospheric Waves

2.1 Introduction

This chapter deals with the mathematical framework of the theory of atmospheric waves for the simple case of a flat, compressible, inviscid atmosphere. The rigorous approach to this would be to start with the full set of governing equations in spherical coordinates on a rotating earth and to treat the flat atmosphere and the non-rotating atmosphere as limiting cases. This method introduces a great deal of mathematical complexity and it will be deferred until Chapter 5 when we deal with atmospheric tides. Because of the long wavelengths that typical tidal oscillations possess, one must solve the equations for them on a spherical, rotating earth. This is not necessary for smaller scale waves and a flat, non-rotating earth provides a useful introduction.

The most heartening aspect of this branch of geophysical fluid dynamics is that it has been possible to identify wave motions by using extremely idealized and simplified models. For example, Rossby (1939, 1940) could identify the planetary scale flows in the atmosphere by using a theory based on an isothermal model of the atmosphere in which it was assumed that there was no wind divergence. Martyn (1950) found that he could explain microbarograph oscillations with a theory based on an inviscid atmosphere. Similarly Hines (1960) managed to interpret the irregular winds in the lower thermosphere in terms of a theory of gravity waves deduced from an isothermal, inviscid collisionless atmosphere, though certain further aspects of these winds had to be analysed by considering dissipative and reflecting mechanisms. Atmospheric tides required a further level of complexity.

The success that these reasonably simple approaches have had has led many authors to apply atmospheric wave theory in a wide variety of situations when it is not always obvious that their mathematical approximations are valid. In particular, there have been many studies of internal atmospheric gravity waves at ionospheric heights. Most of these use the simple windless, isothermal dispersion formula (equation 2.4.3) based on a perturbation theory that linearizes the equations that are used. Now, even though there is experimental evidence which suggests that the wave parameters at these heights are no longer of perturbation magnitude but have amplitudes that can be a considerable fraction of the total quantity, the simple theoretical approach appears to yield useful results. This seems to indicate that atmospheric waves of quite large amplitude are able to retain their sinusoidal form and that the consequences of waves 'breaking', shock wave formation and non-linearities are much less than one would intuitively expect.

The perturbation method is a well-respected method of solving non-linear ordinary differential equations. The differential equations of hydrodynamics are partial, but the general idea of perturbation theory remains applicable. So far, we have been studying the zeroth order, or background solutions for the pressure and temperature variations of an atmosphere composed of an ideal gas. Now it will be assumed that the basic quantities, p, ρ, and $\mathbf{U}$, can be expressed in terms of the zeroth order and first-order terms

$$p = p_0 + p_1 \tag{2.1.1}$$

$$\rho = \rho_0 + \rho_1 \tag{2.1.2}$$

$$\mathbf{U} = \mathbf{U}_0 + \mathbf{U}_1 \tag{2.1.3}$$

where the zero-order terms are taken as being very much larger than the respective first-order terms. There is little doubt that the second-order terms are needed to describe the phenomena properly; however, even the first-order solutions can provide an amazing insight into the physical processes, and by linearizing the basic equations, the first-order solutions provide a mathematically tractable method.

The method may be further simplified by choosing a reference frame that is moving with the neutral wind, so that $\mathbf{U}_0 = 0$, and it will be assumed that only zero-order and first-order perturbations are important. This means, for example, that the product $\rho_1 U_1$, which would appear by substitution into the momentum transfer equation, will be ignored because it is a second-order quantity. Further, only wavelike solutions of the perturbation quantities will be sought, so that*

$$\frac{p_1}{p_0 \mathrm{P}} = \frac{\rho_1}{\rho_0 \mathrm{R}} = \frac{U_{1x}}{\mathrm{X}} = \frac{U_{1y}}{\mathrm{Y}} = \frac{U_{1z}}{\mathrm{Z}}$$
$$= A_0 \exp\left[i(\omega t - K_x x - K_y y - K_z z)\right] \tag{2.1.4}$$

where the P, R, X, Y, Z are called polarization terms. They can be thought of as complex amplitudes which determine the phase relation between the various quantities. I shall deal with them later on. The capital K's are complex wavenumbers which are introduced in order to allow for amplification, or attenuation, in the x, y and z directions. The imaginary part of the wavenumber represents this growth or decay, whereas the real part of K stands for the wavelength of the sinusoidal variation. The real part of the complex wavenumber K will be denoted by the lower case k. x is going to be used to denote a horizontal direction that corresponds to the direction of phase propagation. If this happens to be towards the east then e will be used. Similarly, an n is a coordinate pointing northward.

2.2 Rossby Waves

Geostrophic motions in the earth's atmosphere can exhibit wave motions. We have seen in § 1.4 that for a geostrophic wind $D\mathbf{U}_g/Dt = 0$, so that the wind components stay constant in a reference frame that moves with the fluid's motion $\mathbf{U}_0$. Let us initially take $\mathbf{U}_g = (U_{ge}, 0, 0)$, where e is the x axis directed to the east. If there is now some horizontal pressure disturbance, then a meridional wind U_{gn} will be set up with a pressure gradient extending from a low pressure area L to a high pressure area H. As the motion moves eastward, both U_{ge} and U_{gn} remain constant.

However, the earth's surface is a closed curve. The low pressure area must be followed by a high pressure area which must in turn be followed by a low pressure area. In completing one east-west circuit of the earth's surface there will be regions in which U_{1n} is oppositely

* In case anyone is wondering what happened to the $B\exp[-i(\omega t - K_x x - K_y y - K_z z)]$ term that is necessary for a complete solution, they should flick to § 3.5. In short, we assume that the vertical wavelength we are interested in is very short compared with the depth of the atmosphere, so that there is no downward propagating wave or any other reflected wave.

directed. To an observer on the ground, these will appear to be a periodic sequence of pressure and meriodional wind velocity passing by.

The atmosphere is not quite so simple and geostrophic waves do not actually form like this. If the distance between the high and low pressure area is small, then the rotational effects become important and we have to include the gradient winds. Furthermore, the meridional wind shears can then lead to an unstable cyclone wave. On the other hand, if the distance between the high and low pressure areas is very large, then the geostrophic meridional wind will be very small and meridional variations of the Coriolis parameter become very important. This wave is the Rossby wave.

The geostrophic wind $\mathbf{U}_g$ is non-divergent, which means that $\nabla . \mathbf{U}_g = 0$. This is true for the winds in any incompressible homogeneous atmosphere and follows directly from the equation of continuity. Similarly, the simplest atmospheric model that is capable of generating Rossby waves is that of an inviscid, isothermal non-divergent incompressible atmosphere. This was the starting point for Rossby's (1939, 1940) original investigation of these long waves and it represents the prototype atmosphere. If the wave motion, $\mathbf{U}_1$ is non-divergent then

$$\nabla . \mathbf{U}_1 = 0 \tag{2.2.1}$$

so that if a plane wave exists it will have wavenumber $\mathbf{K}$, that satisfies

$$\mathbf{K} . \mathbf{U}_1 = 0 \tag{2.2.2}$$

so that the wave motion will be transverse and any stable motion for which K is purely real will be linearly polarized. The real part of the complex wavenumber K will be denoted by k. Vertical motions will be ignored.

If we are to investigate Rossby waves on a plane, then we need to include the variation of the Coriolis parameter, $f = 2\Omega_e \cos \theta$, with latitude. This may be regarded as a linear function $f = f_0 + \beta n$ where $\beta = \partial f / \partial n$. The plane in which the Coriolis parameter varies like this is called the beta plane and we shall choose a set of Cartesian axes on the beta plane where the x-axis is in the direction of the wave's phase propagation and the y-axis is perpendicular to it and directed northwards. The x- and y-axes need not correspond to zonal and meridional directions respectively, because both Rossby waves and planetary waves—which are the generalized form of

Rossby waves on a spherical surface—can have phase propagation that is oblique to the axis of rotation as well as phase propagation perpendicular to the axis of rotation.

By having chosen x and y in the longitudinal and transverse directions, the wave components are

$$U_{1x} = 0$$
$$U_{1y} = \mathrm{Y} \exp i(\omega t - kx). \tag{2.2.3}$$

For many purposes it is advantageous to express the horizontal wind field in terms of its rotational properties. Since we know that for any fluid, if

$$\mathbf{U} = \mathbf{\Omega} \times \mathbf{r}$$

then

$$\mathbf{\Omega} = \tfrac{1}{2}\nabla \times \mathbf{U} \tag{2.2.4}$$

the curl of the wind speed is a measure of the wind rotation that is called the vorticity. We shall use the symbol ζ for the vorticity, where

$$\zeta = \nabla \times \mathbf{U}. \tag{2.2.5}$$

Just as we are concerned only with the vertical component of the Coriolis parameter, so we shall be primarily interested in the vertical component of the vorticity which will be denoted by ζ_z. Since the Coriolis parameter f is the vorticity possessed by a fluid at rest with respect to the rotating earth, the absolute vertical vorticity of a fluid element is

$$\zeta_a = \zeta_z + f.$$

For the wave motion we are dealing with

$$\zeta_z = (\nabla \times \mathbf{U}_1)_z = \frac{\partial \mathbf{U}_{1y}}{\partial x}.$$

We now have to find the equation that describes the changes in vorticity. We start with the equation of motion

$$\frac{DU_e}{Dt} = -\frac{\partial p}{\partial e} + fU_n; \qquad \frac{DU_n}{Dt} = -\frac{\partial p}{\partial n} - fU_e$$

and then take the curl of the equation of motion to give

$$\frac{D}{Dt}\left(\frac{\partial U_n}{\partial e}-\frac{\partial U_e}{\partial n}\right) = -f\left(\frac{\partial U_e}{\partial e}+\frac{\partial U_n}{\partial n}\right) - U_n\frac{\partial f}{\partial n} - U_e\frac{\partial f}{\partial e}.$$

Now since $\partial f/\partial t = 0$ let us add $\partial f/\partial t$ to the left-hand side of this equation so that

$$\frac{D}{Dt}(\zeta_z+f) = -f\nabla\,.\,\mathbf{U}$$

For an incompressible atmosphere

$$\nabla\,.\,\mathbf{U} = 0$$

so that

$$\frac{D}{Dt}(\zeta_z+f) = 0 \tag{2.2.6}$$

which shows that absolute vorticity is conserved. If we expand (2.2.6)

$$\begin{aligned} \frac{\partial \zeta_z}{\partial t} + U_{0x}\frac{\partial \zeta_z}{\partial x} + U_{0y}\frac{\partial \zeta_z}{\partial y} + \beta^1 U_{1y} \\ = \frac{\partial}{\partial t}\left(\frac{\partial U_{1y}}{\partial x}\right) + U_{0x}\frac{\partial}{\partial x}\left(\frac{\partial U_{1y}}{\partial x}\right) + \beta^1 U_{1y} \\ = 0 \end{aligned} \tag{2.2.7}$$

where

$$\beta^1 = \frac{\partial f}{\partial y}.$$

Thus, if we substitute (2.2.3) in (2.2.7), we find that the plane waves have a frequency $\omega = U_{0x}k-\beta^1/k = (U_{0e}k-\beta/k)\cos\alpha$ where α is the angle between e and x. We shall use the convention that

$$\beta^1 > 0 \quad \text{and} \quad |\alpha| < \frac{\pi}{2}.$$

The dispersion relation consists of two parts. The first term represents the convection of the zonal wind whereas the β^1/k term implies that the phase propagation of the Rossby wave always has a westward component. In fact this zonal component of phase velocity is independent of the angle α and is always equal to

$$V_{pe} = U_{0e} - \frac{\beta}{k^2}.$$

The longitudinal phase velocity $V_{px} = V_{pe} \cos \alpha$ can best be regarded as a consequence of the westward drift.

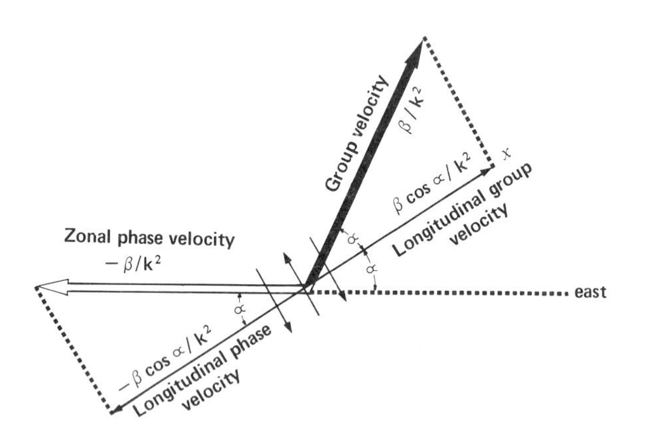

2.1 Rossby wave motion

Platzman (1968) presents an interesting explanation for the westward drift. If we regard the variable Coriolis parameter as a field of east–west lines of force, we can say that by cutting these lines the transverse velocities in the wave create an alternating row of induced vorticities and corresponding induced velocities. The phase propagation that results always has a westward component, an anisotropy that stems from the fact that the gradient of the earth's vorticity (the Coriolis parameter) points northward. The maximum induction effect is attained when the transverse velocities cut the lines of force at the greatest rate. This occurs when the orientation of wave crests is north–south. On the other hand, if the crests are oriented east–west, the induction disappears and the wave cannot move.

Now the dispersion relation for the Rossby wave is

$$\omega = \left(U_{0e}k - \frac{\beta}{k}\right)\cos\alpha$$

where k and α are the polar coordinates of the wave number vector, so the longitudinal component of the group velocity is

$$V_{gx} = \frac{\partial\omega}{\partial k} = \left(U_{0e} + \frac{\beta}{k^2}\right)\cos\alpha$$

and the transverse component is

$$V_{gy} = \left(\frac{1}{k}\right)\left(\frac{\partial\omega}{\partial\alpha}\right) = \left(\frac{\beta}{k^2} - U_{0e}\right)\sin\alpha.$$

The group velocity thus also consists of two parts. One part is the convection of the zonal wind U_{0e}, and the other is the wave group velocity of magnitude β/k^2 and angle 2α (Fig. 2.1).

Rossby waves occur in a divergence-free atmosphere. This will tend to limit their occurrence to altitudes above the boundary layer comprising the lowest kilometre of the atmosphere.

Rossby waves can be divided into two types: free Rossby waves and forced Rossby waves. Free Rossby waves are formed by a random departure from geostrophic equilibrium such as may be occasioned by baroclinic instability, barotropic instability, or thermal perturbations. Forced Rossby waves are produced by departures from geostrophic equilibrium induced by mountains and mountain ranges. Free Rossby waves can be set up by either eastward or westward flows, but forced Rossby waves can only be set up by an eastward wind flow (§ 4.6). If the wind, U_{0x} blows westward, then it will tend to mitigate against the formation of waves. This can be pictured by recalling that, in the northern hemisphere, the Coriolis force tends to deflect the wind to the right of its motion whereas the deflection is to the left in the southern hemisphere. Thus, if we consider a disturbance of the eastward flow at the equator, this will always have a restoring Coriolis force towards the equator and so waves may be set up. On the other hand, westward flows will always have a restoring force directed away from the equator, so that waves are unable to form. The same type of mechanism applies to mid-latitudes as well.

As we have seen, the thermal winds in the upper troposphere are eastward and Rossby waves do play an important role above the Ekman layer. On the other hand, it is possible for the temperature gradients to produce such a large eastward wind that the Rossby wave motions are negligible. This occurs in the tropospheric jet streams of fast-moving eastward winds. In this connection, perhaps one should re-emphasize the fact that Rossby waves are a consequence of the variation of the Coriolis parameter with latitude. If there were no latitude variation of the Coriolis force, then Rossby waves would degenerate into a geostrophic wind: one in which the Coriolis force and the pressure gradient are in equilibrium.

2.3 Isothermal, Windless Case

Having dealt with the Rossby waves let us now examine the theory of the atmospheric waves of shorter period and wavelength which are regarded as 'noise' by meteorologists but which have important ionospheric applications. The analysis of these waves will neglect the Coriolis force and assume a flat earth. The neglect of the earth's curvature is legitimate provided that the horizontal wavelengths do not exceed a few hundred kilometres. The neglect of the Coriolis force does not introduce significant errors when the Rossby number, to be defined in § 4.3, exceeds unity. This is true for waves whose periods are shorter than about 4 hours. Waves of longer period or wavelength need to be dealt with in terms of atmospheric tidal theory.

The basic equations that we need are the equation of motion, the adiabatic equation of state, which also represents energy conservation, and the equation of mass continuity. The equation of motion for an inviscid, isothermal flat atmosphere, when the Coriolis force can be neglected is

$$\frac{D\mathbf{U}}{Dt} = -\frac{1}{\rho}\nabla p + \mathbf{g}$$

where D/Dt represents the Stokes derivative. As previously mentioned there are some caveats to be borne in mind when using the Stokes derivative. In general vector form, the rate of change of any vector $\mathbf{A}$ is

$$\frac{D\mathbf{A}}{Dt} = \frac{\partial \mathbf{A}}{\partial t} + \frac{1}{2}\begin{cases}\nabla(\mathbf{U}\,.\,\mathbf{A}) + (\nabla \times \mathbf{U}) \times \mathbf{A} \\ +(\nabla \times \mathbf{A}) \times \mathbf{U} - \nabla \times (\mathbf{U} \times \mathbf{A}) \\ \qquad + \mathbf{U}(\nabla\,.\,\mathbf{A}) - \mathbf{A}(\nabla\,.\,\mathbf{U})\end{cases}$$

so that

$$\frac{D\mathbf{U}}{Dt} = \frac{\partial \mathbf{U}}{\partial t} + \frac{1}{2}\nabla U^2 + (\nabla \times \mathbf{U}) \times \mathbf{U}.$$

On the other hand, the rate of change of a scalar in a Cartesian coordinate system is given by the operator

$$\frac{D}{Dt} = \frac{\partial}{\partial t} + \mathbf{U}\,.\,\nabla$$

so that we can write the components of the equation of motion in the shorthand form

$$\frac{\partial \mathbf{U}}{\partial t} + \mathbf{U}\,.\,\nabla(\mathbf{U}) = -\frac{1}{\rho}\nabla p + \mathbf{g}. \tag{2.3.1}$$

It must be understood that this is not a correct vector equation but merely a convenient shorthand form for writing the three Cartesian components of the equation of motion.

The other two equations we need are

$$\frac{\partial p}{\partial t} + \mathbf{U}\,.\,\nabla p = c^2\left(\frac{\partial \rho}{\partial t} + \mathbf{U}\,.\,\nabla \rho\right) \tag{2.3.2}$$

which is the adiabatic equation and

$$\frac{\partial \rho}{\partial t} + \nabla\,.\,(\rho \mathbf{U}) = 0 \tag{2.3.3}$$

which is the mass continuity equation.

The solutions to this set of equations will consist of two types of waves—acoustic waves and gravity waves. Both of these types of waves can have variations in the vertical direction and in one horizontal direction. In considering the simplest Rossby wave, only two coordinates—both horizontal—were needed. Similarly, when considering gravity and acoustic waves, it is possible to rotate the horizontal coordinate, which will be denoted by x, so that it lies along the direction of horizontal phase propagation. The vertical component is denoted by z. In this case, when the expressions (2.1.1) to (2.1.4) are substituted into the equations, then the zero-order equations which represent the variations in the background quantities are

$$\frac{DU_{0x}}{Dt} = -\frac{1}{\rho_0}\frac{\partial p_0}{\partial x} \tag{2.3.4}$$

which is the gradient wind equation in the absence of Coriolis forces,

$$\frac{DU_{0z}}{Dt} = -\frac{1}{\rho_0}\frac{\partial p_0}{\partial z} + g \tag{2.3.5}$$

and provided we now assume $\mathbf{U}_0 = 0$,

$$\frac{\partial p_0}{\partial t} = c^2 \frac{\partial \rho_0}{\partial t} \tag{2.3.6}$$

and

$$\frac{\partial \rho_0}{\partial t} = 0. \tag{2.3.7}$$

It may be observed from (2.3.7) that ρ_0 may be treated as being independent of time.

An insight into the theory of atmospheric waves can now be obtained by starting with the isothermal, windless, flat atmosphere:

$$U_{0x} = U_{0z} = \frac{\partial T}{\partial z} = \frac{\partial T}{\partial x} = 0$$

then the first-order perturbed equations of motion, continuity and of state become

$$\rho_0 \frac{\partial U_{1x}}{\partial t} = -\frac{\partial p_1}{\partial x} \tag{2.3.8}$$

$$\rho_0 \frac{\partial U_{1z}}{\partial t} = -\frac{\partial p_1}{\partial z} - g\rho_1 \tag{2.3.9}$$

$$\frac{\partial p_1}{\partial t} + U_{1z}\frac{\partial p_0}{\partial z} = c^2\left(\frac{\partial \rho_1}{\partial t} + U_{1z}\frac{\partial \rho_0}{\partial z}\right) \tag{2.3.10}$$

$$\frac{\partial \rho_1}{\partial t} + U_{1z}\frac{\partial \rho_0}{\partial z} + \rho_0\left(\frac{\partial U_{1x}}{\partial x} + \frac{\partial U_{1z}}{\partial z}\right) = 0. \qquad (2.3.11)$$

The set of equations (2.3.8), (2.3.9), (2.3.10), and (2.3.11) govern the behaviour of atmospheric waves in a flat isothermal atmosphere. They will thus be referred to as the governing equations. By assuming that the perturbed variables have the sinusoidal variations outlined in equation (2.1.4), it is possible to obtain four linear equations in the four variables p_1/p_0, ρ_1/ρ_0, U_{1x} and U_{1z}. These may be expressed in matrix form as

$$\begin{bmatrix} i\omega & 0 & 0 & -iK_x gH \\ 0 & i\omega & g & -iK_z gH - g \\ -iK_x & -\left(\frac{1}{H} + iK_z\right) & i\omega & 0 \\ 0 & \frac{\gamma - 1}{H} & -i\omega\gamma & i\omega \end{bmatrix} \begin{bmatrix} U_{1x} \\ U_{1z} \\ \frac{\rho_1}{\rho_0} \\ \frac{p_1}{p_0} \end{bmatrix} = \mathbf{0}. \qquad (2.3.12)$$

For a non-trivial solution of this equation (2.3.12) the determinant of the coefficient matrix must vanish. Therefore the dispersion relation becomes

$$\omega^4 - \omega^2 c^2 (K_x^2 + K_z^2) + (\gamma - 1) g^2 K_x^2 + i\omega^2 \gamma g K_z = 0. \qquad (2.3.13)$$

An analysis of (2.3.13) reveals that it is not possible for both K_x and K_z to be purely real and non-zero. An attenuation or growth in the wave amplitude must occur in either the vertical or the horizontal directions. In order to continue to deal with reasonably simple cases, I will assume that there is no variation in the amplitude in the horizontal direction, so that K_x will be purely real. In the notation being used, this means $K_x = k_x$.

The real part of the dispersion relation then becomes

$$\omega^4 - \omega^2 c^2 (k_x^2 + k_z^2 - [\mathrm{Im}\,(K_z)]^2) - \omega^2 \gamma g\, \mathrm{Im}(K_z) + (\gamma - 1) g^2 k_x^2 = 0$$

and the imaginary part of the dispersion relation (2.3.13) becomes

$$\omega^2 \gamma g\, \mathrm{Re}\,(K_z) - 2\omega^2 c^2\, \mathrm{Re}\,(K_z)\, \mathrm{Im}\,(K_z) = 0 \qquad (2.3.14)$$

so that, provided the frequency is not zero, either

$$\mathrm{Re}\,(K_z) = 0 \qquad (2.3.15)$$

which means that the vertical wavenumber is purely imaginary; or

$$\mathrm{Im}\,(K_z) = \frac{\gamma g}{2c^2} = \frac{1}{2H}. \tag{2.3.16}$$

The first alternative (2.3.15) means that the waves have no vertical phase variations but only an exponential amplification or decay with height. This is characteristic of surface waves and evanescent waves.

Surface waves are waves whose energy is concentrated at a boundary or discontinuity of some parameter, and they have an exponential variation of amplitude. Evanescent waves, however, can exist independently of a boundary and they can propagate within the body of a fluid, so they are sometimes referred to as body waves*. The generic term 'external waves' has been used to cover both surface and evanescent waves. The external waves, which have no vertical phase variation, contrast with the internal waves that do. Hence the origin of the name internal atmospheric gravity waves. Internal waves and external waves have somewhat different properties, and so § 2.4 will deal with the properties of the internal waves whereas § 2.10 will deal with external waves.

2.4 Internal Waves

The dispersion relation (2.3.13) is a fourth-order equation in ω and, when $\mathrm{Im}\,(K_z) = 1/2H$, there are four separate solutions to the system of equations represented by the matrix in (2.3.12). As the dispersion relation consists of three variables, there are three useful diagrams that may be drawn. Fig. 2.2 gives the ω, k_z dispersion plot for constant k_x contours, Fig. 2.3 gives the ω, k_x diagram for constant k_z contours, and Fig. 2.4 consists of the k_x, k_z plots for constant values of the frequency. The k_x, k_z diagram for a given value of the frequency is known as the propagation surface.

There are two distinct regimes of internal waves on the dispersion diagrams. One regime consists of waves with a frequency greater than ω_a where

$$\omega_a = \frac{\gamma g}{2c} = \frac{c}{2H}. \tag{2.4.1}$$

* Body waves is a generic term covering evanescent waves and cellular waves.

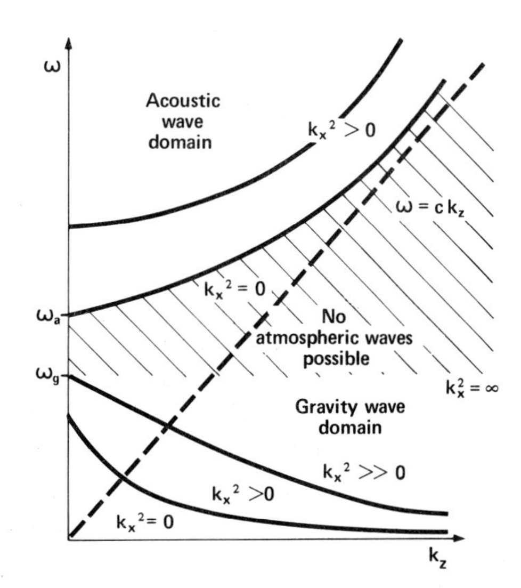

2.2 Dispersion, or diagnostic diagram for constant k_x contours

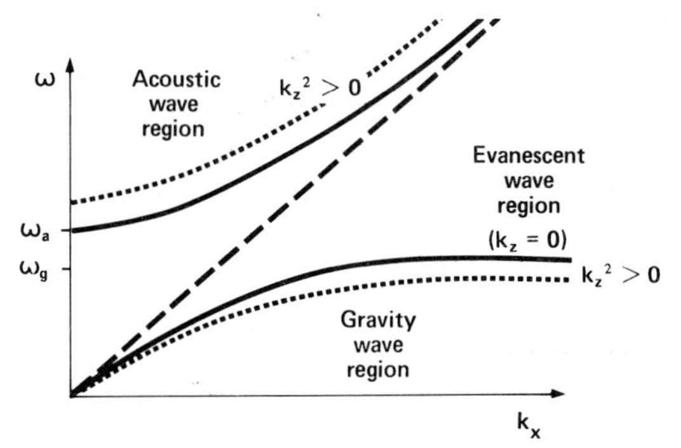

2.3 Dispersion diagram for constant k_z contours

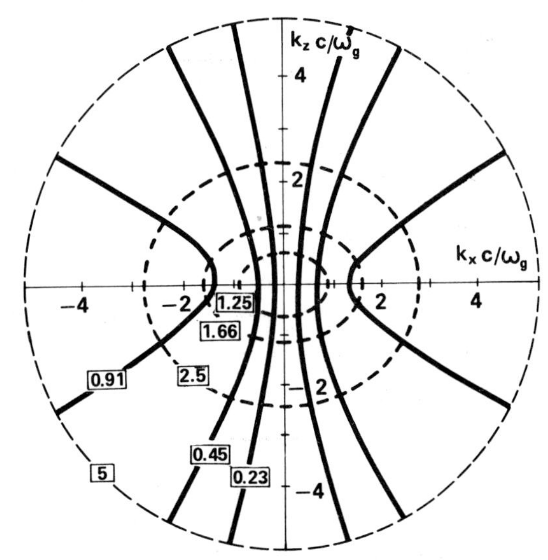

2.4 Constant period propagation surfaces for an isothermal atmosphere. Values of ω/ω_g for each mode are given inside the rectangular boxes. Acoustic waves are represented by dashed lines, gravity waves by unbroken lines (after Hines (1960); used by permission of the National Research Council of Canada)

ω_a is known as the acoustic cut-off frequency and is often just referred to as the acoustic frequency. The waves whose frequency is greater than ω_a are identified as acoustic waves. The other regime consists of low-frequency waves with an upper limit on their frequency of ω_g, the isothermal Vaisala–Brunt frequency given by

$$\omega_g^2 = \frac{(\gamma - 1)g^2}{c^2}. \tag{2.4.2}$$

These low-frequency, long-period waves are the internal atmospheric gravity waves. Then the four solutions that were mentioned earlier correspond to two waves of each type. The two waves of each type do not represent waves with different characteristics, but merely two waves whose phases propagate in opposite direction. Typical values of ω_a and ω_g, as well as their counterparts in a non-isothermal atmosphere, are given in Fig. 2.5.

If the Coriolis term had been retained, then a fifth independent

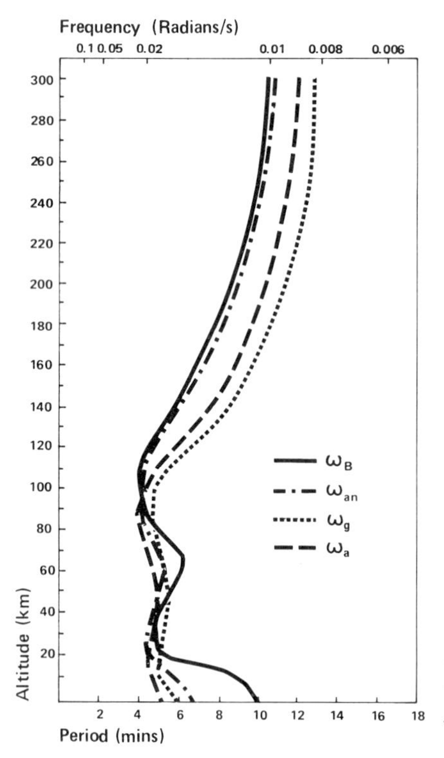

2.5 Values of the isothermal and non-isothermal acoustic cut-off frequencies ω_a, ω_{an} and buoyancy frequencies ω_g, ω_B respectively, for an atmosphere with a high thermospheric temperature

Table 2.1 Comparison between atmospheric waves and magnetoionic radio waves (isothermal atmosphere)

Acoustic waves (isothermal atmosphere)			*Radio waves*		
Acoustic cutoff frequency	$\omega_a^2 = \dfrac{c^2}{4H^2}$	(1*a*)	Plasma frequency	$\omega_N^2 = \dfrac{Ne^2}{\varepsilon_0 m}$	(1*b*)
Vaisala–Brunt frequency	$\omega_g^2 = \dfrac{g}{H}\left(1-\dfrac{1}{\gamma}\right)$	(2*a*)	Gyrofrequency	$\omega_H^2 = \left(\dfrac{e}{m}B_0\right)^2$	(2*b*)
	$X = \left(\dfrac{\omega_a}{\omega}\right)^2$	(3*a*)		$X = \left(\dfrac{\omega_N}{\omega}\right)^2$	(3*b*)
	$Y = \dfrac{\omega_g}{\omega}$	(4*a*)		$Y = \dfrac{\omega_H}{\omega}$	(4*b*)
Refractive index with gravity	$\mu^2 = \dfrac{c^2k^2}{\omega^2} = \dfrac{1-X}{1-Y^2\sin^2\varphi}$	(5*a*)	Refractive index with magnetic field	$\mu^2 = \dfrac{c^2k^2}{\omega^2} = 1 - \dfrac{2X(1-X)}{2(1-X)-Y^2\sin^2\theta \pm [Y^4\sin^4\theta + 4Y^2(1-X)^2\cos^2\theta]^{\frac{1}{2}}}$	
Refractive index with $\omega \gg \omega_g$	$\mu^2 = 1-X$	(6*a*)	Refractive index with $\omega \gg \omega_H$	$\mu^2 = 1-X$	(6*b*)
Vertical propagation ($X < 1$)	$\mu^2 = 1-X$	(7*a*)	Transverse propagation (ordinary wave)	$\mu^2 = 1-X$	(7*b*)
Angle α between wave vector and the ray	$\tan\alpha = -\dfrac{Y^2\sin\varphi\cos\varphi}{Y^2\sin^2\varphi - 1}$	(8*a*)	Angle α between wave vector k and the ray	$\tan\alpha = \dfrac{Y^2\sin\theta\cos\theta}{[Y^4\sin^4\theta + 4Y^2(1-X)^2\cos^2\theta]^{\frac{1}{2}}}$	(8*b*)
Group velocity (V_g) with $Y \ll 1$	$V_g^2 = c^2(1-X)$	(9*a*)	Group velocity with $Y \ll 1$	$V_g^2 = c^2(1-X)$	(9*b*)

solution of the dispersion relation would have emerged. This fifth type of wave is the Rossby wave that has already been dealt with in § 2.2. Because the earth has a preferred direction of rotation, only one Rossby solution exists, so that these waves can only propagate in one direction.

The dispersion relation (2.3.13) has the same significance for atmospheric waves that the Appleton–Hartree equation has in magneto-ionic theory. Presumably some of the readers of this book will be radio physicists whose interest in atmospheric waves stems from the effects such waves have on the ionosphere. For these readers we reproduce, from an article by Davies, Baker and Chang (1969), Tables 2.1 and 2.2 which illustrate the similarities between the dispersion relation for atmospheric waves and the Appleton–Hartree equation for radio waves. The acoustic waves have formulae that

Table 2.2 Dispersion formulae for gravity waves and whistlers

	Gravity wave formulae		*Whistler wave formulae*	
Refractive index	$\mu^2 = \dfrac{X-1}{Y^2\sin^2\varphi - 1}$	(10*a*)	$\mu^2 = \dfrac{X}{\lvert Y\cos\theta\rvert - 1}$	(10*b*)
Limiting refractive index ($X \gg 1$, $Y\sin\varphi \gg 1$)	$\mu^2 \approx \dfrac{X}{Y^2\sin^2\varphi} = \dfrac{\omega_a^2}{\omega_g^2\sin^2\varphi}$	(11*a*)	$\mu^2 = \dfrac{X}{\lvert Y\cos\theta\rvert} = \dfrac{\omega_N^2}{\omega\lvert\omega_H\cos\theta\rvert}$	(11*b*)
Critical propagation angle	$\sin\varphi_{\min} = 1/Y$; $\sin\varphi_{\max} = -1/Y$	(12*a*)	$\cos\theta_{\max} = 1/Y$	(12*b*)
Angle α between wave normal and ray (for long waves)	$\tan\alpha = -\cot\varphi$	(13*a*)	$\tan\alpha \approx 1/2\tan\theta$	(13*b*)
Group velocity for very long waves	$V_g = c\dfrac{\omega_g}{\omega_a}\sin\varphi$	(14*a*)	$V_g = 2c\dfrac{\omega^{\frac{1}{2}}\omega_H^{\frac{1}{2}}}{\omega_N}\cos^{\frac{1}{2}}\theta$	(14*b*)

correspond to those for the transverse propagation of radio waves. The limiting radio case of $\omega_N^2/\omega^2 \gg 1$ and $\omega_H/\omega \gg 1$ represents whistler waves (Helliwell, 1965) and the whistler wave formulae show similarities to the internal atmospheric gravity wave formulae.

There are two differences between gravity waves and whistlers which should be noted. The energy propagation of whistlers tends to be confined to directions about the axis of symmetry which corresponds to the geomagnetic field direction. Energy propagation for gravity waves, however, is confined to directions about the plane perpendicular to the axis of symmetry. As the axis of symmetry for gravity waves is the vertical, their energy propagation is confined to directions about the horizontal plane.

A further distinction between gravity waves and magneto-ionic waves in general is that gravity waves, though anisotropic, are not gyrotropic. There is no double refraction—no splitting into differently polarized components with different phase speeds—and, in the absence of atmospheric rotation effects, the oscillatory motion in any given mode will be confined to the vertical plane of propagation.

The dispersion relation for internal waves when $\text{Im}\,(K_z) = 1/2H$ may be rewritten in terms of ω_a and ω_g and the real parts of the wave-numbers as

$$k_z^2 = k_x^2\left(\frac{\omega_g^2}{\omega^2} - 1\right) + \frac{(\omega^2 - \omega_a^2)}{c^2} \tag{2.4.3}$$

or

$$k_x^2 = \frac{\omega^2}{c^2}\left[\frac{\omega^2 - \omega_a^2 - c^2k_z^2}{\omega^2 - \omega_g^2}\right] \tag{2.4.4}$$

provided that $k_z \neq 0$ and the notation

$$k_x = \text{Real}\,(K_x)$$
$$k_z = \text{Real}\,(K_z)$$

has been used.

Rearranging (2.4.3) into the form

$$\frac{k_x^2}{l^2} + \frac{k_z^2}{m^2} = 1$$

where

$$l^2 = \left(\frac{\omega^2 - \omega_a^2}{\omega^2 - \omega_g^2}\right)\frac{\omega^2}{c^2}$$

and

$$m^2 = \frac{(\omega^2 - \omega_a^2)}{c^2}$$

puts it into the form of the equation for a general conic.

In the real atmosphere, $\gamma < 2$, so that ω_a is always greater than ω_g. In fact, below 200 km altitude the diatomic gases N_2 and O_2 predominate, so that $\gamma = 1{\cdot}4$. Above 400 km monatomic gases predominate, whence $\gamma = 1{\cdot}67$.

The propagation surfaces are plotted in Fig. 2.4. For acoustic waves, $l^2 > 0$ and $m^2 > 0$ and the constant period contours are ellipses. For gravity waves, $l^2 > 0$ but $m^2 < 0$, so that the constant period contours are now hyperbolae.

When $\omega_a > \omega > \omega_g$, the equation (2.4.3) becomes

$$\frac{k_x^2}{|l^2|} + \frac{k_z^2}{|m^2|} = -1$$

which has no values of k_x and k_z that satisfy it and are both real. The only waves that actually do exist in this region are evanescent waves with $k_z = 0$. However, if $k_z = 0$, then $\mathrm{Im}(K_z)$ does not have to be $1/2H$ and therefore equation (2.4.3) need no longer apply for evanescent waves. It should not be thought, however, that evanescent waves are limited to the frequencies $\omega_a > \omega > \omega_g$. Examination of Fig. 2.3 shows that evanescent waves of all frequencies can exist, but that internal waves cannot exist for frequencies $\omega_a > \omega > \omega_g$.

2.5 Constant Horizontal Wind

The next step in my analysis is to consider the atmospheric waves formed on a flat, non-rotating atmosphere when the effects of wind and temperature variations are allowed for. However, in order to leave the problem easily solvable, the background wind vector will be assumed to be a constant wind in the direction of horizontal phase propagation, so that

$$|\mathbf{U}| = U_{0x} = \text{Constant.}$$

One would then intuitively guess that this will produce a Doppler shifted frequency

$$\Omega = \omega - k_x U_{0x}$$

so that Ω should be substituted for ω in the dispersion relation. That this is indeed so will be proved in this section.

The first-order equations become

$$\rho_0 \frac{DU_{1x}}{Dt} = -\frac{\partial p_1}{\partial x}$$

which may be rewritten as

$$\frac{DU_{1x}}{Dt} = -\frac{RT}{M}\frac{\partial}{\partial x}\left(\frac{p_1}{p_0}\right) \tag{2.5.1}$$

$$\rho_0 \frac{DU_{1z}}{Dt} = -\frac{\partial p_1}{\partial z} - g\rho_1.$$

But because

$$\begin{aligned}\frac{\partial}{\partial z}\left(\frac{p_1}{p_0}\right) &= \frac{1}{p_0}\frac{\partial p_1}{\partial z} + \frac{g\rho_0 p_1}{p_0^2} \\ &= \frac{1}{p_0}\frac{\partial p_1}{\partial z} + \frac{g p_1 M}{p_0 RT}\end{aligned}$$

so

$$\frac{DU_{1z}}{Dt} = -g\left(\frac{\rho_1}{\rho_0} - \frac{p_1}{p_0}\right) - \frac{RT}{M}\frac{\partial(p_1/p_0)}{\partial z}. \tag{2.5.2}$$

The continuity equation gives

$$\frac{\partial \rho_1}{\partial t} + U_{0x}\frac{\partial \rho_1}{\partial x} + U_{1z}\frac{\partial \rho_0}{\partial z} + \rho_0\frac{\partial U_{1x}}{\partial x} + \rho_0\frac{\partial U_{1z}}{\partial z} = 0.$$

At this stage it is useful to define an operator

$$\hat{Q} = \frac{\partial}{\partial t} + U_{0x}\frac{\partial}{\partial x}$$

$$= \frac{D}{Dt} - U_{1x}\frac{\partial}{\partial x} - U_{1z}\frac{\partial}{\partial z}.$$

Then for a first-order perturbation

$$\hat{Q}U_{1x} = \frac{DU_{1x}}{Dt}$$

$$\hat{Q}U_{1z} = \frac{DU_{1z}}{Dt}.$$

Now it will be assumed that both the unperturbed temperature and the unperturbed density have no horizontal variation, so that

$$\hat{Q}\left(\frac{\rho_1}{\rho_0}\right) + \frac{\partial U_{1x}}{\partial x} + \frac{\partial U_{1z}}{\partial z} - \left(\frac{gM}{RT} + \frac{1}{T}\frac{\partial T}{\partial z}\right)U_{1z} = 0 \qquad (2.5.3)$$

because

$$\frac{1}{\rho_0}\frac{\partial \rho_0}{\partial z} = -\left[\frac{Mg}{RT} + \frac{1}{T}\frac{\partial T}{\partial z}\right]$$

provided that M and R remain constant. Similarly the first-order perturbed adiabatic equation becomes

$$\frac{\partial p_1}{\partial t} + U_{0x}\frac{\partial p_1}{\partial x} + U_{1z}\frac{\partial p_0}{\partial z} = c^2\left[\frac{\partial \rho_1}{\partial t} + U_{0x}\frac{\partial \rho_1}{\partial x} + U_{1z}\frac{\partial \rho_0}{\partial z}\right]$$

so that

$$\hat{Q}\left(\frac{p_1}{p_0}\right) - \frac{g\rho_0}{p_0}U_{1z} = \frac{\rho_0 c^2}{p_0}\left[\hat{Q}\left(\frac{\rho_1}{\rho_0}\right) - U_{1z}\left(\frac{gM}{RT} + \frac{1}{T}\frac{\partial T}{\partial z}\right)\right].$$

This equation may be rewritten using (2.5.3) as

$$\hat{Q}\left(\frac{p_1}{p_0}\right) + \gamma\left(\frac{\partial U_{1x}}{\partial x} + \frac{\partial U_{1z}}{\partial z}\right) - \frac{gM}{RT}U_{1z} = 0. \qquad (2.5.4)$$

There are now four equations in four unknowns and it is possible to solve these. In the special case of an isothermal atmosphere it was possible first of all to linearize the equations by substituting $-iK_z$ for $\partial/\partial z$, $-iK_x$ for $\partial/\partial x$ and $i\omega$ for $\partial/\partial t$ and then express the linearized equations in matrix form. It would be incorrect to follow the same procedure in an atmosphere with varying temperature. The result obtained would be equivalent to assuming that the scale height was constant, whilst including a non-zero dT/dz from equation (2.5.3).

Therefore, in solving these equations, particular care has to be taken during the manipulations to ensure that the derivatives of the quantities varying with height are included. When this is done, one obtains four second-order differential equations. Alternatively, it is possible to express the parameters in sets of coupled first-order differential equations.

To solve the equations, an exponential variation in the perturbed variables will once again be used. Taking U_{1z} as an example, let

$$U_{1z} = \mathrm{W}(z) \exp\left[i(\omega t - k_x x)\right] \tag{2.5.5}$$

so that

$$\hat{Q} = i(\omega - U_{0x} K_x) \tag{2.5.6}$$

is then linearized. Once again horizontal attenuation or growth of the wave will be neglected, so that K_x may be treated as solely real and will be again denoted by k_x. Then the Doppler-shifted frequency Ω is

$$\begin{aligned} \Omega &= \omega - U_{0x} K_x \\ &= -i\hat{Q} \end{aligned} \tag{2.5.7}$$

and $\partial/\partial x$ will also be a linear operator

$$\frac{\partial}{\partial x} = -ik_x.$$

The procedure to be used in finding the differential equation for U_{1z} is as follows:

STEP 1: Find p_1/p_0 from (2.5.1)

$$\frac{p_1}{p_0} = -\frac{i\hat{Q}}{k_x H g} U_{1x}. \tag{2.5.8}$$

STEP 2: Eliminate p_1/p_0 from (2.5.4)

$$\left(\frac{i\hat{Q}^2}{k_x H g} + i\gamma k_x\right) U_{1x} = \left(\gamma \frac{\partial}{\partial z} - \frac{1}{H}\right) U_{1z}. \tag{2.5.9}$$

STEP 3: Eliminate p_1/p_0 from (2.5.2)

$$\hat{Q}U_{1z} = -g\frac{\rho_1}{\rho_0} + \left(g - gH\frac{\partial}{\partial z}\right)\left(-\frac{i\hat{Q}U_{1x}}{k_x Hg}\right)$$

$$= -g\frac{\rho_1}{\rho_0} - \frac{i\hat{Q}}{k_x H}U_{1x} + \frac{i\hat{Q}}{k_x}\left(\frac{\partial U_{1x}}{\partial z} - \frac{U_{1x}}{H}\frac{dH}{dz}\right). \tag{2.5.10}$$

STEP 4: Eliminate ρ_1/ρ_0 from (2.5.3) and (2.5.10)

$$\left[\frac{\hat{Q}^2}{g} + \frac{1}{H} + \frac{1}{T}\frac{\partial T}{\partial z} - \frac{\partial}{\partial z}\right]U_{1z}$$
$$= \left[\frac{-i\hat{Q}^2}{gk_x H} - \frac{i\hat{Q}^2}{k_x gH}\frac{dH}{dz} - ik_x + \frac{i\hat{Q}^2}{gk_x}\frac{\partial}{\partial z}\right]U_{1x}. \tag{2.5.11}$$

Equations (2.5.9) and (2.5.11) comprise a set of first-order coupled differential equations. They may be solved by substituting (2.5.9) and the derivative of (2.5.9) into (2.5.11) to yield a total differential equation for W

$$\frac{d^2\mathrm{W}}{dz^2} - \frac{1}{H}\left[1 + \frac{k_x^2 c^2}{k_x^2 c^2 - \Omega^2}\frac{dH}{dz}\right]\frac{d\mathrm{W}}{dz}$$
$$+ \mathrm{W}\left[\frac{k_x^2}{\Omega^2}\left\{\frac{(\gamma-1)g^2}{c^2} + \frac{g}{H}\frac{dH}{dz}\right\} + \frac{\Omega^2}{c^2} - k_x^2 + \frac{k_x^2 c^2}{k_x^2 c^2 - \Omega^2}\cdot\frac{1}{\gamma H^2}\frac{dH}{dz}\right]$$
$$= 0. \tag{2.5.12}$$

Similar methods may be used to yield somewhat more complicated second-order differential equations for U_{1x}, p_1/p_0 and for ρ_1/ρ_0. In the isothermal case, all the differential equations become identical and, for example, the equation for W becomes

$$\frac{d^2\mathrm{W}}{dz^2} - \frac{1}{H}\frac{d\mathrm{W}}{dz} + \left[\frac{k_x^2}{\Omega^2}\omega_g^2 + \frac{\Omega^2}{c^2} - k_x^2\right]\mathrm{W} = 0. \tag{2.5.13}$$

If it is now assumed that

$$W(z) = A_0 Z \exp(-iK_z z)$$

then (2.5.13) becomes (2.3.13) with the Doppler-shifted frequency substituted for ω. This agrees with our initial intuitive expectation.

Let us digress and consider what happens if $\Omega = 0$ when $\omega \neq 0$. In this case the group velocity is zero, which means that the atmospheric wave is either reflected or absorbed. Examination of equation (2.5.12) shows that as $\Omega \to 0$ the perturbation amplitude $W \to 0$ so that the wave is absorbed. Hines and Reddy (1967) and Booker and Bretherton (1967) have examined the condition $\Omega = 0$ in the more general case when there are wind shears present. They conclude that the condition $\Omega = 0$ represents absorption of the wave. The place at which this occurs is known as a critical layer or critical level and is analogous to an infinite quantum mechanical potential barrier through which penetration is impossible. If we consider the motion of a wave packet propagating upwards, then the wave packet will approach the critical level for the dominant frequency of the packet, but the wave packet will not reach the critical level in any finite time because the group velocity tends to zero as the packet approaches the critical layer. The wave packet is therefore neither reflected nor transmitted and is absorbed.

A second-order differential equation of the form

$$\frac{d^2W}{dz^2} + f(z)\frac{dW}{dz} + r(z)W = 0 \tag{2.5.14}$$

can be reduced by the standard transformation

$$W = w \exp\left[-\tfrac{1}{2}\int^z f(z)\, dz\right] \tag{2.5.15}$$

to give

$$\frac{d^2w}{dz^2} + q^2 w = 0 \tag{2.5.16}$$

where

$$q^2 = r - \frac{1}{4}f^2 - \frac{1}{2}\frac{df}{dz}. \tag{2.5.17}$$

Equation (2.5.16) is the usual form of the wave equation. The function has an oscillatory type solution when $q^2 > 0$ and an exponential or hyperbolic type solution when $q^2 < 0$. The parameter q may, with some reservations, be interpreted as the vertical component of the wave number. The reservations arise because this procedure will yield a different vertical wave number for each of the parameters U_{1x}, U_{1z}, p_1/p_0, and ρ_1/ρ_0. This point will be considered again in the discussion on wave reflections in § 3.7.

The f(z) term in (2.5.12) arises from the variation of the atmospheric density. The unperturbed pressure has the simple form

$$p(z) = p_g \exp\left\{-\int_0^z \frac{dz}{H}\right\} \tag{2.5.18}$$

but the density in the non-isothermal case assumes the somewhat different form

$$\rho(z) = \rho_g H_g H^{-1} \exp\left\{-\int_0^z \frac{dz}{H}\right\} \tag{2.5.19}$$

where p_g, ρ_g, and H_g are the values of these quantities at $z = 0$. In fact f(z) is

$$\mathrm{f}(z) = \frac{d}{dz}\ln\left(\frac{\rho}{b^2}\right) \tag{2.5.20}$$

where $b^2 = \Omega^2/c^2 - k_x^2$ and then in this notation

$$\mathrm{r}(z) = b^2 - \frac{k_x^2 g}{\omega^2}\left(\mathrm{f}(z) + \frac{g}{c^2}\right). \tag{2.5.21}$$

Provided that the medium varies very slowly, then the higher order derivatives $(dH/dz)^2$ and d^2H/dz^2 may be ignored and the vertical wavelength treated as remaining constant. In this case, one may use the isothermal dispersion relation for internal waves with the substitutions in Table 2.3. The substitutions for ω and ω_g are simple enough. In the non-isothermal case, ω_a is the non-zero value of Ω when $k_x = k_z = 0$. However, a direct substitution into (2.5.12) reveals that ω_a remains unchanged in a non-isothermal atmosphere. This often used assumption is incorrect. In fact we must turn to (2.5.17) and rewrite it as

$$q^2 = \mathrm{r}_0 + \Delta - \omega_{an}^2/c^2$$

where

$$\Delta = \frac{k_x^2 c^2}{k_x^2 c^2 - \Omega^2} \cdot \frac{1}{\gamma H^2} \frac{dH}{dz}$$

$$\mathrm{r} = \mathrm{r}_0 + \Delta$$

and ω_{an} is the acoustic cut-off in the non-isothermal case. Since $\Delta \to 0$ as $k_x \to 0$ and higher order derivatives of H are neglected, then

$$\omega_{an}^2 = \frac{c^2}{4}\mathrm{f}^2 + \frac{c^2}{2}\frac{\mathrm{df}}{dz}$$

$$= \frac{c^2}{4H^2} + \frac{c^2}{2H^2}\frac{dH}{dz}$$

$$= \omega_a^2 + \frac{\gamma g}{2T}\frac{dT}{dz}.$$

The substitutions in Table 2.3 show that it is possible for the Vaisala–Brunt frequency to exceed the acoustic cut-off frequency in a non-isothermal atmosphere. Using the parameters in Table 2.3, this occurs when

$$\frac{dT}{dz} > \frac{(2-\gamma)}{2\gamma}\frac{gM}{R}.$$

Table 2.3 Substitutions in the dispersion relation

Isothermal atmosphere		*Slowly varying atmosphere*
ω	$\rightarrow$	$\Omega = \omega - \mathbf{U}_0 . \mathbf{k}$
ω_g	$\rightarrow$	$\omega_B = \sqrt{\omega_g^2 + \frac{g}{T}\frac{\partial T}{\partial z}}$
ω_a	$\rightarrow$	$\omega_{an} = \sqrt{\omega_a^2 + \frac{\gamma}{2}\frac{g}{T}\frac{\partial T}{\partial z}}$

Thus, if we assume $M = 29$ and $\gamma = 7/5$, then the Vaisala–Brunt frequency exceeds the acoustic cut-off when the temperature gradient exceeds 7·3°K/km. This temperature gradient then provides a rough criterion for a slowly varying atmosphere. If $dT/dz > 7$°K/km, then it is no longer slowly varying and more accurate solutions to the differential equations should be sought, especially for those waves where $\Omega \approx \omega_{an} \approx \omega_{B}$.

Finally, let us use (2.5.9) to derive the vorticity.

$$\zeta_y = \frac{\partial U_{1x}}{\partial z} - \frac{\partial U_{1z}}{\partial x} \tag{2.5.18}$$

of the wave. Because $\partial U_{1z}/\partial x = -ik_x U_{1z}$, this immediately gives

$$\zeta_y = \frac{ik_x}{\Omega^2 - k_x^2 c^2}\left[c^2 \frac{\partial^2 U_{1z}}{\partial z^2} - g\frac{\partial U_{1z}}{\partial z} + (\Omega^2 - k_x^2 c^2) U_{1z}\right]. \tag{2.5.19}$$

There are two special cases when $\zeta_y = 0$. The first is for the case $g = 0$

$$\Omega^2 = c^2(k_x^2 + k_z^2)$$

which corresponds to acoustic waves in a fluid of constant density. The second case is when U_{1z} varies as $\exp(k_x z)$ and $\Omega^2 = k_x g$. (The deep water waves to be discussed in § 4.2.) In these cases the bracket vanishes and $\zeta_y = 0$.

2.6 Wind Divergence Equation

It is occasionally advantageous, especially in meteorological applications, to find the differential equation that describes the wind divergence, rather than the one that describes the vertical velocity component. The wind divergence χ is given by

$$\chi = \frac{\partial U_{1x}}{\partial x} + \frac{\partial U_{1z}}{\partial z} \tag{2.6.1}$$

and it will be necessary to use the y component of the vorticity, ζ_y:

$$\zeta_y = \frac{\partial U_{1x}}{\partial z} - \frac{\partial U_{1z}}{\partial x} \tag{2.6.2}$$

and we define $\nabla^2 = \partial^2/\partial x^2 + \partial^2/\partial z^2$ as a two-dimensional Laplacian, then if one realizes that

$$\nabla^2 U_{1z} = \frac{\partial^2 U_{1z}}{\partial x^2} + \frac{\partial^2 U_{1z}}{\partial z^2} = \frac{\partial \chi}{\partial z} - \frac{\partial \zeta_y}{\partial x}$$

it is possible to write the equations (2.5.1)–(2.5.4) in matrix form in the isothermal case

$$\hat{Q}\begin{bmatrix}\chi \\ \zeta_y \\ \dfrac{\rho_1}{\rho_0}-\dfrac{p_1}{p_0} \\ \dfrac{RT}{M}\nabla^2\left(\dfrac{p_1}{p_0}\right)\end{bmatrix} = B\begin{bmatrix}\chi \\ \zeta_y \\ \dfrac{\rho_1}{\rho_0}-\dfrac{p_1}{p_0} \\ \dfrac{RT}{M}\nabla^2\left(\dfrac{p_1}{p_0}\right)\end{bmatrix} \tag{2.6.3}$$

where

$$B = \begin{bmatrix} 0 & 0 & -g\dfrac{\partial}{\partial z} & -1 \\ 0 & 0 & g\dfrac{\partial}{\partial x} & 0 \\ \gamma-1 & 0 & 0 & 0 \\ -\left[\dfrac{\gamma RT}{M}\nabla^2 - g\dfrac{\partial}{\partial z}\right] & -g\dfrac{\partial}{\partial x} & 0 & 0 \end{bmatrix}. \tag{2.6.4}$$

The dispersion relation is given by $|\hat{Q}I-B|$ where I is the 4×4 unit matrix, and in the non-isothermal case the differential equation for χ may be found in the same way as in the previous section.

Lamb (1945, p. 551) derived this equation in the non-isothermal windless case and found that, when periodic variations in χ are assumed, then

$$\frac{\partial^2\chi}{\partial z^2}-\frac{1}{H}\left(1-\frac{dH}{dz}\right)\frac{\partial\chi}{\partial z} + \left[\frac{\omega^2}{c^2}-k_x^2+\frac{k_x^2}{\omega^2}\left(\frac{g}{H}\frac{dH}{dz}+\frac{(\gamma-1)g^2}{c^2}\right)\right]\chi = 0 \tag{2.6.5}$$

where of course Ω replaces ω in the non-windless case. The sign discrepancy between (2.6.5) and Lamb's formula arises because Lamb takes his vertical coordinate pointing downwards. Martyn (1950) has extended equation (2.6.5) to include the case of arbitrarily varying vertical background winds and temperatures. He also shows that the quantities U_{1x}, U_{1z} and p_1/p_0 are related to the wind divergence by the relations

$$\frac{iG}{k_x c^2} U_{1x} = \left(\Omega^2 - \frac{g}{H} + \frac{g}{H}\frac{dH}{dz}\right)\chi + g\frac{\partial \chi}{\partial z} \tag{2.6.6}$$

$$\frac{G}{c^2} U_{1z} = \left(k_x^2 g - \frac{\Omega^2}{H} + k_x \Omega \frac{\partial U_{0x}}{\partial z}\right)\chi + \Omega^2 \frac{\partial \chi}{\partial z} \tag{2.6.7}$$

and

$$\frac{iG}{\gamma \Omega}\frac{p_1}{p_0} = \left(\Omega^2 - \frac{g}{H} + \frac{g}{H}\frac{dH}{dz}\right)\chi + g\frac{\partial \chi}{\partial z} \tag{2.6.8}$$

where

$$G = k_x^2 g^2 + \left(g k_x \frac{\partial U_{0x}}{\partial z} - \Omega^3\right)\Omega - \frac{\Omega^2 g}{H}\frac{dH}{dz}\ . \tag{2.6.9}$$

These relations may be derived by manipulating the governing equations in much the same manner that was used in § 2.5.

2.7 Energy Densities and the Lagrangian

Given a Lagrangian density $\mathscr{L}$ that is a function of the generalized coordinates q, then D'Alembert's principle of least action leads to Lagrange's equation (Goldstein, 1950)

$$\frac{D}{Dt}\left(\frac{\partial \mathscr{L}}{\partial \dot{q}_j}\right) - \frac{\partial \mathscr{L}}{\partial q_j} = 0 \tag{2.7.1}$$

where q_j is the jth component of q. $\dot{q}$ represents Dq/Dt so that

$$\frac{\partial}{\partial t}\cdot\frac{\partial \mathscr{L}}{\partial(\partial q_j/\partial t)} + \frac{\partial}{\partial x}\cdot\frac{\partial \mathscr{L}}{\partial(\partial q_j/\partial x)}$$

$$+\frac{\partial}{\partial y}\cdot\frac{\partial \mathscr{L}}{\partial(\partial q_j/\partial y)} + \frac{\partial}{\partial z}\cdot\frac{\partial \mathscr{L}}{\partial(\partial q_j/\partial z)} - \frac{\partial \mathscr{L}}{\partial q_j} = 0. \tag{2.7.2}$$

Tolstoy (1963) points out that

$$\mathscr{L} = \frac{1}{2}\rho\left[\left(\frac{\partial\xi}{\partial t}\right)^2 + \left(\frac{\partial\eta}{\partial t}\right)^2 + \left(\frac{\partial\zeta}{\partial t}\right)^2\right] - \frac{1}{2}\lambda\varepsilon^2 + \frac{1}{2}g\zeta^2\left(\frac{d\rho}{dz}\right) + \rho g\zeta\varepsilon \tag{2.7.3}$$

where ξ, η, ζ are the x, y, z, components of the displacement of the fluid element from its equilibrium position and ρ is the density. This means

$$U_{1x} = \frac{\partial\xi}{\partial t}; \qquad U_{1y} = \frac{\partial\eta}{\partial t}; \qquad U_{1z} = \frac{\partial\zeta}{\partial t}$$

and

$$\varepsilon = \frac{\partial\xi}{\partial x} + \frac{\partial\eta}{\partial y} + \frac{\partial\zeta}{\partial z}. \tag{2.7.4}$$

The first term of equation (2.7.3) is the kinetic energy density and the remaining terms represent the scalar potential functions from which may be derived the conservative forces that act. The second term represents the elastic potential energy, where $\lambda = \rho c^2$ is the bulk modulus and ε is the incremental volume change or strain. The third term represents the gravitational potential energy in a gradated incompressible fluid. The last term represents the potential energy associated with the displacement of a compressible fluid.

Thus if one applies the Lagrange equation of motion (2.7.1) to the Lagrangian for the coordinates ξ, η, ζ

$$\left.\begin{aligned} \rho\frac{\partial^2\xi}{\partial t^2} - \frac{\partial}{\partial x}\lambda\varepsilon + \rho g\frac{\partial\zeta}{\partial x} &= 0 \\ \rho\frac{\partial^2\eta}{\partial t^2} - \frac{\partial}{\partial y}\lambda\varepsilon + \rho g\frac{\partial\zeta}{\partial y} &= 0 \\ \rho\frac{\partial^2\zeta}{\partial t^2} - \frac{\partial}{\partial z}\lambda\varepsilon - \rho g\left(\frac{\partial\xi}{\partial x} + \frac{\partial\eta}{\partial y}\right) &= 0. \end{aligned}\right\} \tag{2.7.5}$$

These equations are separable, provided ρ, λ are functions of z only. In localized areas on the surface of a planet, the Coriolis and centrifugal forces may be incorporated directly into the gravitational forces by using the geopotential altitude, so that one needs to add to (2.7.5) the components obtained by expanding

$$2\rho\begin{vmatrix} \mathbf{1}_x & \mathbf{1}_y & \mathbf{1}_z \\ \dfrac{\partial\xi}{\partial t} & \dfrac{\partial\eta}{\partial t} & \dfrac{\partial\zeta}{\partial t} \\ \Omega_{Ex} & \Omega_{Ey} & \Omega_{Ez} \end{vmatrix}$$

where $\mathbf{1}_x$, $\mathbf{1}_y$, $\mathbf{1}_z$ are the x, y, z, unit vectors and $\mathbf{\Omega}_E$ the angular velocity of the planet.

In the case of horizontal plane waves in the absence of Coriolis forces

$$\left.\begin{aligned} \rho\frac{\partial^2\xi}{\partial t^2}-\frac{\partial}{\partial x}\lambda\varepsilon+\rho g\frac{\partial\zeta}{\partial x} &= 0 \\ \rho\frac{\partial^2\zeta}{\partial t^2}-\frac{\partial}{\partial z}\lambda\varepsilon-\rho g\frac{\partial\xi}{\partial x} &= 0 \end{aligned}\right\} \quad (2.7.6)$$

if it is assumed that the x-axis lies in the direction of motion. Tolstoy (1963), in an appendix, proves that equations (2.7.6) are equivalent to the first-order perturbation form of Euler's equation of fluid dynamics. The main advantage in expressing the equations in the Lagrangian form is that they do not require the introduction of thermodynamical concepts, and they allow the actual particle motion for each class of wave motion to be directly determined.

By assuming that both ξ and ζ are proportional to $\exp[i(\omega t-k_x x)]$ then

$$\varepsilon=\frac{\partial\xi}{\partial x}+\frac{\partial\zeta}{\partial z}=-ik_x\xi+\frac{\partial\zeta}{\partial z}$$

and then by assuming that λ and ρ are functions of z only, (2.7.6) becomes

$$\left.\begin{aligned} -\rho\omega^2\xi+\lambda k_x^2\xi+\lambda ik_x\frac{\partial\zeta}{\partial z}-\rho gik_x\zeta &= 0 \\ \rho\omega^2\zeta+\frac{\partial\lambda}{\partial z}\left(-ik_x\xi+\frac{\partial\zeta}{\partial z}\right)+\lambda\left(-ik_x\frac{\partial\xi}{\partial z}+\frac{\partial^2\zeta}{\partial z^2}\right)-\rho gik_x\xi &= 0. \end{aligned}\right\} \quad (2.7.7)$$

Since $\lambda = \rho c^2$, (2.7.7) gives

$$\xi = -ik_x \frac{c^2 \dfrac{\partial \zeta}{\partial z} - g\zeta}{k_x^2 c^2 - \omega^2} \tag{2.7.8}$$

which is exactly equivalent to (2.5.9). Substituting into (2.7.7) gives a second-order differential equation of the form

$$\begin{aligned} &\frac{\partial^2 \zeta}{\partial z^2} + \left\{\frac{d}{dz} \ln (\rho/b^2)\right\} \frac{\partial \zeta}{\partial z} \\ &+ \left(b^2 - \frac{k_x^2}{\omega^2} g \frac{d}{dz} \ln \left(\frac{\rho}{b^2}\right) - \frac{k_x^2}{\omega^2} \frac{g^2}{c^2}\right) \zeta = 0. \end{aligned} \tag{2.7.9}$$

where $b^2 = \omega^2/c^2 - k_x^2$. This second-order differential equation is equivalent to (2.5.12). If the speed of sound is assumed to be a constant (i.e. an isothermal atmosphere) then $d[-\ln (b^2)]/dz = 0$ so that the equation becomes

$$\frac{\partial^2 \zeta}{\partial z^2} + \left\{\frac{d (\ln \rho)}{dz}\right\} \frac{\partial \zeta}{\partial z} + \left(\frac{\omega^2}{c^2} - k_x^2 + \frac{k_x^2 \omega_B^2}{\omega^2}\right) \zeta = 0 \tag{2.7.10}$$

because the Vaisala–Brunt frequency

$$\omega_B^2 = -g\left[\frac{d(\ln \rho)}{dz} + \frac{g}{c^2}\right]. \tag{1.4.5}$$

Now (2.7.10) can be transformed into the wave equation

$$\frac{\partial^2 h}{\partial z^2} + k_z^2 h = 0$$

provided

$$\zeta = h\rho^{-\frac{1}{2}}$$

and

$$\begin{aligned} k_z^2 = \frac{\omega^2}{c^2} - k_x^2 + \frac{k_x^2}{\omega^2} \omega_B^2 - \frac{1}{4}\left(\frac{d(\ln \rho)}{dz}\right)^2 \\ -\frac{1}{2} \frac{d^2(\ln \rho)}{dz^2} \end{aligned}$$

then if ρ is an exponential function of z

$$\rho = \rho_g \exp\left(-\frac{z}{h}\right)$$

then ω_B^2 is a constant, $d^2(\ln \rho)/dz^2 = 0$ and

$$k_z^2 = \frac{\omega^2}{c^2} - k_x^2 + \frac{k_x^2}{\omega^2}\omega_B^2 - \frac{1}{4H^2}$$

which is the same as equation (2.4.3)

The Lagrangian density can also be used as the starting point for an attack on the fluid-mechanical energy-momentum tensor (Jones, 1971). This tensor formulation provides single generalized expressions for the conservation of energy, momentum and angular momentum and enables one to calculate the energy and momenta exchange between the wave and the background.

2.8 Polarization Relations

The time has come to return to the complex quantities P, R, X, Y and Z that were defined in equation (2.1.4). These terms give the phase difference between the sinusoidal pressure variations, the density variations, and the components of the velocity variations. To find these coefficients in the non-isothermal case, it is necessary to use algebraic manipulations, in much the same way as was done in § 2.5. For example, equation (2.5.8) gives the phase between p_1/p_0 and U_{1x}, so that in terms of the polarizations

$$\frac{\mathrm{P}}{\mathrm{X}} = \frac{\Omega}{k_x g H}. \tag{2.8.1}$$

It is rather cumbersome to calculate the other ratios by this method. However, if the governing equations can be linearized and expressed in matrix form, then it is a reasonably simple matter to find the polarization relations. Equation (2.3.12) showed that this could be done in an isothermal atmosphere. Let us now move forward one more step and consider a 'pseudo-thermal' atmosphere (Beer, 1972c). The characteristics of this atmosphere are that the atmosphere is taken as isothermal ($H = T =$ constant) except when the temperature gradient appears explicitly in one of the governing equations, as it does in equation (2.5.3). The results that are obtained offer some insight into the way that temperature gradients affect the isothermal polarization relations, but should not be regarded as being exact solutions.

Thus, to find the air motions when the Coriolis forces are neglected, the two-dimensional equation*

$$\begin{bmatrix} i\Omega & 0 & 0 & -iK_x gH \\ 0 & i\Omega & g & -iK_z gH - g \\ -iK_x & -iK_z - \frac{1}{H} - \frac{1}{T}\frac{\partial T}{\partial z} & i\Omega & 0 \\ -i\gamma K_x & -i\gamma K_z - \frac{1}{H} & 0 & i\Omega \end{bmatrix} \begin{bmatrix} \mathrm{X} \\ \mathrm{Z} \\ \mathrm{R} \\ \mathrm{P} \end{bmatrix} = \mathbf{0} \qquad (2.8.2)$$

can be used. Since the set of equations represented by (2.8.2) are linear and homogeneous, an infinite number of linearly related solutions exist. This means that waves of any amplitude satisfy the equations, provided that the perturbation approximation—that the first-order quantities are small—remains valid. The wave amplitudes in the atmosphere are determined by source terms which may be expected to render the equations inhomogeneous, so that the equations that have been derived in this chapter will be strictly valid only in regions distant from the atmospheric wave source.

To find the polarization relations, which give the ratios of the four unknown variables, it is necessary to treat one of the variables as known and then to use the other three equations. There are four different ways of doing this, all of which will ultimately give the same answer. The easiest combination is to treat Z as known and to solve

$$\begin{bmatrix} i\Omega & 0 & -iK_x gH \\ -iK_x & i\Omega & 0 \\ -i\gamma K_x & 0 & i\Omega \end{bmatrix} \begin{bmatrix} \mathrm{X} \\ \mathrm{R} \\ \mathrm{P} \end{bmatrix} = \mathrm{Z} \begin{bmatrix} 0 \\ iK_z + \frac{1}{H} + \frac{1}{T}\frac{\partial T}{\partial z} \\ i\gamma K_z + \frac{1}{H} \end{bmatrix} \qquad (2.8.3)$$

* It should be realized that (2.8.2), in the isothermal windless case, and (2.3.12) do not have the same elements in the last row of the 4 × 4 matrix, because the other form of the continuity equation was chosen this time. This illustrates the fact that the value of a determinant is unchanged when one multiplies a row (or column) by a constant, γ in this case, and adds the elements so formed to any other row (or column).

for P/Z, R/Z and X/Z. By Cramer's rule, the denominators of the three solutions are always the determinant of the coefficient matrix, so that it is proportional to Z

$$Z \propto \Omega^3 - c^2 K_x^2 \Omega \tag{2.8.4}$$

Then

$$X \propto \Omega c^2 K_x K_z - i\Omega g K_x \tag{2.8.5}$$

$$R \propto \Omega^2 K_z - i\Omega^2 \left(\frac{\gamma g}{c^2} + \frac{1}{T}\frac{\partial T}{\partial z}\right) + iK_x^2 \left[g(\gamma - 1) + \frac{c^2}{T}\frac{\partial T}{\partial z}\right] \tag{2.8.6}$$

$$P \propto -i\Omega^2 \gamma g/c^2 + \Omega^2 \gamma K_z \tag{2.8.7}$$

and it may be shown that if $\chi = DA_0 \exp[i(\omega t - K_x x - K_z z)]$ that

$$D \propto -\Omega g K_x^2 - i\Omega^3 K_z. \tag{2.8.8}$$

Let us now check to see how good an approximation the pseudo-thermal atmosphere is to a real non-isothermal atmosphere. Equations (2.8.5) and (2.8.7) do actually reproduce the exact form (2.8.1), so that P and X are accurate. From (2.5.9) the exact form

$$\frac{X}{Z} = \frac{K_x g + ic^2 K_x K_z}{i(\Omega^2 - c^2 K_x^2)} \tag{2.8.9}$$

shows that Z is accurate. Equation (2.8.6) which gives R is, however, lacking some terms incorporating the temperature gradient. Nevertheless, equation (2.8.6) provides a simply derivable and useful approximation.

Following the success of the pseudo-thermal approach in the treatment of the polarization relations, one may be tempted to use it to derive the dispersion relation. In this case, however, the pseudo-thermal treatment is unsatisfactory. It yields a second-order differential equation

$$\frac{d^2W}{dz^2} - \frac{1}{H}\frac{dW}{dz} + \left[\frac{k_x^2}{\Omega^2}\omega_B^2 + \frac{\Omega^2}{c^2} - k_x^2 - \frac{g}{c^2 T}\frac{\partial T}{\partial z}\right] W = 0 \tag{2.8.10}$$

in which ω_B is always greater than ω_{an} and the difference $\omega_B^2 - \omega_{an}^2$ remains constant. Both the isothermal treatment with ω_B substituted for ω_g and the pseudo-thermal treatment mask many of the complexities of the dispersion relation in a real atmosphere.

The actual motion of the air parcels is given by the relation (2.8.9) between X and Z. For high-frequency acoustic waves $\omega/K_x \rightarrow c$ and so $Z \rightarrow 0$, so that these waves are longitudinal. The air parcel orbits for gravity waves can be found by letting the terms in c^2 dominate in the expression for Z. This is a good approximation when the compressibility effects are small, as they are for gravity waves.

Therefore, neglecting the exponential spatial variation

$$\left.\begin{aligned} U_{1z} &\propto -A_0 c^2 K_x^2 \Omega \exp(i\Omega t) \\ \text{and} \quad U_{1x} &\propto (A_0 \Omega c^2 K_x K_z - i\Omega g K_x) \exp(i\Omega t). \end{aligned}\right\} \tag{2.8.11}$$

By measuring phases relative to U_{1z} it is possible to write $U_{1z} = \mathrm{K} \cos \Omega t$ and so

$$U_{1x} = \mathrm{K}\left[\frac{\left(\Gamma - \dfrac{g}{c^2}\right)}{k_x}\right] \sin \Omega t - \frac{k_z}{k_x} \cos \Omega t. \tag{2.8.12}$$

where $\Gamma = \mathrm{Im}(K_z)$. By eliminating the time,

$$\left(U_{1x} + \frac{k_z}{k_x} U_{1z}\right)^2 = \left[\frac{\left(\Gamma - \dfrac{g}{c^2}\right)}{k_x}\right]^2 (1 - U_{1z}^2). \tag{2.8.13}$$

Now in an isothermal atmosphere

$$\omega_a / c = \Gamma$$

so that, by using the dispersion relation in the form (2.4.3) with ω_{B} substituted for ω_g and once again noting that $\Omega^2 \ll c^2 k_x^2$, then the equation for the particle velocities may be written as

$$U_{1x}^2 + U_{1z}^2\left(\frac{\omega_{\mathrm{B}}^2}{\Omega^2} - 1\right) + \frac{2k_z}{k_x} U_{1x} U_{1z} = \left[\frac{\left(\Gamma - \dfrac{g}{c^2}\right)}{k_x}\right]^2. \tag{2.8.14}$$

The equations for the particle orbits will be of the same form, because the particle displacements (the ξ and ζ of §2.7) will be given by the time integrals of the particle velocities.

Equation (2.8.14) is the equation of an ellipse when $\Omega < \omega_{\mathrm{B}}$. The motion becomes purely vertical in the long wavelength limit when $k_x \to 0$ and so $X \to 0$ and also in the short wavelength limit at the resonant Vaisala–Brunt frequency when $\Omega \to \omega_B$ and $k_x \to \infty$.

For purely horizontal propagation ($k_z = 0$), when $\Omega = \omega_B/\sqrt{2}$ then the air particle orbits are circular, and so the gravity waves of this frequency are circularly polarized. For $\omega_B > \Omega > \omega_B/\sqrt{2}$, the motion is elliptical with a longer vertical axis than a horizontal one, whereas for low frequencies, $\Omega < \omega_B/\sqrt{2}$, the horizontal axis is longer.

Fig. 2.6 illustrates typical air-parcel orbits for gravity waves superimposed on the propagation diagram (Fig. 2.4). Midgley and

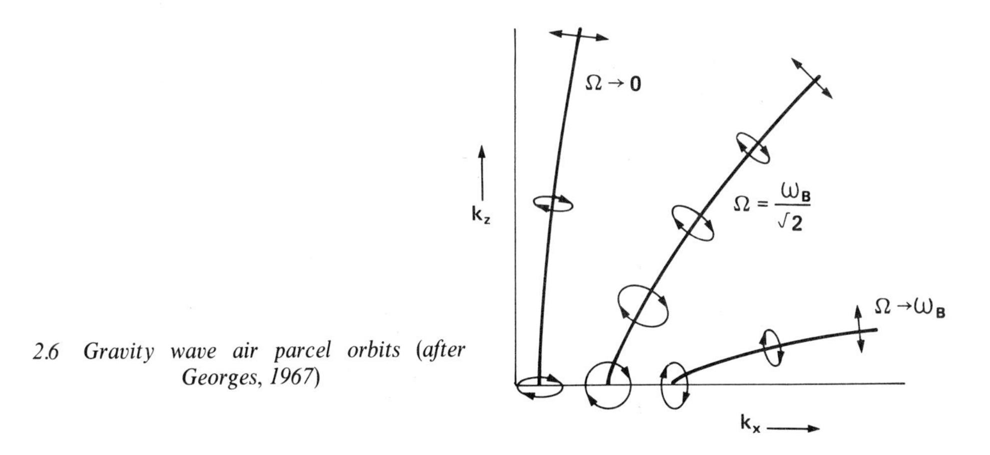

2.6 Gravity wave air parcel orbits (after Georges, 1967)

Liemohn (1966) have also discussed the motion of air packets in isothermal non-dissipative atmospheres and they have produced diagrams like Fig. 2.6 for evanescent and internal waves.

The introduction of the earth's rotation into the equations of motion would require the solution of a three-dimensional problem, since different x and y dependences of the forces would be introduced. Eckart (1960, p. 124) has shown that on a flat earth when $\omega_B > 2\Omega_E$, where Ω_E denotes the local angular frequency of the earth's rotation, then internal atmospheric waves cannot propagate if their frequency $\Omega < 2\Omega_E$. On earth, this means that in a plane atmosphere there will be a low-frequency cut-off at 12 hours (see § 5.1). There is no paradox in the co-existence of this important result and long-period (>24 hour) Rossby waves. Rossby waves are a result of Coriolis parameter variations that arise from the earth's curvature and may be evanescent in the vertical. On a rotating sphere, the cut-off occurs when $\Omega = 2\Omega_E \cos\theta$, where θ is the co-latitude.

In the rather unusual case when $\omega_B < 2\Omega_E$—termed subcritical stability by Eckart—internal gravity waves exist in a plane atmosphere provided $2\Omega_E > \Omega > \omega_B$, though, for waves of large horizontal wavelength, waves with $\Omega < \omega_{an}$ will not exist. The elliptical air parcel orbits are no longer in a vertical plane. Tolstoy (1963) calls these waves gyroscopic waves and he finds that there is only one permissible direction of propagation.

Complex Polarization Terms

We can easily see that most of the polarization terms are complex quantities. This enables one to find the phase relation between any two wave parameters. Any complex number, $A+iB$ say, can be written in exponential form as $C \exp(i\phi)$, where $C = \sqrt{(A^2+B^2)}$ and $\tan\phi = B/A$. For example, since

$$\frac{X}{Z} = \frac{(c^2K_z - ig)K_x}{(\Omega^2 - c^2K_x^2)}$$

if K_x is wholly real and if $K_z = k_z + i/2H$, then the phase difference between X and Z, which corresponds to the phase difference between U_{1x} and U_{1z}, is given by $\tan^{-1}[(\gamma-2)/(2\gamma k_z H)]$. For evanescent waves, when $k_z = 0$, the two motions are exactly 90° out of phase, so that the particle motion is circularly polarized.

2.9 Phase and Group Velocity

The phase and group velocities for Rossby waves were discussed in §2.2. For gravity and acoustic waves, one can use the dispersion diagrams (Figs. 2.2 and 2.3) to analyse qualitatively the vertical and horizontal phase and group velocities respectively. The most significant observation to be made comes from Fig. 2.2, where it may be seen that, for gravity waves, the vertical phase and group velocities are oppositely directed. Even though the vertical phase velocity $\omega/k_z > 0$, the vertical group velocity $\partial\omega/\partial k_z < 0$. Similarly, if $\omega/k_z < 0$, then $\partial\omega/\partial k_z > 0$. The horizontal gravity wave phase and group velocities are, however, in the same direction.

The group velocity represents the direction of energy flow (except in highly dispersive media), whereas the phase velocity is the observed movement of the wave peaks and troughs. Thus, if the energy that sustains internal atmospheric gravity waves has propagated upwards, then observations of the waves themselves would show a downward movement. This appears to explain the downward drift of travelling ionospheric disturbances. Internal gravity waves that are launched in the lower atmosphere propagate to ionospheric heights. At these heights, the wind variations that accompany the waves have become quite large and these winds redistribute the ionization into what is observed as an ionospheric irregularity which moves with the phase velocity of the wave. Radar observations of the irregularity then show a downward drift.

To calculate the phase and group velocities we must use the dispersion relation (2.3.13). Evanescent waves will be discussed in the next section, so that only internal waves will be considered here. Then the horizontal phase velocity in the isothermal case is

$$V_{px} = c\left[\frac{\Omega^2 - \omega_B^2}{\Omega^2 - \omega_a^2 - c^2 k_z^2}\right]^{\frac{1}{2}} \tag{2.9.1}$$

directly from (2.4.4) and the horizontal group velocity is

$$\begin{aligned} V_{gx} &= \frac{c(\Omega^2 - \omega_B^2)^{\frac{3}{2}}(\Omega^2 - \omega_a^2 - c^2 k_z^2)^{\frac{1}{2}}}{\omega_B^2(\omega_a^2 + c^2 k_z^2 - 2\Omega^2) + \Omega^4} \\ &= \frac{c^2 V_{px}(\Omega^2 - \omega_B^2)}{\Omega^2 V_{px}^2 - \omega_B^2 c^2}. \end{aligned} \tag{2.9.2}$$

Similar equations could be obtained for the vertical phase and group velocity. For internal waves this is hardly necessary at this stage. The important result in this case is that the vertical phase and group velocities of gravity waves are oppositely directed, whereas for acoustic waves the vertical group velocity can never exceed the vertical phase velocity.

For gravity waves, their horizontal phase and group velocities are always less than the speed of sound. As $\Omega \to 0$, they both approach the same limit

$$V_{px} = V_{gx} = c\omega_B/\sqrt{\omega_a^2 + c^2 k_z^2}$$

so that, in an isothermal atmosphere, very low-frequency waves have both phase and group velocities that move at less than 0·9 c.

For acoustic waves, both V_{px} and V_{gx} approach the sound speed in the high-frequency limit. This is to be expected. However, as $\Omega \to \omega_a$, the horizontal phase and group velocities of the acoustic waves depend on k_z. The true phase velocity (in the direction of the

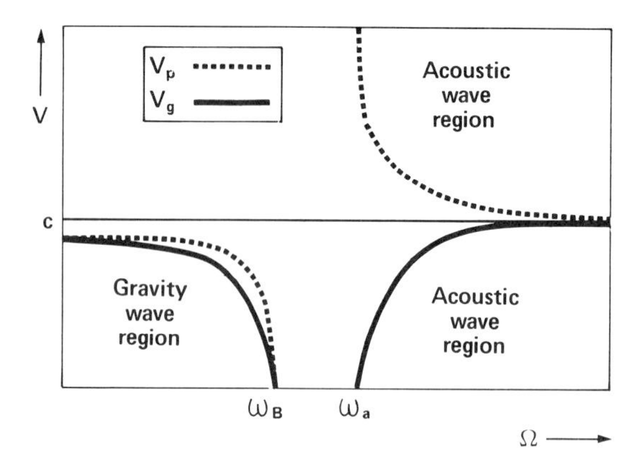

2.7 Internal wave phase and group velocities (after Georges, 1967)

wave normal) becomes infinite and the group velocity approaches zero as $\Omega \rightarrow \omega_a$. Fig. 2.7 illustrates the domains of the group and phase velocities in the (V, Ω) plane.

The true phase velocity V_p is given, in terms of the trace velocities V_{px} and V_{pz}, as

$$\frac{1}{V_p^2} = \frac{1}{V_{px}^2} + \frac{1}{V_{pz}^2} \tag{2.9.3}$$

and is illustrated in Fig. 2.8. V_{px} and V_{pz} are not actually components of V_p because

$$k^2 = k_x^2 + k_z^2$$

so that $\lambda^2 \neq \lambda_x^2 + \lambda_z^2$ but instead

$$\frac{1}{\lambda^2} = \frac{1}{\lambda_x^2} + \frac{1}{\lambda_z^2}. \tag{2.9.4}$$

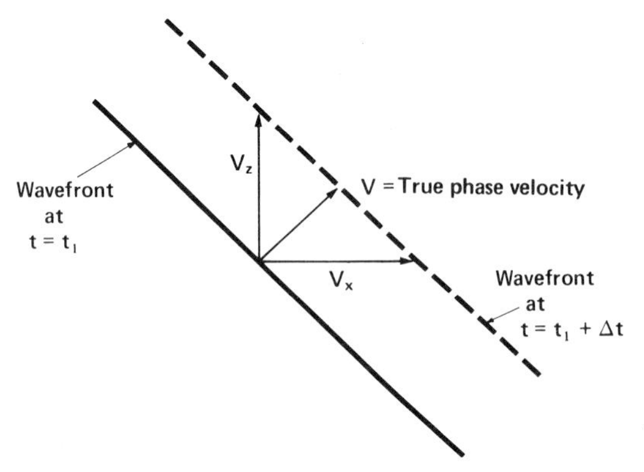

2.8 The relation between the trace velocities and the true velocity

Relations (2.9.3) and (2.9.4) are extremely important when microbarograph records are being analysed. It is common to use an array of three microbarographs placed at the vertices of a right-angled triangle. The time delay of a correlated record gives the horizontal trace velocities. To obtain V_{px} the two velocities have to be combined by using an equation of the form (2.9.3).

2.10 External Waves

External waves may be divided into two types on the basis of energy storage. For any wave to exist there must be two or more types of energy storage, between which the energy of the wave may be repeatedly exchanged. The term surface wave is then reserved for those waves in which one of these means of energy storage exists only at a surface or discontinuity of the medium. At the interface between two different fluids, there will be a discontinuity in density; the surface waves that form are called Helmholtz waves. The capillary waves formed by the restoring force of the surface tension at the surface of a liquid are a type of Helmholtz wave. On the other hand, if the discontinuity is in some parameter other than the density (e.g. temperature, wind speed), then the surface waves that form may be referred to as boundary waves.

Evanescent waves are waves that exist within the body of the fluid rather than merely at a surface. The best-known example of the evanescent wave is the Lamb wave, which is the non-dispersive wave that satisfies the acoustic relation:

$$\omega = ck_x \tag{2.10.1}$$

at all frequencies in an isothermal atmosphere. Lamb waves are a direct consequence of the lower boundary condition imposed by the rigid earth's surface. Like all evanescent waves, most of the energy for the Lamb wave is concentrated near the edge of the discontinuity comprising the boundary. For this reason, Garrett (1969) has coined the term edge wave to describe the equivalent of Lamb waves in non-isothermal atmospheres containing winds.

Cellular waves are standing waves that are formed by the reflections at two fixed boundaries. If the boundaries are taken to be at altitudes $z = 0$ and $z = h$, then, because the wave function must be zero everywhere at these two altitudes, the wave function for the cellular waves that are formed is

$$A_0 \sin\left(\frac{2\pi}{h}z\right) \exp i[\omega t - k_x x - i\,\mathrm{Im}\,(K_z)z].$$

A non-contradictory nomenclature for atmospheric waves with no vertical phase variation is shown on the endpapers of this book. Internal waves are included in this diagram because it is possible for internal waves—which are characterized in an isothermal atmosphere by $\mathrm{Im}\,(K_z) = 1/2H$—to have $k_z = 0$. This class of internal waves marks the boundaries of the internal wave regions on the dispersion diagrams.

The inclusion of boundary conditions, or source terms* means that there is no longer a region representing evanescent waves on the dispersion diagram, and the criterion that $k_z = 0$ is represented by a line on the dispersion diagram. One line that appears frequently has the equation

$$\Omega^2 = gk_x\sqrt{\gamma - 1} \tag{2.10.2}$$

in an isothermal atmosphere, or more generally

$$\Omega^2 = c\omega_B k_x \tag{2.10.3}$$

in a non-isothermal atmosphere. This equation resembles the equation for deep water surface gravity waves ($\Omega^2 = gk_x$) and so the wave represented by (2.10.3) will be referred to as the characteristic surface wave.

The characteristic surface wave always separates the acoustic wave domain from the gravity wave domain, the reason for this being that equation (2.10.3) represents the equation for an atmospheric wave with infinite impedance. Because of this it is impossible for a gravity wave in one region of the atmosphere to couple into an acoustic wave in a region of the atmosphere with different properties. Similarly the evanescent wave region can be divided into non-interacting acoustic and gravity wave portions so that an evanescent wave may interact with an internal wave of the same type.

The characteristic surface wave is elliptically polarized. By using the results of § 2.8, it may be seen that for all evanescent waves Z is

* See the discussion on coupling in § 3.8.

purely real and X is purely imaginary, so that evanescent waves are elliptically polarized—with one exception.

The exception is the Lamb wave. Lamb waves have no vertical variation, so that $Z = 0$ at all heights. Thus

$$\omega/k_x = c \tag{2.10.4}$$

and Lamb waves may then be recognized as being the body acoustic waves propagating in the medium. Because Z = 0, they are linearly polarized. Substituting into the dispersion relation (2.3.13) we find that for Lamb waves

$$\mathrm{Im}\,(K_z) = \frac{g}{2c^2}[\gamma \pm (\gamma - 2)] \tag{2.10.5}$$

Two types of Lamb waves exist both of which have polarization relations given by $P = \gamma$, $R = 1$, $X = c$ and $Z = 0$. Of these two solutions, Lamb waves of the second type, which satisfy $\mathrm{Im}\,(K_z) = g/c^2$ are the waves extant below an energy source. They do not exist if the energy flow is completely upwards. One important property of the remaining Lamb waves is that the total wave energy density, $\mathscr{E}$ in an infinite column of air is finite and in an isothermal atmosphere (neglecting potential energy)

$$\begin{aligned} \mathscr{E} &\propto \tfrac{1}{2}\rho_0 U_{1x}^2 \\ &\propto \int_0^\infty \exp\,[(\gamma - 2)g/c^2]\,dz \\ &\propto c^2/(\gamma - 2)g \end{aligned} \tag{2.10.6}$$

provided $\gamma < 2$.

The curves (2.10.2) and (2.10.4) intersect in the $\omega - k_x$ plane (Fig. 2.9) at the Vaisala–Brunt frequency, when

$$k_x = \sqrt{\gamma - 1}g/c^2 \tag{2.10.7}$$

so that it may be possible for there to be an energy interaction between the characteristic surface waves and Lamb waves. However, one may immediately rule out direct interaction between the linearly polarized Lamb waves and the elliptically polarized characteristic surface waves.

Pitteway and Hines (1965) display a version of the propagation

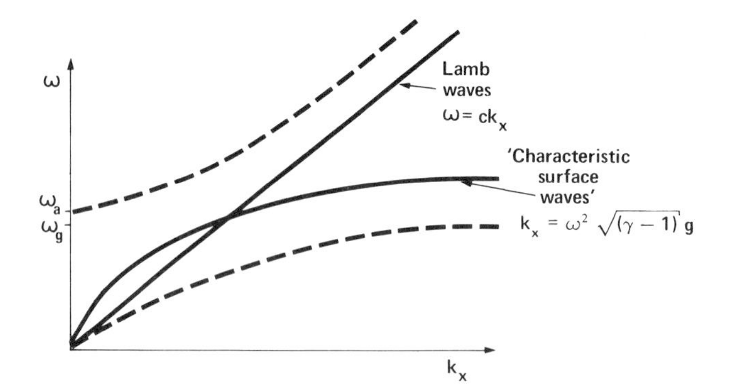

2.9 The dispersion diagram for Lamb waves and surface waves

diagram which is extremely useful for comparing the properties of internal and external waves. They define the refractive indices

$$n_x = ck_x/\Omega$$

$$n_z = ck_z/\Omega \qquad (2.10.8)$$

and plot constant Ω curves in the (n_x^2, n_z^2) plane (Fig. 2.10). The constant Ω curves in an isothermal atmosphere become straight lines. The region below the n_z^2 axis corresponds to imaginary n_z, which represents evanescent waves. The line joining (1, 0) and (0, 1) represents $\Omega = \infty$ and corresponds to acoustic waves in the high frequency limit. No waves—internal, external or body waves—can exist in the shaded region of Fig. 2.10, just as no waves are possible in the shaded region of Fig. 2.2. The refractive index plots are extremely useful for dealing with isothermal atmospheres. They do however mask certain properties of non-isothermal atmospheres

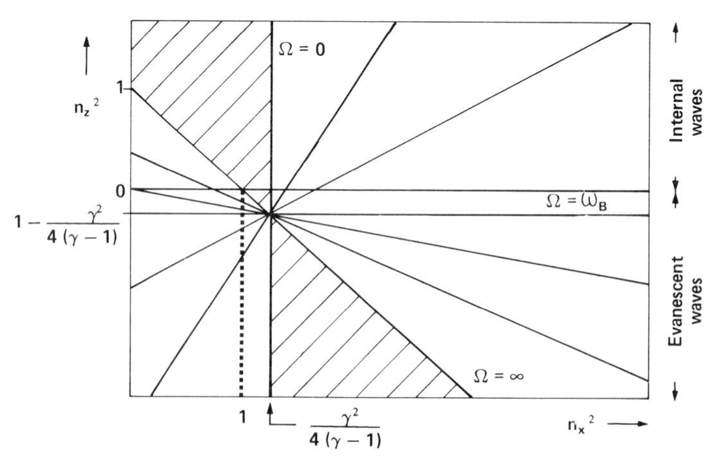

2.10 Constant period curves in the squared refractive index domain (after Georges, 1967 and Pitteway and Hines, 1965; the latter used by permission of the National Research Council of Canada)

(Johnston, 1967) and, in that case, wave vector plots of k_x against k_z are to be preferred (see § 7.3).

One may easily see that for isothermal atmospheres, as $\Omega \to 0$, all waves have the same horizontal phase velocity, which is

$$V_{px} = 2c\sqrt{\gamma-1}/\gamma$$
$$\doteqdot 0{\cdot}9c \qquad \text{if } \gamma = 1{\cdot}4.$$

As Ω increases, the horizontal phase velocity of internal waves decreases. The horizontal phase velocity of evanescent waves varies in a more complicated manner that is depicted in Fig. 2.11. The

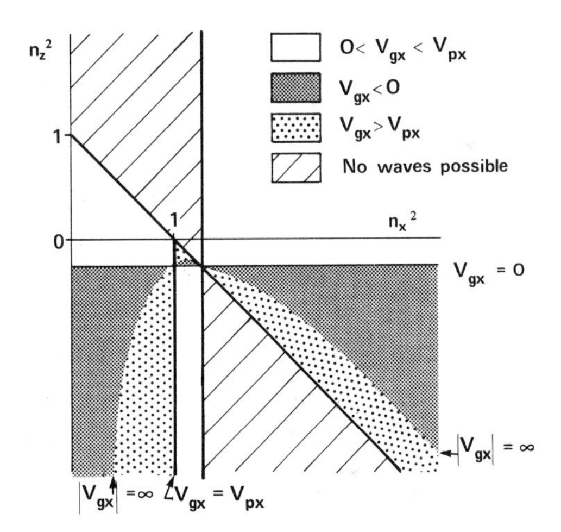

2.11 Horizontal phase and group velocity relationship in the squared refractive index domain (after Pitteway and Hines, 1965; used by permission of the National Research Council of Canada)

phase and group velocities for isothermal atmospheric waves may be written

$$V_{px} = c/n_x \tag{2.10.9}$$

$$V_{gx} = \frac{V_{px}(\Omega^2 - \omega_B^2)}{\left(\dfrac{\Omega^2}{n_x^2} - \omega_B^2\right)}. \tag{2.10.10}$$

Then one may see that for certain waves (which turn out to all be evanescent), $V_{gx} > V_{px}$ and that V_{gx} can become infinite. This occurs when

$$\Omega^2 = ck_x\omega_B \tag{2.10.11}$$

which is the characteristic surface wave. Its vertical amplification is

$$\operatorname{Im}(K_z) = \frac{\gamma g \pm \sqrt{\gamma^2 g^2 + 4k_x^2 c}}{2c^2}. \tag{2.10.12}$$

The physical interpretation of the result when $V_{gx} \rightarrow \infty$ is still somewhat unclear. However, it may be possible to draw an analogy with Lamb's famous fish-line problem (Lamb, 1945, p. 468). When an obstacle moves in still water, waves form in front of the obstacle. These bow waves remain stationary relative to the obstacle, so that the bow waves have a phase velocity equal to the obstacle's velocity, The group velocity of the bow waves must, however, exceed the obstacle's velocity in order to transfer the energy to a point in front of the obstacle. This would indicate that the evanescent waves of large group velocity are related to the movement of high impedance atmospheric 'obstacles' such as meteors, aeroplanes, etc. When the group velocity becomes infinite then the possibility of instability exists. This is discussed in the book by Clemmow and Dougherty (1969, Chapter 6) and is exemplified by the material in § 7.3.

No consensus has emerged on the importance of evanescent waves in the upper atmosphere, though they are liable to be of some consequence at meteorological heights. In the troposphere, Scorer (1953) has shown that evanescent waves may be generated by a wind shear with a gradual variation of the horizontal wind velocity in the vertical direction. At ionospheric heights, the evidence for the existence of evanescent waves is more dubious. Fogle and Haurwitz (1966) suggested that noctilucent clouds may be surface waves on a temperature discontinuity at the mesopause, and Jones and Maude (1965, 1972) and Thome (1966) present some evidence for their existence in the *E* and *F* regions respectively.

Lamb and edge waves, on the other hand, only exist above the stratosphere as forced oscillations (Lindz and Blake, 1972). Over 99 per cent of the energy of an edge wave is located below 110 km (Garrett, 1969). However, it is apparent that most of the distinct part of the microbarograph observations of distinct pressure pulses is due to edge waves (Bretherton, 1969b; Garrett, 1969). This includes the pulses generated by such events as the 1883 Krakatoa eruption, which seems to have prompted Lamb's original work (Lamb, 1945, p. 553), the 1908 Siberian meteorite impact, the 1964 Alaskan earthquake, and the artificial detonation of large nuclear weapons in the atmosphere.

2.11 Boundary Waves

Let us find the waves formed at the boundary between two isothermal layers of the atmosphere with different densities and temperatures. Let the origin be in the interface and let the subscript l refer to the lower layer ($z < 0$) and the subscript u refer to the upper layer ($z > 0$).

At the boundary, continuity of pressure of the vertical component of velocity is required. The pressure and velocity on either side of the boundary cannot simply be equated at $z = 0$, because the boundary is free to move. Even though this motion is arbitrarily small, there is a zero-order gradient of pressure in the vertical direction (dp_0/dz) and the boundary condition must be applied so as to account for this.

The appropriate conditions are

$$\left[\left(\frac{Dp}{Dt}\right)_l = \left(\frac{Dp}{Dt}\right)_u\right]_{z=0} \tag{2.11.1}$$

and

$$\left[\left(\frac{DU_z}{Dt}\right)_l = \left(\frac{DU_z}{Dt}\right)_u\right]_{z=0} \tag{2.11.2}$$

where

$$\frac{Dp}{Dt} = \frac{\partial p_1}{\partial t} + U_{1z}\frac{\partial p_0}{\partial z}$$

and

$$\frac{DU_z}{Dt} = \frac{\partial U_{1z}}{\partial t}$$

when we assume that the background parameters are in a steady state.

Applying (2.1.4) to (2.11.1) reveals that

$$\frac{\rho_l c_l^2(i\omega^2 K_{zl} + K_x^2 g)}{\rho_u c_u^2(i\omega^2 K_{zu} + K_x^2 g)} = \frac{A_u}{A_l} \tag{2.11.3}$$

where ρ_l and ρ_u refer to the densities in the respective half space at $z = 0$ and applying (2.1.4) to (2.11.2) gives

$$\frac{\omega^2 - c_l^2 K_x^2}{\omega^2 - c_u^2 K_x^2} = \frac{A_u}{A_l}. \qquad (2.11.4)$$

The wavenumbers K_{xu} and K_{xl} have been replaced by K_x since the phase fronts match at the boundary.

The dispersion equation (2.3.13) may be applied in each layer to find K_{zl} and K_{zu}. If we assume that the surface waves are external and $K_z = i\Gamma$, then in each layer

$$\Gamma = \frac{1}{2H} \pm \sqrt{\left(\frac{1}{2H}\right)^2 - \frac{\omega^2}{c^2} + K_x^2\left(1 - \frac{\omega_g^2}{\omega^2}\right)} \qquad (2.11.5)$$

where $\omega_g^2 = (\gamma - 1)g^2/c^2$.

Equating (2.11.3) and (2.11.4) reveals that

$$\frac{\rho_l c_l^2}{\rho_u c_u^2} \frac{(K_x^2 g - \omega^2 \Gamma_l)}{(K_x^2 g - \omega^2 \Gamma_u)} = \frac{\omega^2 - c_l^2 K_x^2}{\omega^2 - c_u^2 K_x^2} \qquad (2.11.6)$$

provided no background winds exist. The presence of background horizontal winds can be taken into account by substituting Ω_l and Ω_u respectively for ω, where $\Omega = \omega - U_0 K_x$.

If $c_l = c_u = c$ at the boundary then (2.11.6) reduces to

$$\rho_l(K_x^2 g - \omega^2 \Gamma_l) = \rho_u(K_x^2 g - \omega^2 \Gamma_u).$$

Thus the characteristic equation for surface gravity waves over an incompressible homogeneous half space is obtained by taking $c = \infty$, $H_l = \infty$ so that $\Gamma_l = K_x$ and provided $K_x \gg 1/2H_u$, then $\Gamma_u = -K_x$ (choosing the signs so that energy decays from the boundary) and the phase velocity of the waves V_{px}, is given by

$$V_{px}^2 = \left(\frac{\omega}{K_x}\right)^2 = \frac{g}{K_x} \frac{(\rho_l - \rho_u)}{(\rho_l + \rho_u)}. \qquad (2.11.7)$$

This is the well-known equation for surface waves at the interface between two incompressible fluids such as oil and water. When $\rho_l \gg \rho_u$, such as at air–water interfaces, this equation further reduces to

$$V_{px}^2 = g/k_x \qquad (2.11.8a)$$

giving the dispersion relation for surface gravity waves in deep water

$$\omega^2 = gk_x. \tag{2.11.8b}$$

This equation is essentially unaffected by the compressibility or density stratification of the lower half space, but is dependent upon an abrupt density discontinuity between the two layers. However, in a compressible lower layer there is a second solution which corresponds to the Lamb wave (Tolstoy, 1963). In this case it is not necessary to invoke the lower boundary condition, since the Lamb wave is an acoustic wave guided by the boundary. There is no interaction between the surface wave mode and the acoustic (Lamb) wave mode because the surface wave is circularly polarized and irrotational whereas the Lamb wave is linearly polarized and non-irrotational.

Density discontinuities are, however, unlikely to exist in the atmosphere, except at the lower boundary. There is thus good reason to suppose that surface gravity waves obeying (2.11.8) extend into the atmosphere above the oceans, but their amplitude will be too small to be of any meteorological significance.

A more realistic atmospheric boundary condition is to take $\rho_l = \rho_u$ at $z = 0$. It can be shown that in an isothermal atmosphere $\Gamma = 1/2H$ corresponds to constant energy flow at all heights. In order for the energy carried by the boundary wave to be finite, Γ must be less than $1/2H$ above the boundary, and greater than $1/2H$ below. That is, energy flow is to be concentrated at the boundary. In the expression for Γ then, the minus sign is used in the upper region and the plus sign in the lower region.

In the low frequency limit, provided $(1/2H) > K_x$ then

$$\Gamma_l = \frac{1}{H_l} - \frac{(\gamma-1)}{\gamma}\frac{gK_x^2}{\omega^2}$$

and

$$\Gamma_u = \frac{(\gamma-1)}{\gamma}\frac{gK_x^2}{\omega^2}$$

so

$$c_l^2\left((2\gamma-1)gK_x^2 - \frac{\omega^2\gamma}{H_l}\right)(\omega^2 - c_u^2K_x^2)$$
$$= c_u^2 gK_x^2(\omega^2 - c_l^2K_x^2) \tag{2.11.9}$$

or in terms of V_{px}

$$c_l^2\left[(2\gamma-1)g-\frac{V_{px}^2\gamma}{H_l}\right][V_{px}^2-c_u^2]$$
$$=c_u^2 g[V_{px}^2-c_l^2] \qquad (2.11.10)$$

which gives a quadratic in V_{px}^2:

$$V_{px}^4\gamma^2 g - V_{px}^2\{c_l^2(2\gamma-1)g+c_u^2(\gamma^2-1)g\}$$
$$+c_u^2 c_l^2(2\gamma-2)g = 0.$$

Provided $c_u > c_l$ then the two solutions for V_{px}^2 are approximately

$$V_{px}^2 = \frac{2c_l^2}{\gamma+1}\left\{1-\frac{(2\gamma-1)}{(\gamma^2-1)}\frac{c_l^2}{c_u^2}\right\} \qquad (2.11.11)$$

and

$$V_{px}^2 = \frac{\gamma^2-1}{\gamma^2}c_u^2\{1+(\gamma+1)^{-2}c_l^2/c_u^2\}. \qquad (2.11.12)$$

Basically there are two possible types of boundary wave: one which moves at almost the sound speed of the lower layer and one which moves at almost the sound speed of the upper level. In a general n-layer model of the atmosphere the equation describing the phase velocity will be of degree $2(n-1)$. The number of boundary waves possible is then also $2(n-1)$.

2 Background Reading

Eckart, C. (1960) *Hydrodynamics of oceans and atmospheres*, Pergamon Press, New York.

Tolstoy, I. (1963) The theory of waves in stratified fluids including the effects of gravity and rotation, *Reviews of Modern Physics*, **35**, 207.

Lighthill, M. J. (1966) Dynamics of rotating fluids: a survey, *Journal of Fluid Mechanics*, **26**, 411.

3

Waves in Real Atmospheres

3.1 Introduction

In the previous chapter, the properties of certain atmospheric waves were discussed when these waves propagate in highly idealized atmospheres. When we come to consider these waves in real atmospheres, then the energy dissipating processes that occur in the real atmosphere due to its viscosity are of prime importance. The viscous effects arise both from the diffusion of energetic molecules and from turbulence. Of comparable effect to viscosity will be the energy losses due to thermal conduction.

Even in the simple case of an atmosphere composed of a single constituent ideal gas, the complexities introduced by the inclusion of viscosity require special methods of solution. The further problems that could be introduced by considering the atmosphere as a mixture of gases and water vapour, apt to be both sources and sinks of neutral and charged particles, lie outside the intended scope of this book and at present are still fruitful fields for research.

3.2 Turbulence

Turbulent flow is characterized by the presence of random oscillations of velocity which produce irregularities in the path of a particle. Turbulence also involves the notions of dissipation and dispersion in addition to random fluctuations in velocity. The kinetic energy of the fluid due to the fluctuations in velocity will eventually be dissipated as heat. The fluctuations in velocity also produce mixing and are different in nature from regular wave motions. It is the fluid's stability which determines whether a given sample is turbulent.

It is not simple to differentiate between turbulence and waves, and there is likely to be no clear-cut distinction between them. For example, when turbulence exists in or near a stratified fluid, some of the energy is transformed into waves. On the other hand, turbulence can grow at the expense of the energy in waves either by wave-breaking phenomena or simply by extracting energy from the strain rates produced by waves. When energy is constantly flowing between the two kinds of motion by non-linear effects, then the energy cannot be clearly identified as belonging to either kind.

The analysis of turbulence is based on the postulate of the existence of a range of scales of turbulent motion. Energy is being continually passed from larger to smaller scale motions. A lower limit is eventually reached when the energy has been passed down to fluctuations that are too small to permit the formation of still smaller eddies. At this point the energy is directly converted into the random motion of the molecules by viscosity.

The average properties of the small-scale components of any turbulent motion are determined uniquely by the kinematic viscosity of the fluid, η, and the average rate of dissipation of energy per unit mass of fluid ($\overline{\mathscr{R}}$). There is also a sub-range of the small eddies within which the average properties are determined only by $\overline{\mathscr{R}}$. This conceptual framework is known as the universal equilibrium theory or as the Kolmogoroff similarity theory (Batchelor, 1953, Chapter 6).

While it is the largest eddies which obtain their energy directly from large-scale ordered motions such as atmospheric winds, it is the smallest eddies which convert their energy into heat through the viscous interactions. The energy-containing eddies are of intermediate size. Their characteristic velocity U_c and diameter L_c are basic parameters in the mathematical description of turbulence. If U_c and L_c are known, then

$$\overline{\mathscr{R}} = \frac{U_c^3}{L_c}.$$

This is the rate at which energy is extracted from the atmospheric winds and converted into heat. The diameter of the smallest eddies L_{min} and the mean lifetime $\overline{\mathscr{T}}_{\mathrm{min}}$ of the smallest eddies are given by

$$L_{\mathrm{min}} = \left(\frac{\eta^3}{\overline{\mathscr{R}}}\right)^{\frac{1}{4}}$$

and

$$\overline{\mathscr{T}}_{\mathrm{min}} = \left(\frac{\eta}{\overline{\mathscr{R}}}\right)^{\frac{1}{2}}.$$

The flow of air in the earth's atmosphere differs from the flow of fluids in straight pipes. Some of the earliest investigations on laminar and turbulent flow were performed by Reynolds, who examined the flow in pipes in terms of the Reynolds number. In geophysics there are two Reynolds numbers. The first is the ordinary one

$$\mathrm{R_e} = \frac{U_f L}{\eta}$$

where U_f is the flow velocity, L a typical length, which is the pipe diameter in the case of motion in pipes, and η is the kinematic viscosity which is defined in terms of the coefficient of viscosity μ, (defined in equation (3.3.1) and the density as

$$\eta = \frac{\mu}{\rho}.$$

The other Reynolds number is

$$\mathrm{R_g} = \frac{\omega_{\mathrm{B}} L^2}{\eta}$$

where ω_{B} is the Vaisala–Brunt frequency. In laboratory experiments $\mathrm{R_g} \ll \mathrm{R_e}$ and the latter will dominate the phenomena. For large-scale or moderately large-scale phenomena in geophysical fluids, $\mathrm{R_g} \gg \mathrm{R_e}$ and the former will dominate. Consequently, it is most improbable that the results of laboratory investigations of turbulence will have much relevance to turbulence in the air or oceans, because the viscous stresses introduced by the solid walls of pipes will be lacking in the large-scale motions.

The Reynolds number $\mathrm{R_g}$, though of rather dubious relevance as a measure of the onset of geophysical turbulence, is certainly pertinent to the extinction of turbulence at the small-scale end of the spectrum. Spectral components for which $\mathrm{R_g} \approx 1$ will be strongly damped.

A quantity that indicates whether turbulent motions will persist or decay in a density gradated fluid is the Richardson number defined by

$$\mathrm{R_i} = \frac{\omega_{\mathrm{B}}^2}{\left(\dfrac{\partial U}{\partial z}\right)^2}.$$

This number provides a measure of the stabilizing influence of gravity, modified by temperature gradients, in comparison with the destabilizing effects of wind shears. It has applications in the theory of convective motions, in instabilities in shearing slows, and, by implication, in the generation of turbulence.

Imagine the situation in which an initially very sharp shear exists, such that the Richardson number falls below a critical value and turbulence is generated. The turbulence will broaden the transition zone, raise the Richardson number, and eventually bring it to its critical value when a stationary state can be established. For turbulence to be generated without immediate quenching, however, the Reynolds number associated with its largest eddies must exceed unity and this in turn implies that the Reynolds number derived from the characterizing velocity and scale of the shear zone must exceed unity. As a limiting case, the shear will just succeed in producing turbulence if its Richardson number is just critical and its Reynolds number is comparable to unity.

The above argument gives the necessary conditions for the onset and decay of turbulence. It is unlikely, however, that they are sufficient conditions. For example, Menkes (1961) finds that the critical Richardson number is related to the ratio of the shear scale to the scale height, L/H, and, as L/H increases, the critical value decreases until at $L = H/2$ the Richardson number must vanish or become negative for an instability to occur. This is apparent in very deep streams, which can have layers of uniform R_i less than $\frac{1}{4}$ and yet be stable. This implies that the smaller scale modes are most important in the generation of turbulence. The actual value of the critical Richardson number when L/H is small is still being disputed. The Miles–Howard theorem states that

$$R_i \leqslant \tfrac{1}{4}$$

is a necessary condition for instability in an incompressible fluid. The same criterion also applies to compressible fluids (Warren, 1968; Chimonas, 1970a). The original criterion of Richardson (1920) that instability occurs when

$$R_i \leqslant 1$$

seems to be more relevant to the maintenance of turbulence already in existence than to the onset of turbulence from a non-turbulent state (Hines, 1971).

The Richardson number is unable to provide an explanation for the turbopause. In the atmosphere it is almost everywhere greater than unity, and even in the layer 105 to 109 km where it falls below unity it still exceeds $\frac{1}{2}$. This suggests that wind shear is not the sole turbulence generator in the atmosphere. Lindzen (1968a) found that the diurnal tide could become unstable and he used a Richardson number based on the tidal temperatures and velocities as well as the background atmospheric temperatures and velocities. The unstable diurnal tide led to turbulence and Lindzen and Blake (1971) showed that molecular viscosity and conductivity cause the diurnal tide's amplitude to decay above a certain well-defined height—108 km in their case—and that this height corresponds to the turbopause.

Tchen (1970) studied various methods of generating turbulence in the atmosphere. They are characterized by the variation of the energy of the turbulence as a function of the wavenumber of the Fourier component of the turbulent motion. The results for the categories studied were as follows:

(*a*) *Turbulence due to viscous dissipation*
In this case, the viscosity removes enough energy from the wave to produce turbulence. In the absence of wind shears, the spectral energy distribution, F is proportional to k^{-7} which corresponds to the Heisenberg law. In the presence of wind shear however, $F \propto k^{-1}$. Neither of these distributions is found to occur in the atmosphere because the dominant effect of gravity must be included.

(*b*) *Turbulence due to buoyant instability*
When the Vaisala–Brunt frequency becomes imaginary, the atmosphere is in convective instability. In the absence of wind shear, the energy spectrum distribution obeys the Kolmogoroff law $F \propto k^{-\frac{5}{3}}$, whereas, when the instability occurs in the presence of wind shear (the Kelvin–Helmholtz instability) $F \propto k^{-3}$. Both these spectral distributions have been observed in the atmosphere, though the $F \propto k^{-3}$ spectrum can also be produced by the instability of amplified gravity waves (Hodges, 1967). These results are summarized in Table 3.1.

Turbulence is also believed to be responsible for the energy exchange between the planetary waves of different wavenumbers that are found to exist in the earth's upper troposphere. However,

Table 3.1 Energy spectral distributions for various forms of turbulence

Wind shear	Viscous dissipation	$F = \frac{3}{\sqrt{\pi}} \frac{\bar{\mathscr{R}}}{(\partial U/\partial z)} k^{-1}$
	Thermal instability	$F = \frac{1}{3}(\partial U/\partial z)^2 k^{-3}$
No shear	Viscous dissipation	$F = \frac{\pi}{6}(\bar{\mathscr{R}}/\eta^2)^2 k^{-7}$
	Thermal instability	$F = (32/9\pi)^{\frac{1}{3}} \bar{\mathscr{R}}^{\frac{2}{3}} k^{-\frac{5}{3}}$
	Non-linear wave amplification	$F = (\text{const})\omega_{\mathrm{B}}^2 k^{-3}$

since planetary waves basically comprise two-dimensional motions in a fluid under hydrostatic equilibrium, the theory of isotropic three-dimensional turbulence is no longer adequate. Theoretical determinations of the structure of two-dimensional isotropic turbulence yield two regimes of planetary scale turbulence. For planetary waves of high wavenumber the energy transfer determines the spectral distribution law as being of the Kolmogoroff $k^{-\frac{5}{3}}$ type. However, for the low wavenumber planetary waves—which are generally forced waves produced by the major mountain ranges or by the continents—the transfer of mean square vorticity (enstrophy) leads to a spectral distribution in which the kinetic energy per unit scalar wavenumber varies as k^{-3} (Charney, 1971).

3.3 Viscosity

Viscosity, from a molecular point of view, is a momentum transfer phenomenon that occurs within a gas in which a mean velocity gradient exists. The molecules that have a larger momentum will diffuse into regions of lower momentum as a result of their random thermal motion, whilst molecules from the region of less momentum will move into regions of higher momentum. This tendency to equalize the momenta can be described either by a coefficient of momentum diffusion or by a coefficient of viscosity.

The general form of the diffusion equation for a quantity Q is

$$\text{Flux of } Q = D \times \text{divergence of } Q \text{ per unit volume}$$

where D is the coefficient of diffusion. If we take Q to be the momentum, then the momentum flux is the viscous force per unit area, or pressure, being exerted by the diffusing particles.

Consider the simple one-dimensional case when a gas is moving in the x direction with a uniform velocity U_x that increases in the z direction. There will be a transfer of momentum from the faster moving particles to the slower ones, so that the viscous force F_z acting on an area A is given in terms of the diffusion coefficient for momentum by

$$\frac{F_z}{A} = D\frac{\partial(\rho U_x)}{\partial z}.$$

The coefficient of viscosity, μ, of the gas is defined in this simple case by the equation

$$\frac{F_z}{A} = \mu\frac{\partial U_x}{\partial z} \tag{3.3.1}$$

so that, in a body of uniform density, the momentum diffusion coefficient and the coefficient of viscosity are related by $D = \mu/\rho$. The coefficient of viscosity thus has dimensions of N-s/m^2 and the diffusion coefficient is expressed in m^2/s.

Simple kinetic theory yields the viscosity coefficient in terms of the mean molecular speed $\bar{v}$ as

$$\mu = \frac{1}{3}\frac{m\bar{v}}{\sqrt{2}\sigma_m}$$

where m is the mass of a molecule and σ_m the cross-section of the molecule. Since

$$\bar{v} = \sqrt{8kT/\pi m}$$

the coefficient of viscosity of an ideal gas is independent of both the pressure and density and is only a function of temperature.

If we now turn to the more complicated situation where U_x is no longer constant and a variable U_y and U_z has to be included, then we have to replace the simple scalar stress F_z/A by a tensor which shall be denoted by $\underline{\underline{\boldsymbol{\sigma}}}$. The components which are derived in Lamb's *Hydrodynamics* (but misprinted in the final equation) are

$$\boldsymbol{\sigma}_{xx} = -p-\frac{2}{3}\mu\chi+2\mu\frac{\partial U_x}{\partial x}$$

$$\boldsymbol{\sigma}_{yy} = -p-\frac{2}{3}\mu\chi+2\mu\frac{\partial U_y}{\partial y}$$

$$\boldsymbol{\sigma}_{zz} = -p-\frac{2}{3}\mu\chi+2\mu\frac{\partial U_z}{\partial z}$$

so that the total pressure

$$p = -(\boldsymbol{\sigma}_{xx}+\boldsymbol{\sigma}_{yy}+\boldsymbol{\sigma}_{zz})/3$$
$$= -\mathrm{Trace}\,(\underline{\boldsymbol{\sigma}})/3$$

where $\chi = \nabla \,.\, \mathbf{U}$ is the velocity divergence. The off-diagonal components of the stress tensor are

$$\boldsymbol{\sigma}_{yz} = \mu\left(\frac{\partial U_z}{\partial y}+\frac{\partial U_y}{\partial z}\right) = \boldsymbol{\sigma}_{zy}$$

$$\boldsymbol{\sigma}_{zx} = \mu\left(\frac{\partial U_x}{\partial z}+\frac{\partial U_z}{\partial x}\right) = \boldsymbol{\sigma}_{xz}$$

$$\boldsymbol{\sigma}_{xy} = \mu\left(\frac{\partial U_y}{\partial x}+\frac{\partial U_x}{\partial y}\right) = \boldsymbol{\sigma}_{yx}.$$

Since the ith component of the viscous force per unit volume is given by the divergence of the stress tensor,

$$\frac{\partial \boldsymbol{\sigma}_{ix}}{\partial x}+\frac{\partial \boldsymbol{\sigma}_{iy}}{\partial y}+\frac{\partial \boldsymbol{\sigma}_{iz}}{\partial z}$$

the equation of motion of a viscous fluid which has a constant coefficient of viscosity can be represented by

$$\rho\frac{D\mathbf{U}}{Dt} = \rho g - \nabla p + \frac{\mu}{3}\nabla\chi + \mu\nabla^2\mathbf{U}$$

where χ is again the velocity divergence. This equation is a bastardized representation, valid only in Cartesian coordinates, which represents the three-component equations. A form of the equation of motion that is invariant under a change of coordinate system is

$$\rho\frac{D\mathbf{U}}{Dt} = \rho g - \nabla p + \mu[\tfrac{4}{3}\nabla(\nabla \,.\, \mathbf{U}) - \nabla\times(\nabla\times\mathbf{U})] \qquad (3.3.2)$$

where D/Dt can be expanded by using the expressions given in §2.3. This is the Navier–Stokes equation.

When both sides of the Navier–Stokes equation are divided by the gas density, then it is advantageous to define the coefficient of kinematic viscosity as

$$\eta = \frac{\mu}{\rho}. \qquad (3.3.3)$$

The coefficient of kinematic viscosity is thus the same as the momentum diffusion coefficient and it also has units of m^2/s.

Molecular nitrogen, N_2, has a coefficient of molecular viscosity $\mu = 1{\cdot}73 \times 10^{-5}$ N-s/m^2, which is close to the value of μ for air at ground level, which experiments reveal to be $1{\cdot}86 \times 10^{-5}$ N-s/m^2. As the density of air at 10^5 N/m^2 and 300° K is 1·16 kg/m^3, we obtain a coefficient of kinematic molecular viscosity of $1{\cdot}60 \times 10^{-5}$ m^2/s for air at ground level. However, meteorological experiments (as opposed to the measurements of the viscosity of air in pipes) show that the viscosity of the air near the ground is greater than the value of the molecular viscosity. This is because the momentum carriers for the atmosphere below the turbopause are not the small-scale molecules, but are the turbulent atmospheric eddies whose scales are of the order of a metre.

In the lowest few kilometres of the atmosphere, the eddy viscosity depends on both the topography and the background winds. Estimates of the eddy viscosity are made by measuring the distribution of wind velocity with height and they provide the values for η_{eddy} given in Table 3.2.

Table 3.2 η_{eddy} **at the ground**

Above land:	Eddy kinematic viscosity
Strong winds	6·2 m^2/s
Moderate winds	5·0 m^2/s
Light winds	2·8 m^2/s
Above sea	0·077 to 0·69 m^2/s

For a Maxwellian gas, the coefficient of molecular viscosity is independent of the gas pressure or density and varies as $T^{\frac{1}{2}}$, where T is the neutral gas temperature. Dalgarno and Smith (1962) have found that, for atomic oxygen, the coefficient of kinematic viscosity is given by

$$\eta = \frac{3{\cdot}34 \times 10^{-7} T^{0{\cdot}71}}{\rho} \text{ m}^2/\text{s} \tag{3.3.4}$$

where ρ is measured in kg/m^3. Fig. 3.1 gives the variation of the kinematic viscosity with height. Below the turbopause at 105 km, it is the eddy viscosity that dominates.

If η is constant, as it is in the lowest 40 km of the atmosphere, then it is not possible to use the Navier–Stokes equation (3.3.2), because

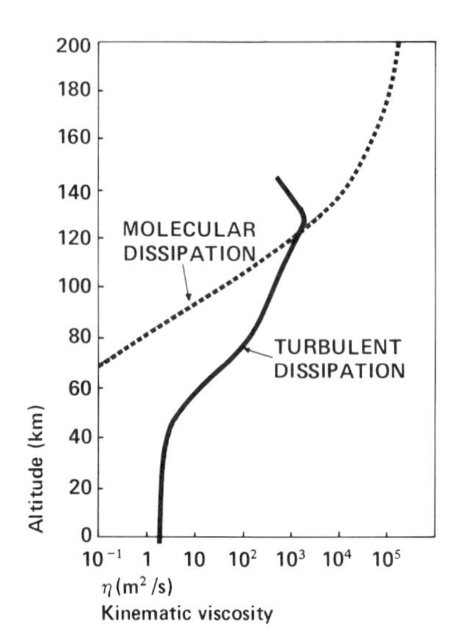

3.1 Atmospheric kinematic viscosity including turbulent dissipation (after Midgley and Liemohn, 1966)

if η is constant in the atmosphere, then μ varies, so that a more complicated form must be used for the viscous force density:

$$(\nabla \,.\, \mu\nabla)\mathbf{U}+\tfrac{1}{3}\nabla(\mu\nabla \,.\, \mathbf{U})+([\nabla\mu]\times\nabla)\times\mathbf{U} \qquad (3.3.5)$$

(Howarth, 1953, p. 48).

The complete hydrodynamic equations for a viscous fluid involve not only the addition of a term like (3.3.5) to the force equation. To be completely accurate, viscosity must be included in the energy equation as well, which becomes (Lamb, 1945, p. 579)

$$\frac{Dp}{Dt}-c^2\frac{D\rho}{Dt}=(\gamma-1)\{\nabla \,.\, [\underline{\underline{\boldsymbol{\sigma}}} \,.\, \mathbf{U}]-\mathbf{U} \,.\, [\nabla \,.\, \underline{\underline{\boldsymbol{\sigma}}}]\} \qquad (3.3.6)$$

where $\underline{\underline{\boldsymbol{\sigma}}}$ is the stress tensor, which can be written as

$$\boldsymbol{\sigma}_{ij}=\mu\left(\frac{\partial U_i}{\partial x_j}+\frac{\partial U_j}{\partial x_i}-\frac{2}{3}\delta_{ij}\nabla \,.\, \mathbf{U}\right)-\delta_{ij}p \qquad (3.3.7)$$

where

$$\left.\begin{aligned}\delta_{ij}&=1 \quad \text{if } i=j\\ \delta_{ij}&=0 \quad \text{if } i\neq j\end{aligned}\right\} \qquad (3.3.8)$$

To avoid the complications that arise from the inclusion of the viscous term in the energy equation it is generally assumed that the viscous dissipation is small so that the waves remain nearly adiabatic. In this case, one obtains the modified dispersion relation for internal waves in an isothermal atmosphere (Pitteway and Hines, 1963)

$$\Omega(\Omega - i\eta\phi)\left\{\Omega^2 - \left(\frac{4i\eta\Omega}{3} + c^2\right)\phi\right\} + k_x^2\left(g + \frac{i\omega\eta}{H}\right)\left\{g(\gamma-1) - \frac{2i\omega\eta}{3H}\right\}$$
$$= 0 \qquad (3.3.9)$$

where $\phi = k_x^2 + k_z^2 + 1/4H^2$.

This equation may be regarded as a quadratic in ϕ for a given value of k_x and thus is a quartic in k_z. Viscosity therefore introduces extra roots to the dispersion equation, and so leads to a generalized form of the viscous waves that occur in hydrodynamics.

In the upper atmospheric situation, when μ is a constant and the kinematic viscosity varies with height as $\exp(z/H)$, it is no longer possible to assume that $p_1/(p_0\mathrm{P}) = \rho_1/(\rho_0\mathrm{R}) = U_{1x}/\mathrm{X} = U_{1z}/\mathrm{Z}$, because each of the wave parameters now experiences different attenuations, and so it is not possible to describe these by a single dispersion equation.

One can obtain some assessment of these dissipative effects by equating the mean energy content per unit mass of the atmosphere with the product of the period and the mean rate of energy dissipation per unit mass.

The rate of dissipation of energy per unit mass, through the action of viscosity, is given by (3.3.6).

$$\mathscr{R} = (\nabla \cdot [\underline{\underline{\sigma}} \cdot \mathbf{U}] - \mathbf{U} \cdot [\nabla \cdot \underline{\underline{\sigma}}])/\rho \qquad (3.3.10)$$

where $\mathbf{U}$ is the real velocity vector. The complex representation of the velocity can be used in calculating the mean value of $\mathscr{R}$ represented by $\overline{\mathscr{R}}$

$$\overline{\mathscr{R}} = (\nabla \cdot [\underline{\underline{\sigma}} \cdot \mathbf{U}^*] - \mathbf{U}^* \cdot [\nabla \cdot \underline{\underline{\sigma}}])/\rho \qquad (3.3.11)$$

where $\underline{\underline{\sigma}}$ is still defined by (3.3.7) and the star denotes the complex conjugate. Hines (1960) has derived the full expression for $\overline{\mathscr{R}}$ by assuming an exponential form for the velocities and by using the polarization relation. He also showed that for high-frequency sound waves

$$\overline{\mathscr{R}} \approx \frac{2\eta}{3}(k_x^2 + k_z^2)\mathbf{U}_1 \cdot \mathbf{U}_1^* \qquad (3.3.12)$$

and for the low-frequency internal gravity waves

$$\overline{\mathscr{R}} \approx \eta(k_x^2+k_z^2)^2 \frac{U_{1x}U_{1x}^*}{2k_z^2} \tag{3.3.13}$$

in the asymptotic limit $k_z^2 \gg (1/2H)^2$.

The energy content per unit mass of the atmosphere $\mathscr{E}$ is the sum of the kinetic energy due to atmospheric motion, the gravitational potential energy, and the thermodynamic potential energy for adiabatic compressions, so that

$$\mathscr{E} = gz + \frac{U^2}{2} + (\gamma-1)^{-1}\frac{p}{\rho} \tag{3.3.14}$$

as measured with respect to a stationary element of atmosphere at an arbitrary height, $z = 0$. The excess due to the wave perturbation is then*

$$\mathscr{E}_1 = gz_1 + \frac{U_1^2}{2} + (\gamma-1)^{-1}\left[\frac{p}{\rho} - \frac{p_0}{\rho_0}\right] \tag{3.3.15}$$

which, in the high-frequency case, gives a mean excess energy of

$$\overline{\mathscr{E}}_1 \approx \frac{U_1^2}{2} \tag{3.3.16}$$

and it yields in the asymptotic case ($k_z^2 \gg (1/2H)^2$) for internal gravity waves

$$\overline{\mathscr{E}}_1 \approx \left[\frac{(\gamma-1)g^2k_x^2}{2\omega^2k_z^2c^2}\right]U_{1x}U_{1x}^* \tag{3.3.17}$$

The importance of dissipation can now be investigated by explicitly selecting those modes in which

$$\frac{2\pi\overline{\mathscr{R}}}{\omega} = \overline{\mathscr{E}}_1. \tag{3.3.18}$$

This approach suffers somewhat from its logical inconsistency. The condition (3.3.18) is one of gross dissipation, yet the formulae on which its derivation was based assumed that the viscous dissipation was small. Nevertheless, the results that are obtained by using this

* I have followed Hines' (1960) treatment. The wave energy is a second-order perturbation and there are a number of different choices available for the wave energy. See Eckart (1960, p. 53).

rather simple approach appear to be experimentally verifiable (Hines, 1964c; Zimmerman, 1964). The results indicate that there will be a minimum vertical wavelength and a minimum horizontal wavelength below which viscous dissipation completely damps the wave. More complete calculations have been done by Pitteway and Hines (1963) and Midgley and Liemohn (1966), who essentially confirm this result. They find, however, that for an internal gravity wave represented by its values of k_x and ω, there is a height at which the amplitude of the wave is maximized. This arises from the competing influences of the exponential growth of amplitude with height (due to the constancy of energy flux, in the presence of a density decrease) and the depletion of the energy flux due to viscous dissipation. Fig. 3.2 gives the parameters of the internal gravity wave of maximum

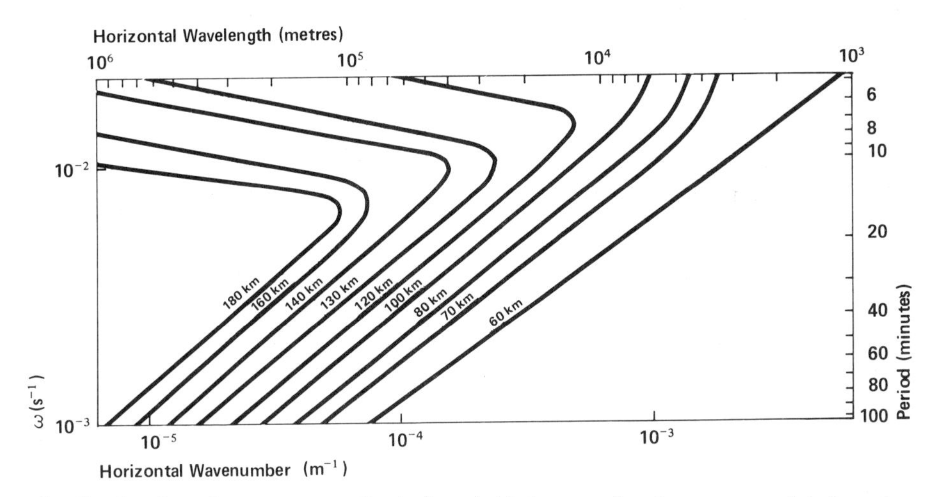

3.2 Height of maximum wave amplitude for a 'white' source (i.e. Source strength independent of frequency)

amplitude at various ionospheric heights if the originating spectrum of waves is assumed to have equal amplitudes.

The height of complete quenching may be found by using (3.3.18) along with (3.3.17), (3.3.13) and the dispersion relation, so that the condition for total energy dissipation of low-frequency waves in an isothermal atmosphere is

$$\frac{2\pi\eta k_x^2 \omega_g^2}{\omega^3} = 1 \qquad (3.3.19)$$

when $k_z^2 \gg (1/2H)^2$. This expression may be rewritten in terms of k_z as

$$2\pi\eta k_z^2 = \omega\left(1 - \frac{\omega^2}{\omega_g^2}\right) \qquad (3.3.20)$$

which yields a maximum value for k_z^2 when $\omega = \omega_g/\sqrt{3}$. This maximum wavenumber corresponds to a minimum vertical wavelength of

$$\lambda_z|_{\min} = \left[\frac{3^{\frac{3}{2}}.4\pi^3}{\omega_g}\right]^{\frac{1}{2}} \eta^{\frac{1}{2}} \text{ metres.} \qquad (3.3.21)$$

In the E region, where the approximations on which (3.3.21) is based are true, $\gamma = 1{\cdot}4$ and $g = 9{\cdot}4$ m/s^2, so that the minimum wavelength becomes

$$\lambda_z|_{\min} = 20\eta^{\frac{1}{2}}H^{\frac{1}{4}} \text{ metres.}$$

This value will change with height since the kinematic viscosity is inversely proportional to the gas density. As the height increases, $\lambda_z|_{\min}$ increases. The values shown in Fig. 3.1 then indicate that the smallest possible value for the vertical wavelengths of internal atmospheric gravity waves is 125 metres. These waves would occur at the mesopause where the scale height is also a minimum, and they possess the same horizontal wavelength and have periods of eight minutes.

Pitteway and Hines (1963) have graphed the corrected form of equation (3.3.19) when $|k_z|$ may actually be smaller than $|1/2H|$. Their results are depicted in Fig. 3.3, and the appropriate value for η may be found from Fig. 3.1.

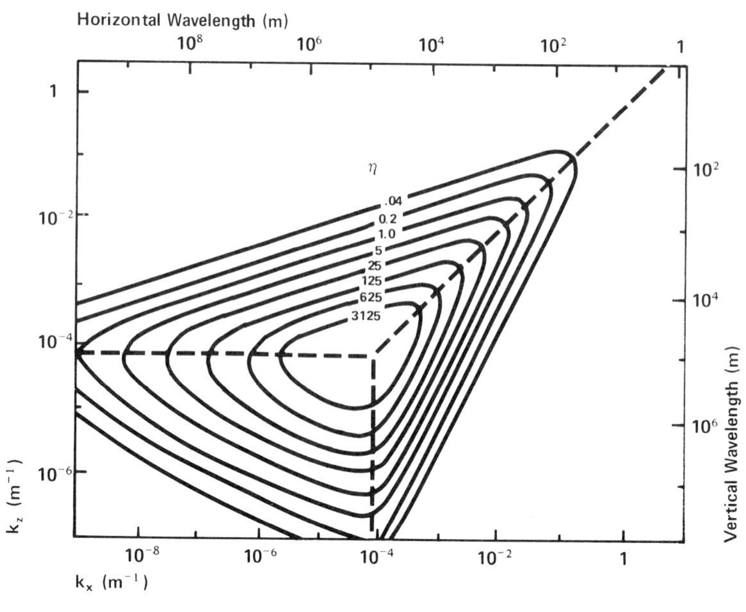

3.3 Points outside the triangular contours correspond to internal waves which are heavily damped in the vertical direction by loss of energy due to viscous damping (after Pitteway and Hines, 1963; used by permission of the National Research Council of Canada)

Tchen (1970) claims that, below the turbopause, it is not sufficient to assume an $H^{\frac{1}{4}}$ dependence for the minimum irregularity size and to use a kinematic viscosity deduced from the eddy diffusion (Fig. 3.4). He shows that, in the region of turbulent dissipation, the minimum scale size will be proportional to the scale height. The rate of energy dissipation may be found by employing the Kolmogoroff similarity theorem. If the energy-containing eddies have characteristic velocity V_c and diameter L_c then

$$\bar{\mathscr{R}} \sim V_c^3/L_c \sim \omega_B^3 L_c^2$$

when the wave is being broken up to generate higher harmonics, and then it is the potential energy gL which is being dissipated in a time of the order of ω_B^{-1}. Hence the energy balance of the dissipation of the wave energy by the turbulence is

$$\omega_B g L_c = \omega_B^3 L_c^2$$

$$\therefore \quad \lambda_z|_{\min} = L_c = \frac{g}{\omega_B^2} \sim H.$$

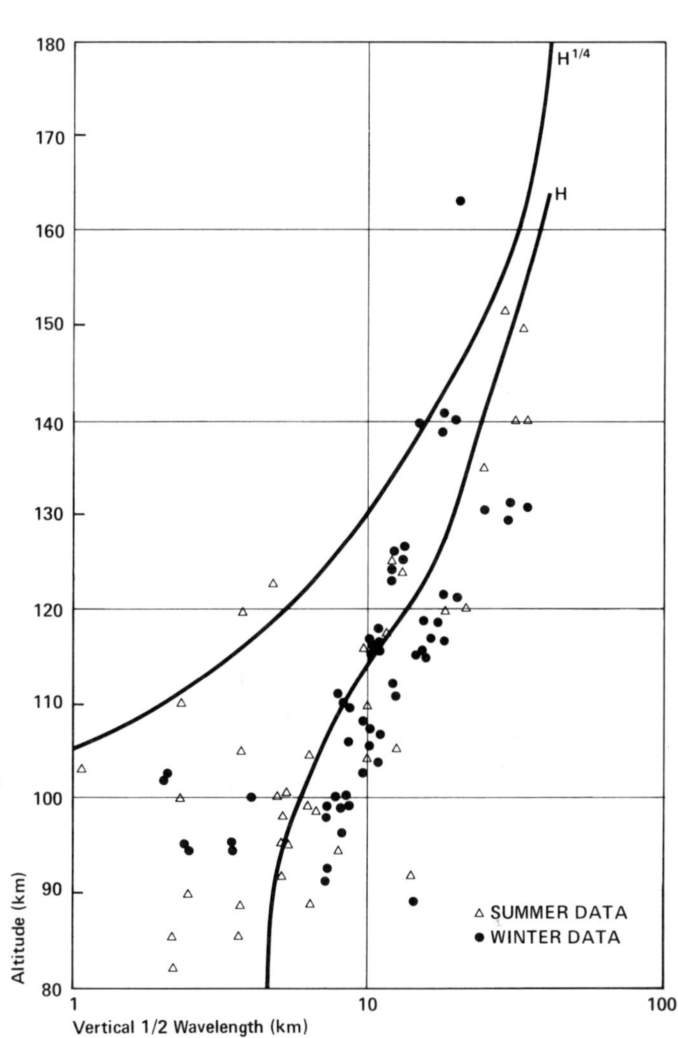

3.4 *Minimum vertical scale of internal gravity waves (after Tchen, 1970)*

Viscosity causes a partial reflection of incident gravity waves (Yanowitch, 1967; Volland, 1969a, c). This result complicates somewhat the choice of an upper boundary condition for the atmosphere, but it should not be unexpected. Viscosity acts as a momentum diffusion term describing the pressure. Momentum variations will then become manifest as pressure oscillations through the coupling due to viscosity. The same type of thing is true for the thermal conductivity and we shall treat the problem a bit more fully in § 3.6.

3.4 Acoustic Wave Attenuation

The frequency range of acoustic frequencies which can propagate in the atmosphere has a lower limit that is determined by the acoustic cutoff frequency ω_{an}, given by

$$\omega_{an}^2 = \left(\frac{\gamma g}{2c}\right)^2 + \frac{\gamma}{2}\frac{g}{T}\frac{\partial T}{\partial z}. \tag{3.4.1}$$

The upper limit is imposed by absorption.

Absorption occurs when molecular collisions are not frequent enough to transfer wave energy effectively. When a molecule is set in motion by the wave and it does not collide again for an appreciable part of a wave cycle, some of its energy is lost to the thermal particle motion. The damping effect due to heat conduction in particular increases rapidly as the length of the mean free path of the air particles approaches the length of the sound waves. On account of this, waves with shorter lengths (higher frequency) are more severely damped than waves of greater wavelength.

Stokes (1845) developed the first successful theory of sound wave attenuation. This was extended by Kirchhoff (1868) to include the effects of thermal conductivity as well as of viscosity. In more recent times it has become evident that sound absorption cannot be adequately described in terms of classical hydrodynamic concepts and it has become necessary to adopt a microscopic point of view of fluids and to consider the molecular interactions (Herzfeld and Litovitz, 1959).

The absorption of energy from sound waves is associated with a time lag relative to the acoustic pressure. This lag depends on the characteristic time or relaxation time required for the viscous stresses associated with relative fluid particles velocities to tend to equalize these velocities, and for heat conduction to occur between high pressure (high temperature) and low pressure (low temperature) regions. When the pressure and density variations are in phase, then no energy is lost from the wave.

The classical Stokes–Kirchhoff formula for the amplitude attenuation coefficient or absorption factor for sound waves is

$$\Lambda = \left(\frac{4}{3}\eta + \frac{\gamma - 1}{C_p}\frac{\mathcal{K}}{\rho}\right)\frac{\omega^2}{2c^3} \tag{3.4.2}$$

where the amplitude is diminished in the ratio e:1 when a distance Λ^{-1} has been traversed by the wave. $\mathcal{K}$ represents the coefficient of thermal conductivity that is defined in § 3.6. Expressed in terms of the relaxation time τ

$$\Lambda = \frac{\omega^2 \tau}{2c} \tag{3.4.3}$$

provided that $\omega\tau \ll 1$ so that the classical approach remains valid. When $\omega\tau \gg 1$, then molecular thermal relaxation becomes the predominant mechanism. Since μ, $\mathcal{K}$, and c all have the same temperature dependence, a useful form of (3.4.2) for computational purposes is

$$\Lambda = 2{\cdot}0 \times 10^{-5}\frac{\mathrm{f}^2}{p}\,\mathrm{km}^{-1}$$

where f is the frequency in Hz and p the pressure in millibars. This represents the molecular attenuation for infrasonic and audible frequencies.

Attenuation in the amplitudes of particle displacement, acoustic pressure, etc., is frequently expressed in nepers, a natural logarithmic unit corresponding to a reduction in amplitude of 1/e with respect to the initial value. Therefore, Λ has the dimension of nepers/m. The change in intensity level is then given by

$$\text{Change in intensity level } (\Delta IL) = 10 \log \frac{I_x}{I_0} = 10 \log \exp(-2\Lambda x)$$

$$= -8{\cdot}7\Lambda x \text{ decibels}$$

so that $8{\cdot}7\Lambda$ is a measure of the decrease in intensity level expressed in db/m.

Fig. 3.5 illustrates attenuation of low-frequency sound waves in the atmosphere. Diagrams of this type are useful when examining the propagation of sound waves produced by nuclear disturbances originating at arbitrary heights in the atmosphere. Absorption in excess of 10^{-3} db/m (i.e. 1 db/km) provides a convenient demarcation between the existence and non-existence of long-distance sound waves. The upper frequency limit in the lower atmosphere is then

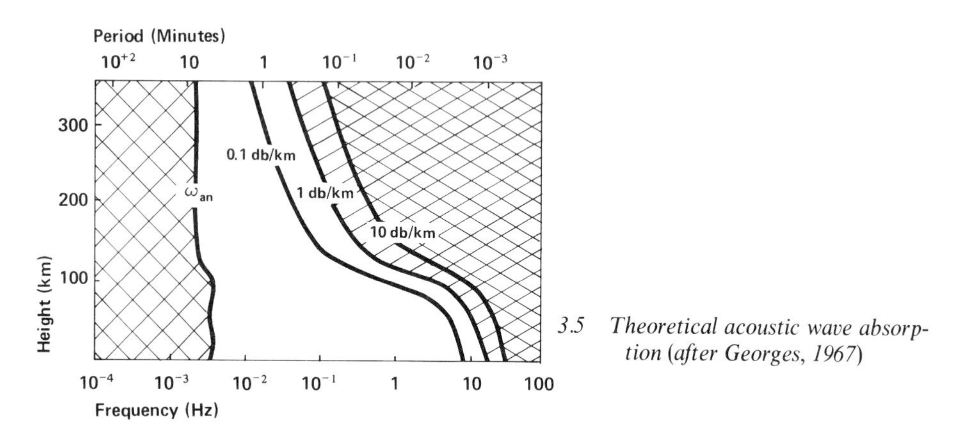

3.5 Theoretical acoustic wave absorption (after Georges, 1967)

imposed by the large-scale turbulent dissipation appropriate to the atmosphere rather than by the molecular dissipation that would be applicable to the laboratory studies of sound attenuation. The classical formula (3.4.2) has been found to be experimentally valid for acoustic waves generated in the laboratory when the molecular viscosity and the molecular thermal conductivity are used (Sivian, 1947).

The attenuation in air at ultrasonic and audible frequencies is heavily influenced by the presence of water vapour. Apparently, the presence of small amounts of water vapour acts as a catalyst which reduces the number of collisions required to excite the vibrational mode of the oxygen molecule. This in turn reduces the relaxation time of the normally unexcited vibrational mode of the oxygen molecules, which gives rise to a large excess attenuation at frequencies between 1 and 100 kHz. It should be recalled that in the high-frequency case ($\omega\tau > 1$), the attenuation and the relaxation time are not related by equation (3.4.3). Fig. 3.6 gives the plot of molecular attenuation against relative humidity for various representative frequencies.

Golitsyn (1965) examined the viscous damping of Lamb waves in the special case when the coefficient of kinematic viscosity remains constant. For high-frequency Lamb waves ($\omega^2 \gg \omega_B^2$), the damping was the same as the damping of ordinary acoustic waves. In the low-frequency case, the ratio of the rate of energy loss to the rate of energy flux in the wave tends towards a constant value of $0{\cdot}62\eta\omega_B^2/c^3$ that is independent of the wave frequency.

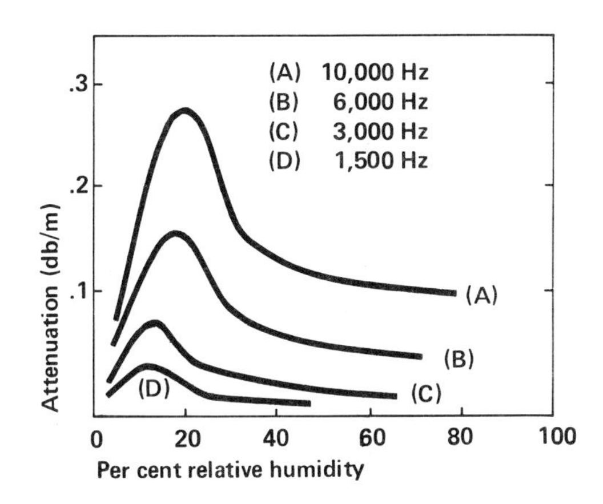

3.6 *Molecular attenuation in air as a function of relative humidity*

3.5 The Radiation Condition

The dispersion relation for the non-dissipative atmosphere contains only terms in k_z^2. Thus the wave properties in this case are unchanged if $-|k_z|$ replaces $+|k_z|$. As the vertical phase propagation and vertical energy propagation of gravity waves and atmospheric tides are oppositely directed, which may immediately be seen from Fig. 2.2, and as a term of the form $\exp\,[i(\omega t - k_z z)]$ has been used, when $k_z > 0$ this corresponds to upward phase propagation and downward energy propagation, whereas $k_z < 0$ is associated with upward propagation of energy. The complete solution to the second-order differential equation would be a linear combination of the two waves

$$A \exp\,(i|k_z|z) + B \exp\,(-i|k_z|z) \tag{3.5.1}$$

where the exponential time and horizontal variation has been dropped.

The identification of the $A \exp\,(i|k_z|z)$ term with an upward travelling wave, and the $B \exp(-i|k_z|z)$ term with a downward travelling one, may also be made by considering the mean rate of transfer of horizontal momentum of the wave or alternatively by introducing a small damping and tracing the wave amplitude back to the energy source (Booker and Bretherton, 1967).

Provided that $|B| < |A|$ then the solution (3.5.1) represents a wave with upward propagating energy. The radiation condition postulates that there are no energy sources at $z = \infty$ and that, in addition $B = 0$, so that there exist no wave reflections. This was implicit in the isothermal non-dissipative calculations, but it is not always applicable in the non-isothermal case or in the case of energy dissipation because, in a medium that does not satisfy the WKB condition (that the properties do not vary substantially over a wavelength), there can be waves partially reflected from the temperature gradients and from the gradients in viscosity and thermal conductivity.

The result of Pitteway and Hines (1963) and Midgeley and Liemohn (1966) predict the decay of waves in the direction of energy propagation, thus supporting the use of the radiation condition in the adiabatic theory that treats the dissipative effects as merely a small variation on the basic wave motion. However, both Midgeley and Liemohn (1966) and Yanowitch (1967) find that, when the vertical wavelength λ_z greatly exceeds the density scale height H, the radiation condition does not hold and a substantial part of the upgoing wave energy is reflected back.

This result agrees with the qualitative predictions of the WKB approximation. Yanowitch (1967), however, obtained quantitative results for internal gravity waves in a plane, non-rotating, infinite, exponentially stratified, incompressible, non-conducting fluid where kinematic viscosity is finite and proportional to $1/\rho$. In addition to an upward propagating wave component of amplitude A_0, there was a reflected component of amplitude $A_0 \exp(-\pi H k_z)$ where k_z is the vertical wavenumber that the wave would have had in the absence of viscosity. The phase of the reflected wave is a complicated function of wave period, molecular viscosity, and horizontal wavelength. Further work by Lindzen (1968b) not only corroborates Yanowitch's result, but indicates that the results depend neither on the assumptions nor on the details of the dissipative mechanism, but on the fact that the dissipation rate varies inversely with the density.

Despite the possibility of reflected waves, the simplicity of the radiation condition endears it to the student of thermospheric dynamics. Fortunately, it appears that even in the presence of dissipative processes the radiation condition, when applied to the thermosphere, yields realistic results (Volland and Mayr, 1972). This view tends to be confirmed by observations of the motion of ionospheric irregularities which are believed to form as a consequence of internal gravity waves.

3.6 Thermal Conduction

Wave damping by thermal conduction is closely related to dissipation by viscosity, since they are both transport effects caused either by molecular diffusion or by turbulent motions. Viscosity is related to momentum transfer within a fluid, whereas the thermal conduction deals with the energy transfer within the fluid. The coefficient of thermal conductivity $\mathscr{K}$ is defined by

$$\mathbf{J} = \mathscr{K}\nabla T \tag{3.6.1}$$

where $\mathbf{J}$ is the heat flowing per unit area per unit time when a temperature gradient, represented by ∇T exists. $\mathscr{K}$ can thus be recognized as the coefficient of heat diffusion. If no other energy flow occurs then the energy equation becomes

$$\frac{Dp}{Dt} - c^2\frac{D\rho}{Dt} = (\gamma-1)\nabla\,.\,(\mathscr{K}\nabla T) \tag{3.6.2}$$

where $\mathscr{K}$ has units of joule/m-s-degree.

For a gas, the viscosity and the thermal conductivity are related by

$$\mathscr{K} = f\mu C_v = \frac{fR}{M(\gamma-1)}\mu \tag{3.6.3}$$

where f is a numerical factor that depends on the nature of the gas and equals unity for an ideal gas in which the molecules can be taken as rigid spheres without mutual repulsion.

For real gases (Dalgarno and Smith, 1962)

$$\begin{aligned} f &= 1{\cdot}87 \quad \text{for } N_2 \\ &= 1{\cdot}91 \quad \text{for } O_2 \\ &= 2{\cdot}57 \quad \text{for } O. \end{aligned} \tag{3.6.4}$$

Equation (3.6.3) is a well-established result for molecular transport processes and consequently is also assumed to be true for the turbulent eddy coefficients.

Pitteway and Hines (1963) show that the damping due to conduction has essentially the same functional form as the damping due to viscosity, provided that $\mathcal{K}$ is proportional to the density. In this case, the modified dispersion relation becomes

$$\omega^4 - \omega^2 c^2 \phi + (\gamma - 1) g^2 k_x^2 = \frac{i\omega \mathcal{K} \phi}{\rho_0 C_v} (\omega^2 - gH\phi) \qquad (3.6.5)$$

which leads to the isothermal dispersion equation

$$\omega^2 = gH\phi = gH\left(k_x^2 + k_z^2 + \frac{1}{4H^2}\right) \qquad (3.6.6)$$

for large values of $\mathcal{K}/\rho_0$. This represents only a single wave sequence one in which $\omega^2 > \omega_a^2/\gamma$ for k_x and k_z real, and which is then the dispersion relation for acoustic waves being propagated by means of isothermal compressions. No internal gravity waves $(\omega < \omega_g)$ can propagate by means of isothermal compressions in an isothermal atmosphere.

When $\mathcal{K}/\rho_0$ is only a weak perturbation, then the damping due to conduction exceeds the damping due to viscosity by the factor f/γ, which is about 1·3 for air and can be treated as unity. Therefore the curves of Fig. 3.3 can be interpreted as representing damping due to conduction. The overall damping will then be the sum of the damping due to conduction and the damping due to viscosity.

In the more genuine upper atmospheric case, when $\mathcal{K}$ remains constant, it is once again possible to determine the height of maximum amplitude relevant to different parameters and to use Fig. 3.2.

Viscosity and Thermal Conduction Waves

The existence of partially reflected gravity waves due to viscosity was pointed out by Yanowitch (1967) and of those due to thermal conductivity by Volland (1969a, c). The aura of mystery surrounding these waves can be slightly cleared if one realizes that both these dissipative processes are also diffusion effects. Viscosity is a measure of momentum diffusion whereas thermal conductivity is a measure of energy (i.e. heat) diffusion. Furthermore, the two coefficients are simply related. Thus, both the atmospheric viscosity waves and thermal conduction waves may be treated as diffusion waves.

The flux equation for the diffusion of the property Q was

$$\frac{\partial^2 \mathbf{Q}}{\partial A \, \partial t} = D\nabla\left(\frac{\partial Q}{\partial V}\right) \tag{3.6.7}$$

where A is the area through which the quantity occupying a volume V flows. It is especially interesting to note that the flux of a scalar quantity is a vector. Also the flux of a vector quantity is a scalar if the diffusion coefficient is a scalar. In order to allow for vector fluxes of vector quantities, a diffusion tensor has to be introduced. It was for this reason that the diffusion treatment of viscosity was only undertaken for the simple two-dimensional case.

If we integrate the diffusion flux equation over the total volume and then take the divergence, we obtain the classical diffusion equation

$$\frac{\partial Q}{\partial t} = D\boldsymbol{\Delta} Q \tag{3.6.8}$$

provided that D remains nicely constant. $\boldsymbol{\Delta}$ is the three-dimensional Laplacian. This equation has given delight to the mathematical physicists of many generations, any of whom will point out that the steady-state case of (3.6.8) is simply Laplace's equation. The solution of Laplace's equation represents a spatial harmonic wave. Hence the existence of viscosity and thermal conduction waves.

3.7 Reflection

The dispersion relation (2.4.3) may be written as

$$k_z^2 = k_x^2\left(\frac{\omega_\mathrm{B}^2}{\omega^2} - 1\right) + \frac{(\omega^2 - \omega_{an}^2)}{c^2}. \tag{3.7.1}$$

Vertical variations in any one of the governing parameters will vary the value of k_z. The most important variations, though, will be those imposed by gradients in the temperature or wind profile. If k_z^2 becomes negative, and remains negative for at least half a wavelength, then a vertically propagating wave will not be transmitted through the region where k_z^2 is negative. The wave is then reflected.

Reflection will occur when

$$k_x^2\left(1-\frac{\omega_B^2}{\omega^2}\right) = \frac{(\omega^2-\omega_{an}^2)}{c^2}. \tag{3.7.2}$$

This has two distinct asymptotes: at low frequencies (tides and gravity waves)

$$\frac{k_x^2\omega_B^2}{\omega^2} = \frac{\omega_{an}^2}{c^2} \tag{3.7.3}$$

and at high frequencies (acoustic waves)

$$k_x^2 = \frac{\omega^2}{c^2}. \tag{3.7.4}$$

Equation (3.7.4) does not represent a reflection condition; it is merely the dispersion relation for high-frequency acoustic waves. These waves are evanescent and always have $k_z = 0$.

For acoustic gravity waves and for short-period gravity waves $\omega \approx \omega_{an}$ and so the reflection condition becomes

$$\omega = \omega_B. \tag{3.7.5}$$

Thus an increasing temperature gradient reflects low-frequency waves. A positive temperature gradient means that ω_B is decreasing with height. Therefore as the wave propagates upwards, internal waves are reflected whose period lies between the Vaisala–Brunt frequency at the ground and the Vaisala–Brunt frequency at the top of the temperature gradient.

In the troposphere, where the temperature gradient is usually negative, an upwards increasing temperature profile is known as a temperature inversion. Gossard and Munk (1954) found that internal gravity waves of periods from 5–10 minutes occurred on ground-based pressure recorders only when there was a temperature inversion present. The gravity waves propagating in the regions of temperature inversions are often made visible by cloud formations—the so-called mackerel sky. At other times the waves may become unstable leading to the onset of billow clouds.

At upper atmospheric heights it is possible to combine the reflection limits (3.7.3) and (3.7.5) with the viscous dissipation limits depicted in Fig. 3.3 in order to arrive at the range of values of k_x and k_z that could exist in the thermosphere. ω_B has a minimum value, corresponding to a period of approximately 8 minutes, at 54 km whereas $c\omega_B/\omega_{an}$ has a minimum at 79 km. These two limits are superimposed on the propagation surface in Fig. 3.7 along with constant-period contours and viscous dissipation limits.

The low-frequency asymptote (3.7.3) corresponds to a horizontal line on Fig. 3.7 at a value of $k_z = \pm 10^{-4}$. This boundary is marked by

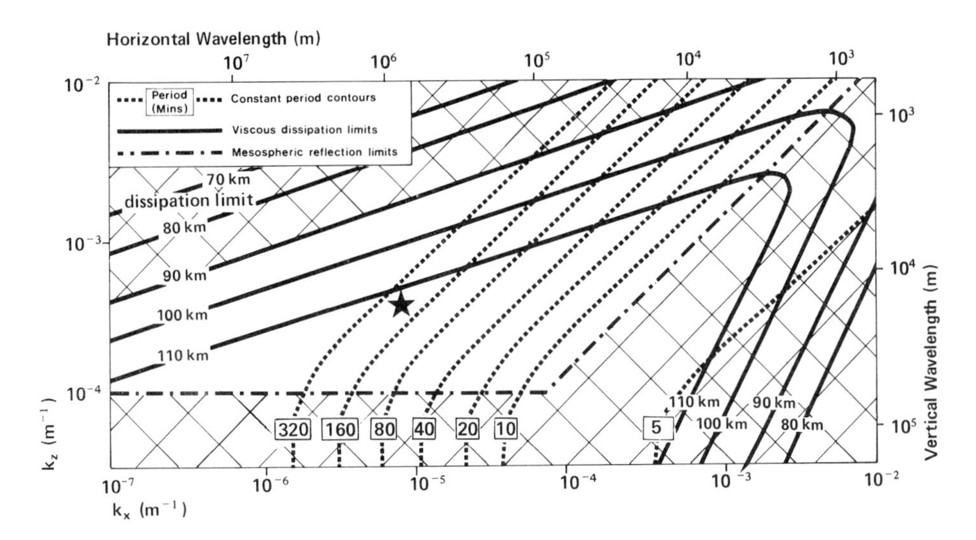

3.7 Composite diagram showing the reflection conditions, the viscous dissipation limits and the constant period contours for gravity waves propagating to the D and E regions from below. The unhatched area represents the spectrum that will exist at 90 km and the star represents the dominant wavelengths

the bottom horizontal boundary of the diagram. Similarly, the unbroken straight diagonal line that is parallel to the dashed constant-period contours represents condition (3.7.5). It is also the 8-minute-period contour. Atmospheric gravity waves propagating vertically upwards can then only reach above 60 km if their wave-numbers fall in the region of Fig. 3.7 that is above $k_z = \pm 10^{-4}$ and to the left of the constant-period contours. At the same time in order to escape dissipation, the wavenumbers must lie within (i.e. below and to the left of) the dissipation boundaries depicted in Fig. 3.7.

The star in Fig. 3.7 represents the dominant mode at 90 km inferred from meteor wind data (Spizzichino, 1969). It lies centrally in the range that is free from severe reflection and severe dissipation. As the waves progress to higher heights, the Vaisala–Brunt period increases, so that the constant period contours appear to move towards the left in Fig. 3.7. At the same time, the dissipation becomes more severe, so that only an extremely small spectrum of tropospherically launched gravity waves could reach the F region. The permissible wavenumbers would of course increase greatly if the gravity wave sources resided above or in the ionosphere.

Let us now return to (3.7.4). This, it was pointed out, does not represent a reflection condition. Nevertheless, it is possible for acoustic waves to be refracted and reflected in the atmosphere. This leads to anomalous sound propagation. For example, during the First World War it was noticed that many people upwind of the firing could hear gunfire at distances 150 to 400 km from the explosion, though people much nearer could not. The explanation—that a region of increasing temperature gradient exists at a large altitude where the rays are then reflected—was the first evidence for the existence of the stratosphere.

The path of an atmospheric wave that is propagating at an angle to the vertical may be computed by dividing the atmosphere into layers of differing refractive index and then applying Snell's law at each interface. For example, in the case of high-frequency sound waves ($\omega/k_x = c$) when horizontal temperature gradients and wind shears can be ignored, if c_1, c_2, etc., are the phase velocities in the first, second, etc., layers (see Fig. 3.8), then from Snell's law

$$\frac{c_1}{\cos\theta_1} = \frac{c_2}{\cos\theta_2} = \frac{c_3}{\cos\theta_3} = \ldots, \text{etc.} \tag{3.7.6}$$

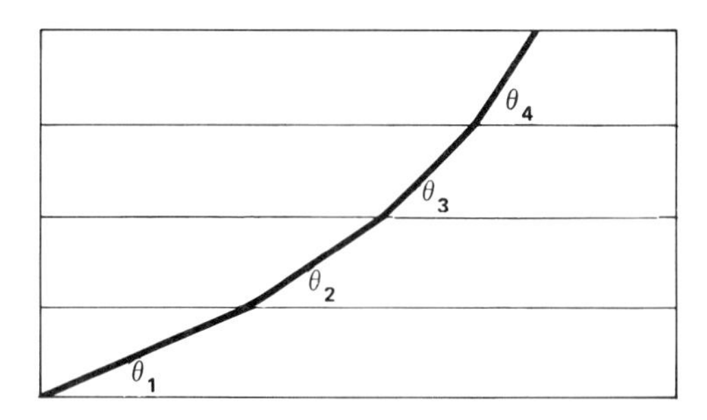

3.8 *Acoustic wave ray tracing in the troposphere*

Using (1.2.12) with $M = 28{\cdot}9$ and $\gamma = 1{\cdot}4$, the speed of sound in the troposphere is

$$c = 20{\cdot}1\sqrt{T}\ \text{m/s} \tag{3.7.7}$$

so that, in the normal troposphere, as c decreases, $\cos\theta_1$ decreases and thus θ_1 increases. All the rays then follow upward curved paths. During periods of temperature inversions, or when the sound wave reaches the tropopause, the rays will start to be bent downwards, and they will eventually be reflected. The thermospheric reflection is illustrated in Fig. 3.9 for an acoustic gravity wave of period 4 minutes.

Though it is a very common technique to divide the atmosphere into a number of isothermal layers and to then trace the path of an

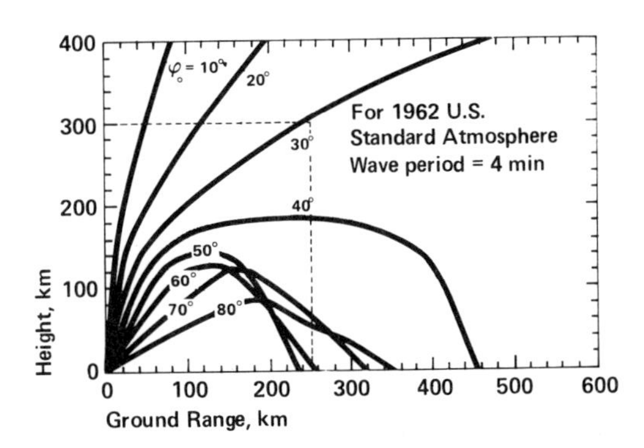

3.9 Acoustic ray paths in a stationary model atmosphere (after Baker and Davies (1969), J. Atmos. Terr. Phys., 31, 1351, Fig. 5)

acoustic or gravity wave ray through each layer, it is not obvious that the solutions that are obtained are necessarily correct (Hines, 1973). There will be partial reflections from the boundary of each layer and these partial reflections will interfere in a manner that may differ from the reflections, if any, from the temperature gradient. For evanescent waves, the use of multilayer techniques is appropriate provided that the medium does not change appreciably over one wavelength. For the low-frequency waves, which have a large wavelength and are not necessarily evanescent, other criteria are needed. In fact, Vincent (1969) has shown that, in order to obtain adequate solutions, more than ten layers are needed in each vertical wavelength.

It is worth noting that the coefficient of the $U'_{1x}, U'_{1z}, \ldots$ term in the differential equation (2.5.14) is proportional to $(1/\rho_0)d\rho_0/dz$. Thus, in proceeding from an isothermal to a real atmosphere it changes from H^{-1} to $H^{-1}(1+dH/dz)$. Thus the correction term is of the order $H^{-1}\,dH/dz$ and not, for example, of the order $k_z^{-1}\,dH/dz$. More specifically, this means that the correction term is independent of the vertical wavelength, and does not necessarily become small if the medium is taken to be 'slowly varying' in the traditional sense (i.e. variations within a wavelength being small). In this case, the criterion that the correction term shall be small, namely that $|dH/dz| \ll 1$, is a property of the medium that is irrespective of the particular atmospheric wave.

However, if a suitable transformation of the differential equation is used, then it is possible to obtain criteria for the solutions in terms of the wavelength. Martyn (1950) employed a variable, ϕ, proportional to $p_0^{\frac{1}{2}}\chi$ and obtained a differential equation

$$\frac{\partial^2\phi}{\partial z^2}+\frac{1}{H}\frac{dH}{dz}\frac{\partial\phi}{\partial z}+q^2\phi = 0 \tag{3.7.8}$$

where

$$q^2 = \left(\frac{\omega_B^2}{\Omega^2}-1\right)k_x^2+\frac{(\Omega^2-\omega_a^2)}{c^2}.$$

A solution to (3.7.8) may be readily found by taking

$$\phi \propto |qH|^{-\frac{1}{2}}\exp\left(-i\int q\, dz\right). \tag{3.7.9}$$

The condition for the validity of this approximation is that

$$\left|\frac{dH}{dz}\right| \ll \left|\frac{q}{H}\right|.$$

In the region where q is zero, then somewhat sophisticated techniques have to be applied to match the boundary conditions within the region of reflection. Examples of this are given in Martyn (1950) and Yamamoto (1957).

Another problem, which has to date not been completely resolved, arises if the reflection condition for a variable other than ϕ is sought. It was mentioned in § 3.5 that in this case the reflection coefficient appears to be different for each parameter. It is not certain just how important this is. For example, if the pressure and density perturbation obey the ideal gas equation

$$p_1 = \rho_1 R T_0$$

then, provided that T_0 increases markedly with height, it would be possible for p_1 to be growing in a region where ρ_1 is damped. This would be equivalent to reflection of the density perturbations at a lower height than of the pressure perturbations. In this case, it appears quite reasonable for each parameter to suffer reflections from a different height. This indicates that it is incorrect to use the reflection condition on ϕ or on U_{1z} to examine microbarograph records, though this is quite often done. Nevertheless, unless the temperature gradient is exceptionally large, the error is not likely to be severe.

The coefficients in the differential equations for the various parameters involve the height derivatives of H and U_{0x}. This implies that, where there is a sudden change of temperature or wind, the equations are not well defined. Once again, this problem may be circumvented by choosing new variables for which the height derivatives of H and U_{0x} are not involved in the coefficients of the respective coupled differential equations, and the wave variables so chosen are then continuous across a boundary where the temperature or wind speed changes suddenly. Pitteway and Hines (1965) take the variables

$$f_1 = \frac{\chi}{\Omega} \tag{3.7.10}$$

and

$$f_2 = \frac{K_x U_{1z}}{\Omega} \tag{3.7.11}$$

to obtain the coupled first-order differential equations

$$\frac{df_1}{dz} = \left(\frac{1}{H} - \frac{gK_x^2}{\Omega^2}\right) f_1 - \left(\frac{\Omega^2}{c^2 K_x} - \frac{g^2 K_x}{c^2 \Omega^2}\right) f_2 \tag{3.7.12}$$

and

$$\frac{df_2}{dz} = K_x \left(1 - \frac{c^2 K_x^2}{\Omega^2}\right) f_1 + \frac{gK_x^2}{\Omega^2} f_2. \tag{3.7.13}$$

The coefficients involve no height derivatives. Other workers (Garrett, 1969; Bretherton, 1969b) use combinations of the pressure and vertical velocity. The boundary conditions, that f_1 and f_2 are continuous, may be shown to be identical with the continuity of the pressure and of the normal component of the velocity. Furthermore, when the second-order equation is formed from (3.7.12) and (3.7.13), then, provided that the second and higher order terms in dH/dz, such as d^2H/dz^2, $(dH/dz)^2$, etc., are neglected, the correction term is once more $H^{-1}\, dH/dz$, as in (3.7.8), and similar methods of solution apply.

3.8 Ducting

Consider a wave that has been reflected from some atmospheric gradient or discontinuity. This wave will be propagating downwards but, upon reaching the ground, will be totally reflected. It is thus possible to establish a duct, or waveguide, for the propagation of atmospheric waves.

The classic example of this is the perfect acoustic waveguide in which sound waves are trapped between two rigid walls or between a rigid wall and a free surface. A waveguide system is dispersive even though the medium in the guide may not be, so that even the high-frequency sound waves, which are normally non-dispersive, will have propagation characteristics that are determined by their frequency. The dispersion in this case is due to interference between up-going and down-going waves. Energy at a given frequency can propagate only at certain angles of incidence without losses due to destructive interference. The condition of constructive interference, giving the dispersion law, has the same form as the diffraction grating equation (Brekhovskikh, 1960). This is called 'geometrical' dispersion to distinguish it from other types of dispersion not due to interferences between reflections. Thus, the dispersion of acoustic waves and of gravity waves is due to a completely different type of process. The dispersion that yields resonant and cut-off frequencies is best referred to as 'structural' dispersion, since it is determined entirely by the natural resonant frequencies of the medium.

Certain boundary conditions need to be satisfied at the waveguide walls. The ground may be considered as a rigid boundary which cannot be displaced and the fluid in contact with it cannot move in the vertical z direction. Thus the boundary condition there is that $U_{1z} = 0$ at the ground. In reality, the ground is a solid in which both longitudinal and transverse sound waves can travel. To allow for this would involve the modulus of rigidity of the earth and thus introduce complexities outside the intended scope of this book.

If the atmosphere is being approximated by a number of isothermal layers, then at the boundary of each layer there is a temperature discontinuity which does not act like a rigid boundary. The boundary conditions in this case are that the pressure and vertical velocity be continuous across the boundary. The quantitative form for these boundary conditions is given in § 2.11.

In boundary value problems of this kind, only certain characteristic values, often called eigenvalues, of the wavelengths and frequencies can satisfy the boundary conditions. The waves which satisfy the boundary conditions are the permissible modes of propagation. Other modes cannot exist. This is in contrast to the continuous spectrum of waves possible in an unbounded medium. We have already met a simple case of a boundary value problem in the case of Lamb waves. In that case, the boundary condition was that the vertical particle velocity vanished everywhere and this gives a dispersion relation that was a line, corresponding to one mode, on the dispersion diagram. In general, ducted internal waves travel slower than the lowest sound speed in the ducting region.

To illustrate the method of analysing the waveguide modes, consider the case of a density gradated isothermal atmosphere with a free surface at an altitude $z = h$ and a rigid bottom at $z = 0$. Then for internal waves

$$\frac{p_1}{p_0 \mathrm{P}} = \frac{U_{1z}}{\mathrm{Z}} = A_0 \exp\left(\frac{z}{2H}\right) \exp\left[i(\omega t - k_x x - k_z z)\right] \qquad (3.8.1)$$

and the boundary condition at the rigid surface: $U_{1z} = 0$ at $z = 0$ implies

$$\frac{p_1}{p_0 \mathrm{P}} = \frac{U_{1z}}{\mathrm{Z}} = A_0 i \exp\left(\frac{z}{2H}\right) \sin k_z z. \qquad (3.8.2)$$

neglecting the harmonic time variation. Applying the free surface boundary condition gives

$$\left(i\omega \mathrm{P} - \frac{\gamma g}{c^2}\mathrm{Z}\right) \sin k_z h = 0$$

so that

$$k_z h + \phi = m\pi \qquad (3.8.3)$$

where

$$\phi = \tan^{-1}\frac{\text{Im}\,(i\omega\text{P}-Z/H)}{\text{Real}\,(i\omega\text{P}-Z/H)} \tag{3.8.4}$$

and $m = 0, 1, 2, \ldots$ gives the mode condition. The relation (2.4.3) describes the structural dispersion of k_z.

A detailed examination of (3.8.4) reveals that $\phi \neq 0$. Thus, in a rather complicated way, we have derived the physically obvious result that evanescent waves will not exist above a rigid surface. The reason that this should be obvious is that if $U_{1z} = 0$ at the bottom boundary then, if the wave is evanescent, U_{1z} must be zero throughout the waveguide, so that evanescent waves cannot exist—except for the Lamb wave which is defined as having no vertical motions.

This precludes the observation of evanescent waves other than the Lamb wave on ground based microbarographs, but it does not necessarily mean that no evanescent waves can exist in the upper atmosphere, or in a waveguide with a non-rigid lower boundary.

The one-layer model that we have used, with one free and one rigid boundary is a realistic approximation for the earth's oceans, but it is not a good choice for the atmosphere. The pioneering work in this field was done by Pekeris (1948a, b), who studied trapped acoustic modes in the atmosphere and in the ocean. His work, and that of Scorer (1950), failed to provide an adequate explanation of the waves generated by explosions in the atmosphere because he and Scorer had inadequate data concerning the upper atmosphere. Their results predicted that waves with periods less than two minutes could not propagate through the atmosphere. In fact, periods as low as 0·5 min have been identified on microbarograph records of waves from nuclear explosions.

There are certain simple temperature variations for which the differential equation that describes the wave parameters is analytically solvable. These solutions involve well-known and tabulated functions such as the Airy function, the Whittaker function, and the Confluent hypergeometric function. However, the number of cases where this can be done is small and, for most realistic profiles, multilayer techniques are required.

Within each layer there is a local ω_{B} and ω_{an}. The energy of internal gravity waves can be ducted by the reflection processes described by equations (3.7.3) and (3.7.5) which depend on ω_B and ω_{an}. Another ducting mechanism could be the result of the bending of the acoustic waves by variations in c. One would expect, by a cursory examination of the ω_B, ω_{an} and c profiles that there would be two distinct ducts for internal gravity waves that correspond to the ω_B maxima in the stratosphere and near the top of the mesosphere. There could also be two waveguides for high-frequency acoustic waves that correspond to the stratospheric and upper mesospheric minima in the speed of sound, as well as several more acoustic-gravity wave ducts due to minima of the acoustic cut-off frequency. Of the last, the most important is probably associated with the thermosphere itself.

Only a limited number of phenomena appear to be capable of generating waves that can travel large distances around the world and still have sufficient magnitude to be recorded on sensitive microbarographs. Multi-megaton nuclear explosions are capable of this, as was the eruption of the volcano Krakatoa in 1883 and the fall of the great Siberian meteorite in 1908, and large numbers of workers have investigated the problem by using various different atmospheric profiles. Fig. 3.10 represents an idealized picture of the principal modes that can exist in a windless atmosphere based upon the work of Pfeffer and Zarichny (1963) Friedman (1966) and Francis (1973).

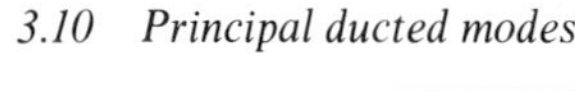

3.10 Principal ducted modes

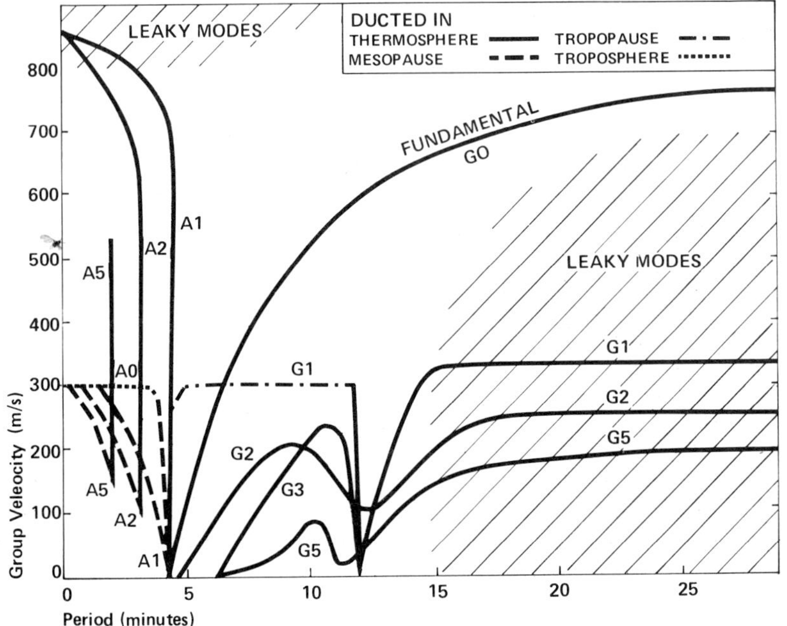

Different modes are important at different heights, depending on where the kinetic energy of the mode is localized, and observations at any one height will reflect those modes that are ducted there. For example, at the troposphere one would expect to observe the second, third and fourth acoustic modes that manifest themselves as waves in the period range from 20 s to 50 s. From 50 s to 100 s, the first acoustic mode dominates and, from 150 to 290 s, the fundamental acoustic mode is most important. The fundamental gravity mode is not at all important at the ground, but the first gravity mode controls the response from 290 s to 12 minute periods. Presumably higher gravity modes control the response at higher periods though there is an element of uncertainty as to which modes dominate. The situation is depicted in Fig. 3.10, which also summarizes the regions in which the various modes are principally excited. The acoustic wave high-frequency limit at around 300 m/s is the tropospheric sound speed, whereas the limit at 850 m/s represents the highest thermospheric sound speed. The low-frequency gravity waves reach group velocities about ninety per cent of this (§ 2.9).

Garrett (1969) has shown that most of the energy of the short-period waves from nuclear explosions lies in the fundamental $A0$ mode, which corresponds to a Lamb wave with a group velocity described by the formula

$$V_{\mathrm{g}} = \bar{c}\left(1 - \frac{3\mathscr{D}\omega^2}{\bar{c}^2}\right)$$

where $\bar{c}$ is the averaged tropospheric sound speed and $\mathscr{D}$ is a dispersion coefficient with a value of the order of 1 km^2.

The energy of certain modes is able to leak out of the waveguide. This was mentioned by Hines (1960) in order to explain the characteristic of travelling ionospheric disturbances in the F region, and is illustrated in Fig. 3.11. These leaky ducts were quantitatively investigated by Friedman (1966) by retaining a complex horizontal

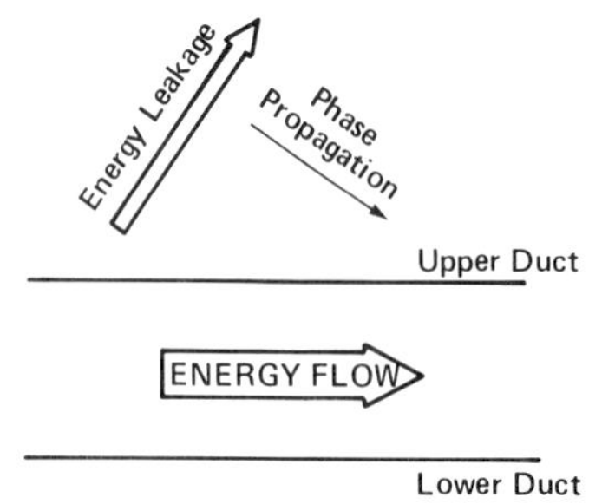

3.11 Leaky Duct model of energy propagation

wavenumber in the analysis. Friedman found that a large number of leaky modes existed with orders between 1 and 35. The degree to which each mode is excited depends on the altitude and spectral content of the source. It is consequently rather unlikely that modes with orders higher than 5 contain significant amounts of energy (Wickersham, 1966). Press and Harkrider (1962) and Pfeffer and Zarichny (1963) found that the first few modes adequately described the ground-based barograph records of nuclear explosions. This is likely to be true for natural sources as well, though a few extra modes become important in the observation of phenomena at ionospheric heights.

Inverse Dispersion

Donn and Shaw (1967), when examining microbarographic oscillations from nuclear explosions, found that, for gravity waves within certain period ranges, there existed inverse dispersion. This meant that the group velocity, V_g, decreased with period in contrast to the normal gravity wave situation, where V_g is an increasing function of period. Inverse dispersion was not predicted in the earlier theoretical models of Press and Harkrider and Pfeffer and Zarichny, but it was shown by Balachandran (1968) that inverse dispersion arises when realistic wind profiles are included in the analysis.

Coupling

In Fig. 3.10 it is apparent that the same mode can be important at different atmospheric heights. This arises through coupling between different modes wherein, for example, a gravity mode in the lower atmosphere transfers part of its energy to the upper atmospheric fundamental mode and, at the same time, the fundamental transfers its energy to the gravity mode. Coupling is more clearly representable on dispersion plots of angular frequency and wavenumber, and it occurs when two modes (represented as lines on the ω, k diagram) intersect. Jones (1970) has shown that modes can couple either through an 'embrace', in which case the energy exchange is gradual, or through a 'kiss', in which case the energy exchange is abrupt. The terms kiss and embrace arise from the appearance of the two forms of coupling on the dispersion diagram. In the kissing modes the two modal curves come together quite suddenly. In the embracing mode the modal curves come together gradually. Jones's results indicate that kissing modes occur for couplings of evanescent modes whereas embracing modes occur for internal wave couplings.

We have already mentioned that the elliptically polarized characteristic surface wave and the linearly polarized Lamb wave are unable to couple because of their different polarizations. However, by introducing a tropospheric source, Jones (1970) showed that coupling could indeed take place with the tropospheric Lamb wave being coupled to the stratospheric characteristic surface wave through a kissing mode (Fig. 3.12b). Similarly, the tropospheric characteristic surface wave is coupled to the upper atmospheric Lamb wave.

Energy coupling in kissing modes is related to the presence of boundary conditions or to the presence of loss or source mechanisms.

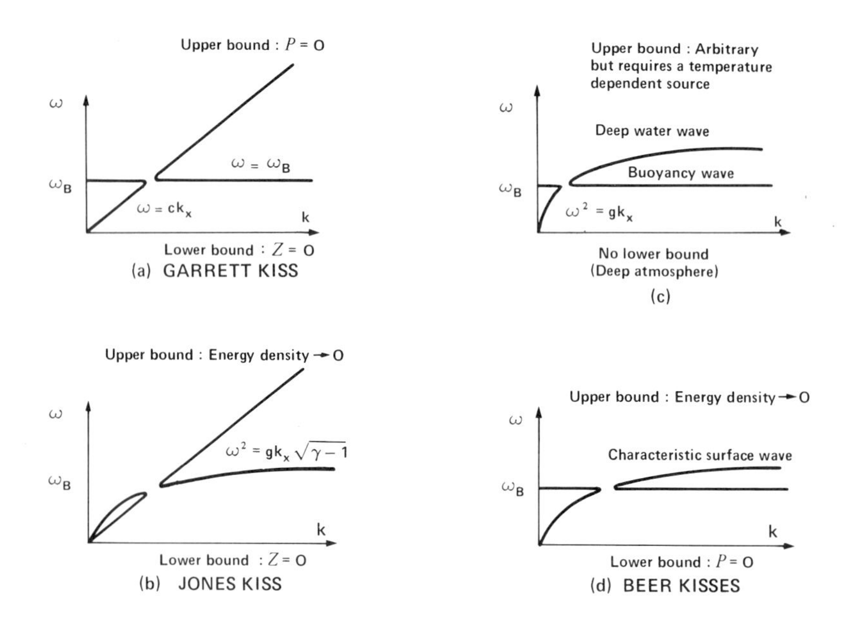

3.12 Kissing modes of energy coupling

Garrett (1969) examined the effect of bounding the atmosphere at a fixed height at which the pressure perturbations everywhere vanished. This immediately introduced a new modal curve at the boundary, namely the buoyancy wave $\omega = \omega_B$, which was coupled to the Lamb wave. Beer (1971) postulated a source term proportional to the atmospheric temperature, and demonstrated that a kissing mode exists between the buoyancy wave and an atmospheric equivalent of the deep water wave, though presumably a more complete analysis would indicate that the coupling was extremely weak. It does, however, emphasize the importance of source terms in producing coupled modes. These couplings are indicated in Fig. 3.12.

The importance of the buoyancy wave, the Lamb wave and the characteristic surface wave arises because they represent waves of characteristic impedance 0, ρc and infinity respectively.

The characteristic impedance, I, of an atmospheric wave in a windless atmosphere may be shown to be (Eckart, 1960, p. 113)

$$\mathrm{I} = \frac{(\omega^4 - \omega^2 \omega_{\mathrm{B}}^2)}{\omega^4 - k_x^2 c^2 \omega_{\mathrm{B}}^2} \rho c \tag{3.8.5}$$

where I is defined as the complex ratio of the pressure perturbations to the velocity perturbations. In the high frequency limit (3.8.5) reduces to $\mathrm{I} = \rho c$ which is the specific acoustic impedance for plane progressive waves.

It thus becomes evident that coupling between various atmospheric modes is intimately related to the impedances of the sources and the boundaries.

3.9 Winds

So far we have considered only constant zonal winds in the atmospheric wave derivations, though we know from the discussion in § 1.4 that there are vertical gradients in both the zonal and meridional winds which will not be identical. These vertical gradients can be combined to yield a horizontal wind in the direction of the atmospheric wave's phase propagation, whose magnitude varies with altitude. There are then two ways to tackle the resulting problem. One method is to solve the second-order differential equation that describes one of the wave parameters, ensuring that both temperature and wind-speed gradients are included. The other method is to divide the atmosphere into layers and to assume a constant wind speed within each layer. Both these methods are strongly reminiscent of the treatment applied to the situation in which there was a vertical temperature gradient. In fact, winds can produce reflections of the waves, which can lead to ducting just as temperature variations do and in the same manner. However, the effects due to wind variation appear to be richer than those due only to temperature variations; for, at the critical layers, the winds act like dissipators rather than reflectors of wave energy. The particular mathematical technique chosen depends on the desired result. It is easier to use the differential equation to find the reflection conditions but more tractable to use multilayer methods to examine the ducting.

Both the relevant equations and the procedure for solving them are outlined in Chapter 2. The straightforward but rather lengthy elimination process eventually yields a second-order ordinary differential equation which is homogeneous when source and loss terms are neglected and is linear when plane wave solutions are sought.

The coefficients of the differential equation are highly complicated functions of the winds, the temperature and the derivatives of the winds and the temperature. This method was employed by Martyn (1950) who gives the explicit equation for χ. For gradients of the neutral wind, which then have a simple analytical form, it is possible to solve the differential equation. However, the number of occasions when the neutral wind gradient is a simple expression is rather limited. The prime examples of mathematically tractable background winds are the tidal winds. In their case, the differential equation can be transformed into Mathieu's differential equation and exact solutions obtained. These solutions indicate that ducting will occur (Liu, 1970). A similar situation arises if we regard the background winds for one particular gravity wave as arising from another gravity wave. This effectively couples the gravity waves of different frequencies and can lead to non-linear interactions between the waves (Spizzichino, 1969, 1970).

A simpler method is also available. We should recall that, to include the effects of constant winds, it was merely necessary to replace ω by $\omega - U_{0x}k_x$ in the equations derived in the windless case. Is it also possible to derive equations in the steady-state case ($\omega = 0$) including the winds and then to replace U_{0x} by $U_{0x} - \omega/k_x$? In practice, the answers that are obtained by doing this are quite adequate for most physical applications (Booker and Bretherton, 1967; Synge, 1933), so that it is immediately possible to extend the Scorer equation, which is derived in Chapter 4, to obtain

$$\frac{d^2 \mathrm{w}}{dz^2} + \left[\frac{\omega_{\mathrm{B}}^2}{(U_{0x} - \omega/k_x)^2} - \frac{U''}{(U_{0x} - \omega/k_x)} - k_x^2\right]\mathrm{w} = 0 \qquad (3.9.1)$$

where

$$U_{1z} = \mathrm{w} \exp\left[i(\omega t - k_x x)\right] \exp\left[-\int \frac{dz}{2H}\right]$$

$$U'' = \frac{d^2(U_{0x})}{dz^2}.$$

If $U''/(U_{0x} - \omega/k_x)$ is small (as it usually is), this equation becomes the low-frequency (Ω) form of (2.5.13). The $(1/H)(d\mathrm{W}/dz)$ term of (2.5.13) is eliminated by the transformation (2.5.17). However, for low-frequency waves $\omega_{\mathrm{B}}^2 k_x^2/\Omega^2 \gg \frac{1}{4}H^2$ so that the $(1/H)(d\mathrm{W}/dz)$ term may be completely ignored. In this case, we have low-(Doppler-shifted)-frequency upward-propagating waves which will be reflected when

$$\Omega = \omega_{\mathrm{B}} \qquad (3.9.2)$$

which is a condition that has already been derived for thermally stratified atmospheres. In this case, however, we are primarily interested in the ducting due to the wind and a notable example of this occurs when a periodic train of lee waves extends downwind from a range of mountains. If ω_{B} and U_{0x} change sharply with height then ducting may occur (see Fig. 3.13). Of course, in that case, equation (3.9.1) becomes inadequate and the full terms that include the temperature variations need to be retained.

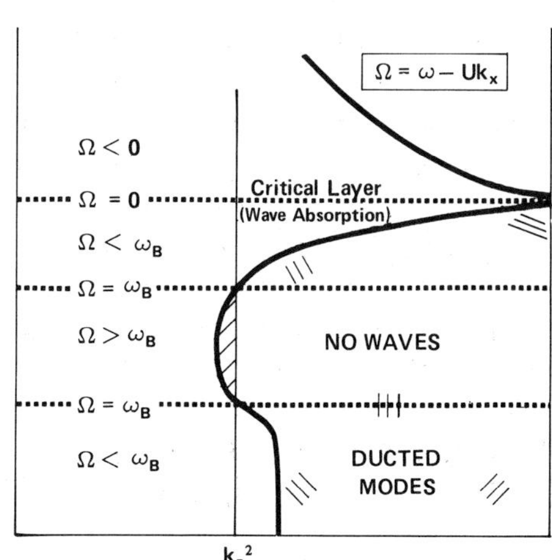

3.13 Wave ducting and critical layer absorption

Another possibility, which has already been briefly mentioned, is critical layer absorption. To examine this phenomenon, let us first find the vertical group velocity V_{gz} for internal waves. It should be recalled that there is no vertical group or phase velocity for evanescent waves because they do not propagate vertically. By differentiating (2.4.3), rearranging the terms, and applying Table 2.3

$$V_{gz} = \frac{\Omega^3 k_z c^2}{\Omega^4 - k_x^2 \omega_B^2 c^2} \tag{3.9.3}$$

so that, as a wave approaches the critical level where $U_{0x} = \omega/k_x$, the vertical group velocity becomes zero. In other words, the wave energy is unable to pass through the critical layer and is absorbed at that altitude. Furthermore, as $\Omega \to 0$, the vertical wavenumber $k_z \to \infty$, so that the vertical scale size of the wave becomes smaller and the rate of viscous dissipation increases. The existence of critical layers in both the lower atmosphere—where jet streams are likely to be major absorbers of gravity waves—and in the upper atmosphere will have a profound effect on the spectrum of gravity waves that can exist at any one height. Critical layers will remove the waves with small horizontal wavelengths because they have low horizontal phase velocities. Another interesting effect has been postulated by Jones and Houghton (1972), who point out that, at some height, the perturbation wind velocity of gravity waves U_{1x}, which is increasing in magnitude exponentially, will have reached a value comparable with the phase velocity of the gravity wave. In this case, the wind perturbation acts as the wave's critical layer through momentum exchange with the mean background wind, and the wave becomes self-destructing.

Dickinson (1970) has analysed the development of critical levels for Rossby waves. These occur in the same way as for gravity waves, but in a rotating system, the critical level occurs when $\Omega = \omega - U_{0x}k_x = f$, the Coriolis parameter. Since the phase velocity of a Rossby wave is always westward, the effects of a critical layer should become most pronounced when a Rossby wave tries to cross the equatorial easterly (westward) upper air winds. Surprisingly, recent numerical simulations of the forced Rossby wave pattern indicate that the presence or absence of equatorial easterlies makes little difference. (Vergeiner and Ogura, 1972). This most probably demonstrates that Rossby wave critical layers are unlikely to play a major role in the general atmospheric circulation, due to the failure of most of the important Rossby waves to have significant meridional components of phase or group velocity, resulting in their failure to reach their critical layer.

3. Background Reading

Infrasonics and atmospheric acoustics, *Geophysical Journal of the Royal Astronomical Society*, **26**, part 1–4. (1971).

4

Waves in the Lower Atmosphere

4.1 Vorticity

It is impossible to discuss the motion of the earth's lower atmosphere without introducing the earth's rotation and it is extremely difficult to discuss rotating fluids adequately without introducing vorticity. The concept arises because any vector field **U** can be uniquely divided into a solenoidal and an irrotational component, $\mathbf{U}_s$ and $\mathbf{U}_i$ respectively, where

$$\nabla \times \mathbf{U}_i = 0$$

and

$$\nabla \cdot \mathbf{U}_s = 0.$$

To determine the velocity of a real fluid completely, it is also necessary to know the value of **U** at the boundaries. The divergence of **U** has been represented by the scalar $\chi = \nabla \cdot \mathbf{U}$ and it represents the rate of change of volume of the fluid. The vorticity ζ is given by

$$\zeta = \nabla \times \mathbf{U}$$

and it represents a rotation of the fluid at an angular velocity of $\zeta/2$.

Provided that the flow is wholly irrotational, then the non-linear terms in the equations of motion can be written as exact differentials, so that

$$U_x\, dx + U_y\, dy + U_z\, dz = -d\phi$$

and $\mathbf{U} = -\nabla\phi$ where the scalar function ϕ is known as the velocity potential.

On the other hand, if the motion is in two dimensions, so that

$U_z = 0$ and the motion is wholly non-divergent, then $U_x\,dy - U_y\,dx$ can be written as an exact differential

$$U_x\,dy - U_y\,dx = -d\psi$$

where ψ is Lagrange's stream function and is related to the velocity components—which are assumed to be functions of x and y only—by

$$U_x = -\frac{\partial\psi}{\partial y}; \qquad U_y = \frac{\partial\psi}{\partial x}$$

or, in plane polar coordinates,

$$U_r = -\frac{1}{r}\frac{\partial\psi}{\partial\theta}; \qquad U_\theta = \frac{\partial\psi}{\partial r}.$$

The equation of continuity reveals that the flow of an incompressible fluid is non-divergent and, if it is also irrotational, then it will satisfy Laplace's equation $\mathbf{\Delta}\phi = 0$, where $\mathbf{\Delta}$ is the three-dimensional Laplacian. The functions ϕ and ψ are connected by the relations

$$\frac{\partial\phi}{\partial x} = \frac{\partial\psi}{\partial y}; \qquad \frac{\partial\phi}{\partial y} = -\frac{\partial\psi}{\partial x}$$

and any complex function

$$f(x+iy) = \phi + i\psi$$

will satisfy the governing relations. The vertical component of the vorticity ζ_z is related to the stream function by

$$\zeta_z = \frac{\partial^2\psi}{\partial x^2} + \frac{\partial^2\psi}{\partial y^2}$$
$$= \nabla^2\psi$$

where ∇^2 is restricted to the two-dimensional Laplacian.

The vorticity equation is derived by taking the curl of both sides of the vector equation of motion. If we assume an isothermal inviscid flow and neglect the Coriolis term, then in Cartesian coordinates

$$\frac{D\mathbf{U}}{Dt} = \frac{\partial\mathbf{U}}{\partial t} + (\mathbf{U}\,.\,\nabla)\mathbf{U} = 0. \qquad (4.1.1)$$

We shall need to use the following vector identities

(1) $(\mathbf{U}\,.\,\nabla)\mathbf{U} = \frac{1}{2}\nabla U^2 - \mathbf{U}\times(\nabla\times\mathbf{U})$

(2) $\nabla\times\nabla\phi = 0$

(3) $\nabla\times(\mathbf{A}\times\mathbf{B}) = (\mathbf{B}\,.\,\nabla)\mathbf{A} - \mathbf{B}(\nabla\,.\,\mathbf{A}) - (\mathbf{A}\,.\,\nabla)\mathbf{B} + \mathbf{A}(\nabla\,.\,\mathbf{B})$

where **A** and **B** are any arbitrary vectors and ϕ is any scalar function. Then, if we take the curl of (4.1.1) and use the vector identities, we obtain

$$\frac{\partial\zeta}{\partial t} + (\mathbf{U}\,.\,\nabla)\zeta + \zeta(\nabla\,.\,\mathbf{U}) - (\zeta\,.\,\nabla)\mathbf{U} - \mathbf{U}(\nabla\,.\,\zeta)$$

$$= \frac{D\zeta}{Dt} + \zeta(\nabla\,.\,\mathbf{U}) - (\zeta\,.\,\nabla)\mathbf{U}$$

$$= 0$$

because

(4) $\nabla\,.\,(\nabla\times\mathbf{U}) = \nabla\,.\,\zeta = 0.$

If we now limit ourselves to two-dimensional flow, then $(\zeta\,.\,\nabla)\mathbf{U} = 0$ and, if we consider an incompressible fluid of constant density, then $\nabla\,.\,\mathbf{U} = 0$, so that the vorticity equation becomes

$$\frac{D\zeta}{Dt} = 0 \tag{4.1.2}$$

and represents the fact that the vorticity of a fluid element is conserved. This equation was used in the discussion of the elementary Rossby wave. In most cases, however, the vorticity equation will not have a right-hand side that equals zero. One common case is two-dimensional flow accompanied by changes in the depth, h, of the fluid due either to the topography of the earth's surface, which is acting as the lower boundary of the fluid, or to the presence of large-scale external-type waves (e.g. tides) on the upper surface. In this case, the continuity of mass gives the two-dimensional wind divergence as

$$\nabla\,.\,\mathbf{U} = -\frac{1}{h}\frac{Dh}{Dt}$$

so that the vorticity equation may be rewritten as

$$\frac{D(\zeta/h)}{Dt} = 0.$$

When the right-hand side of the vorticity equation (4.1.2) does not equal zero, then Bjerknes' theorem points out that the surfaces of constant pressure (isobars) and the surfaces of constant density (isopycnics) are no longer parallel. An atmosphere in which the isopleths of constant density and pressure are parallel is called a barotropic atmosphere, and one in which they are not parallel is known as a baroclinic atmosphere. The tubes caused by the intersection of the isobaric and isopycnic surfaces at fixed unit intervals are called solenoids and represent a measure of the baroclinicity. The intensity of the solenoidal field is a measure of the circulation, Γ, which is defined by the line integral

$$\Gamma = \oint \mathbf{U} \cdot d\mathbf{l}.$$

The vorticity may be related to the circulation by applying Stoke's theorem to obtain

$$\Gamma = \oint_C \mathbf{U} \cdot d\mathbf{l} = \iint_S (\nabla \times \mathbf{U})\, dA$$

$$= \iint_S \zeta \cdot dA$$

where dA is a unit element of the surface bounded by the closed curve C.

4.2 Filtering

Computational Instability

There is still substantial interest in gravity waves in the troposphere. This dates from the late 1940s, when the 'lee waves' that are set up on the leeward side of mountains by the wind blowing over them were successfully interpreted in terms of gravity waves. Gravity waves can also exist in regions of the troposphere where there are no mountains and, when they form, they then show up on microbarograph records. On the other hand, both lee waves and tropospheric gravity waves are quite weak and are of far less interest in the troposphere than the pressure fluctuations and the consequent climatic changes that are generated by the planetary waves in the mid-latitude troposphere. In the tropics, the strength of the planetary waves is greatly reduced and the atmospheric tides predominate (Fig. 1.4).

It is thus apparent that there is a wide variety of solutions to the equations that describe the state of the atmosphere. In order to make the mathematics tractable, the normal procedure is to decide which waves are of interest and to eliminate those terms of the equations that are responsible for the existence of the extraneous solutions. This process is known as filtering and it is the mathematical analogue of the mechanical or electrical filtering done by a microbarograph when it eliminates the unwanted frequencies. The development so far has actually utilized filtering approximations. By neglecting the Coriolis force and the sphericity of the earth, tidal solutions have been filtered out.

Various attempts have been made to solve the dynamic equations of the atmosphere by numerical methods. The pioneering work was done by L. F. Richardson (1922), who attempted a six-hour forecast for the European area. Unfortunately, Richardson was unaware of the importance of wave filtering and his predictions completely failed to agree with the observed developments of the weather. To illustrate the problems involved let us attempt to find finite difference solutions to the wave equation in one dimension

$$\frac{\partial^2 p}{\partial t^2} = V_\mathrm{p}^2 \frac{\partial^2 p}{\partial x^2} \tag{4.2.1}$$

where V_p is the phase velocity of the wave we are dealing with. Both the spatial coordinate, x, and the temporal coordinate, t, are now divided into equal intervals, so that

$$x = n\Delta x \qquad n = 0, 1, 2, \ldots$$

$$t = m\Delta t \qquad m = 0, 1, 2, \ldots$$

and then

$$\frac{\partial p}{\partial x} = \frac{p_{n+1} - p_n}{\Delta x}$$

where p_n represents the value of p when $x = n\Delta x$. The second derivative is then

$$\frac{\partial^2 p}{\partial x^2} = \frac{p_{n+1} - 2p_n + p_{n-1}}{(\Delta x)^2} \tag{4.2.2}$$

and similarly

$$\frac{\partial^2 p}{\partial t^2} = \frac{p_{m+1} - 2p_m + p_{m-1}}{(\Delta t)^2}. \tag{4.2.3}$$

Now, if we take $p = p_0 \exp[i(\omega t - kx)]$, where $\omega/k = V_p$, and substitute it into (4.2.2) and (4.2.3), then (4.2.1) becomes

$$\frac{\exp(i\omega\Delta t) - 2 + \exp(-i\omega\Delta t)}{(\Delta t)^2} = \frac{V_p^2}{(\Delta x)^2}[\exp(ik\Delta x) - 2 + \exp(-ik\Delta x)]$$

so that

$$\sin^2\left(\frac{\omega\Delta t}{2}\right) = \left(V_p \frac{\Delta t}{\Delta x}\right)^2 \sin^2\left(\frac{k\Delta x}{2}\right)$$

We are interested in the development of this atmospheric wave as the time progresses, so that $V_p\Delta t/\Delta x$ must be less than or equal to unity for the resultant solution to represent a well-behaved wave. If $V_p\Delta t/\Delta x$ is greater than unity, then the solution $\omega\Delta t$ becomes imaginary and the solution represents an unstable wave. The wave's instability is, however, solely a consequence of the numerical methods we have used to solve the equation and is therefore referred to as a computational instability. The stability criterion

$$\frac{V_p\Delta t}{\Delta x} < 1$$

is known as the Courant–Friedrichs–Lewy (CFL) criterion.

The onset of computational instability is dependent on the spatial and temporal intervals at which readings are taken and also on the phase velocity of the wave. Let us imagine a network of meteorological sounding stations spaced 100 km apart and taking readings approximately every hour. In this case, $\Delta x/\Delta t \approx 28$ m/s, so that, if the relevant wave's phase velocity is less than 28 m/s, the results obtained will be stable. If the phase velocity exceeds 28 m/s, then computational instability will manifest itself. Now for sound waves, the phase velocity is of the order of 300 m/s; for gravity waves, the horizontal phase velocity is of the order of 100 m/s; and for Rossby waves, it is around 20 m/s. The solutions will therefore be unstable unless sound waves and gravity waves are excluded. The meteorologically least relevant waves are the waves that are responsible for computational instability and hence the acoustic waves and gravity waves are often lumped together under the rather descriptive title of 'meteorological noise'. It is intriguing to notice that computational instability can be avoided either by filtering out the meteorological noise, by greatly reducing the time interval between the readings, or surprisingly by *increasing* the distance between successive readings. This is because sound waves and gravity waves have much shorter wavelengths than the Rossby waves, so that, by choosing stations close together, one is compiling more information about the shorter wavelength waves than is necessary or desirable. Increasing the distance between stations smoothes out the fluctuations due to meteorological noise and accentuates the oscillations arising from the meteorologically relevant waves. Another way of surmounting the problem of computational instability is to increase the complexity of the original equations that are being used by including the frequency-dependent absorption terms that effectively damp out the meteorological noise. This approach, however, increases both the complexity of the problem and the computing time involved in the calculations without any concomitant improvement in physical insight and is not greatly favoured. A review of the recent developments in the computational problems of the finite difference method as it is applied to geophysical problems has been recently published by Shapiro (1970).

Incompressible Fluid Dynamics

The simplest way to eliminate sound waves is to assume that the atmosphere is completely incompressible, in which case γ, the ratio of specific heats, is infinite and the speed of sound becomes infinite. If this assumption is used, but density gradients are still included, then the Vaisala–Brunt frequency is non-zero and gravity waves characterized by density variations can still occur. In this case, $\nabla \cdot \mathbf{U} \neq 0$, and the Vaisala–Brunt frequency is given by

$$\omega_B^2 = -g\frac{d}{dz}(\ln \rho).$$

It is more common to impose the further condition that the density remains constant throughout the body of the atmospheric fluid, so that ω_B becomes zero. This eliminates internal gravity waves.

We now wish to find the height, h, of the equivalent constant-density incompressible atmosphere. It was pointed out in § 1.2 that the scale height H is the height that a column of air would have to be in order to produce the same pressure at the ground that the compressible atmosphere would if the column of air had a constant density equal to the ground value. Therefore, we could intuitively think of h, which is conventionally called the equivalent depth, as being equal to H in an isothermal atmosphere. We shall return to this in the next section.

In the incompressible case with constant mean density, there will be waves at the surface which will produce perturbations in the pressure and the vertical velocity, which are related by

$$\frac{\partial p_1}{\partial t} = \rho g U_{1z}.$$

Substituting this into the equation of motion we get

$$\frac{\partial^2 U_{1z}}{\partial t^2} + g \frac{\partial U_{1z}}{\partial z} = 0 \qquad (4.2.4)$$

at the surface, $z = h$. This is the well-known differential equation for surface waves (Lamb, 1945, p. 364). Furthermore, in the incompressible case, the mass continuity equation shows that the motion is irrotational, so that

$$\nabla . \mathbf{U} = 0$$

or, in terms of the perturbation quantities,

$$\frac{\partial U_{1x}}{\partial x} + \frac{\partial U_{1z}}{\partial z} = 0. \qquad (4.2.5)$$

At this stage, it is possible either to follow the method that has been used previously or to define a velocity potential. The first alternative requires us to take

$$\frac{U_{1x}}{\mathrm{X}} = \frac{U_{1z}}{\mathrm{Z}} = \exp\left[i(\omega t - K_x x - K_z z)\right]$$

in which case, it is possible to use (4.2.4) and (4.2.5) to show that

$$\frac{X}{Z} = -\frac{K_z}{K_x} = -\frac{i\omega^2}{gk_x} \tag{4.2.6}$$

which indicates that the surface waves are circularly polarized. The second alternative defines the velocity potential ϕ, such that

$$\mathbf{U} = -\nabla\phi \tag{4.2.7}$$

so that (4.2.5) becomes

$$\frac{\partial^2\phi}{\partial x^2}+\frac{\partial^2\phi}{\partial z^2} = 0 \tag{4.2.8}$$

and (4.2.4) is

$$\frac{\partial^2\phi}{\partial t^2}+g\frac{\partial\phi}{\partial z} = 0. \tag{4.2.9}$$

By assuming, once again, plane wave solutions for

$$\phi = A \exp [i(\omega t - K_x x - K_z z)],$$

we find that

$$K_z^2 = -K_x^2 \tag{4.2.10}$$

or, in other words, that the only waves that can exist in an incompressible fluid without density gradients are evanescent waves or horizontally decaying internal waves. By assuming the plane-wave solutions, we also find that

$$K_z = \frac{i\omega^2}{g}. \tag{4.2.11}$$

Equations (4.2.10) and (4.2.11) do *not* satisfy the boundary condition that $\partial\phi/\partial z = U_{1z} = 0$. In order to satisfy this condition, we need to take

$$\phi = [A_a \exp(iK_z z)+A_b \exp(-iK_z z)] \exp [i(\omega t - K_x x)]$$

where A, A_a and A_b are all arbitrary constants. Then the boundary condition implies that $A_a = A_b = A/2$, say, and then, because (4.2.10) is still true,

$$\phi = A \cosh (k_x z) \exp [i(\omega t - k_x x)]$$

where we assume no horizontal attenuation, so that by using (4.2.9) at $z = h$, the dispersion relation becomes

$$\omega^2 = gk_x \tanh(k_x h). \qquad (4.2.12)$$

Equation (4.2.12) has two important limiting cases. The first one occurs when $k_x h \ll 1$, so that for a shallow fluid

$$\frac{\omega^2}{k_x^2} = gh \qquad (4.2.13)$$

gives the square of the phase velocity of evanescent waves in a shallow atmosphere. The second limiting case occurs when $k_x h \gg 1$ and then $\tanh(k_x h) = 1$ so that

$$\omega^2 = gk_x. \qquad (4.2.14)$$

This is the dispersion relation for deep-water waves and corresponds to equation (4.2.11). This case has already been discussed in connection with evanescent waves in a compressible atmosphere.

We thus find that the filtering condition of an incompressible atmosphere of depth h filters out both the sound and internal gravity waves completely and filters the external and body waves out at the bottom of the atmosphere. External and body waves can be completely eliminated by postulating a fixed upper boundary condition for the incompressible fluid. On the other hand, if one wishes only to eliminate the acoustic waves whilst retaining the gravity waves, it is possible to use the Boussinesq approximation, in which the atmosphere is taken to be uniform and incompressible in the horizontal direction, but to be compressible and in hydrostatic equilibrium in the vertical direction. Both these criteria may be expressed in terms of the wind velocities and the wind divergence, as we have seen. In the incompressible constant-density case, the mass continuity equation shows that

$$\nabla . \mathbf{U} = 0$$

whereas the adiabatic equation and the equation of motion imply that

$$\frac{DU_{0z}}{Dt} = 0.$$

On the other hand the Boussinesq approximation will allow for solenoidal perturbation velocities, so that the filtering conditions are that

$$\nabla \cdot \mathbf{U}_0 = 0 \quad \text{and} \quad \frac{DU_{0z}}{Dt} = 0.$$

These conditions are identical with the geostrophic wind condition, namely that the background wind is in equilibrium under the action of the pressure gradient forces and the Coriolis force. The geostrophic wind is divergence-free. The e (zonal) and n (meridional) components are given (§ 1.4) by

$$U_{0e} = -\frac{1}{\rho f}\frac{\partial p}{\partial n}$$

and

$$U_{0n} = \frac{1}{\rho f}\frac{\partial p}{\partial e}$$

so that, in a Boussinesq fluid, $\nabla \cdot \mathbf{U}_0 = 0$ in the geostrophic approximation. Because it is only the background winds that are in geostrophic balance, this condition is known as the quasi-geostrophic approximation. Temporary departures from hydrostatic equilibrium lead to oscillations about a mean state with periods of minutes or more, departures from geostrophic equilibrium lead to oscillations with periods of several hours or more. The quasi-geostrophic approximation permits these departures, though it needs to be emphasized that the geostrophic approximation is invalid at low latitudes. Eliassen and Kleinschmidt (1957, p. 93) deal extensively with the quasi-geostrophic approximation and they point out that, if one uses the geostrophic approximation as a filtering method in a finite-difference method, then there is another CFL criterion to avoid computational instability, namely that the time intervals Δt should be related to the mesh size Δx by

$$\Delta t < \frac{\Delta x}{(\sqrt{2}U_0)}$$

where U_0 is the maximum geostrophic wind velocity which occurs.

The quasi-geostrophic equations form a consistent system of equations that can be used to predict meteorological developments. However the geostrophic approximation is a rather crude one and the quasi-geostrophic equations do not always give sufficient accuracy.

In an attempt to obtain better filtering approximations, Herbert (1971) shows that there are two families of sound filtering conditions that correspond to the equations

$$\left(\frac{D}{Dt}\right)^m U_{0z} = 0$$

and

$$\left(\frac{D}{Dt}\right)^m \nabla \, . \, \mathbf{U}_0 = 0 \qquad \text{where } m = 0, 1, 2, \ldots$$

As we have already seen, $m = 0$ gives the quasi-geostrophic approximation and $m = 1$ gives the quasi-hydrostatic one (i.e. the Boussinesq approximation in the vertical direction). Higher orders of m yield further equations, some of which may be recognized as well-known approximations (e.g. the Richardson equation). Herbert points out that the higher order filtering systems are necessary if non-linear atmospheric wave systems are to be studied.

4.3 Planetary Waves

Planetary waves, of which the well-known Rossby waves are a special case, occur in rotating fluids. Rossby waves are the idealizations of planetary waves on to a plane. When motions are referred to axes which rotate, then the Coriolis and the centrifugal forces must be taken as acting on the fluid. The importance of the Coriolis force in relation to the inertial forces is given by the Rossby number which is defined as

$$\frac{U}{L\Omega_E}$$

where U is a representative velocity, measured relative to the rotating axes and L is a measure of the distance over which the velocity varies appreciably. When $U/L\Omega_E \gg 1$, Coriolis forces are likely to cause only a slight modification of the flow pattern, but when $U/L\Omega_E \ll 1$, the effects of the Coriolis force are likely to be dominant. We are mainly interested in the in-between situation, when $U/L\Omega_E \sim 1$.

A simplified theory of the Rossby wave was outlined in § 2.2. This neglected the changes in vorticity induced by the changes in depth of the fluid. For a barotropic atmosphere the zonal phase velocity of a Rossby wave measured relative to the ground is (Charney, 1947)

$$V_{pe} = \frac{U_{0e} - \beta\lambda^2/4\pi^2}{1 + f^2\lambda^2/(4\pi^2 gH)} \tag{4.3.1}$$

where λ is the wavelength and H is the scale height derived from the mean temperature of the fluid. In all practical situations where the β plane approximation is valid

$$\frac{f^2\lambda^2}{4\pi^2 gH} \ll 1$$

so that the zonal phase velocity is

$$V_{pe} = U_{0e} - \frac{\beta\lambda^2}{4\pi^2}$$

as derived in § 2.2.

There will be a wavelength λ_s, called the stationary wavelength that determines the apparent direction of motion of the Rossby wave. If $\lambda > \lambda_s$, then the wave will appear to move westward and, if $\lambda < \lambda_s$, then the wave will appear to move eastward. If the background wind is directed westward (i.e. an easterly wind), then no stationary

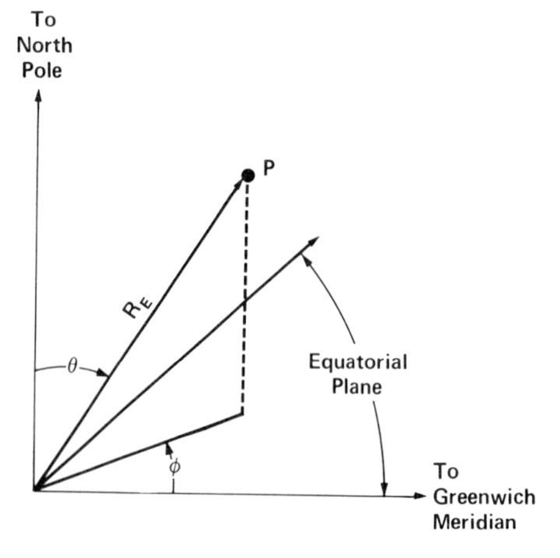

4.1 Spherical coordinate system

waves are possible and all Rossby waves propagate westward. The stationary wavelength is given by setting $V_{pe} = 0$ in (4.3.1) to yield

$$\lambda_s = 2\pi\sqrt{\frac{U_{0e}}{\beta}} \tag{4.3.2}$$

however, because

$$\beta = \frac{\partial f}{\partial n} = \frac{\partial(2\Omega_E \cos\theta)}{-\partial(R_E\theta)}$$

$$= \frac{(2\Omega_E \sin\theta)}{R_E} \tag{4.3.3}$$

then

$$\lambda_s = 2\pi\sqrt{\frac{U_{0e}R_E}{(2\Omega_E \sin\theta)}} \tag{4.3.4}$$

where R_E is the radius of the earth. Table (4.1) gives values of the stationary wavelength.

A Rossby wave in the earth's atmosphere must be spatially periodic around the circumference of the earth, so that

$$2\pi R_E \sin\theta = m\lambda \tag{4.3.5}$$

where $m = 1, 2, \ldots$.

Table 4.1 The stationary wavelength in kilometers as a function of latitude and mean eastward wind speed

U_{0e} / $90° - \theta$	4 m/s	8 m/s	12 m/s	16 m/s	20 m/s
30°	2822	3990	4888	5644	6310
45°	3120	4412	5405	6241	6978
60°	3713	5252	6432	7428	8304
Latitude / U_{0e}	14·4 km/hr	28·8 km/hr	43·2 km/hr	57·6 km/hr	72·0 km/hr

$\Omega_E = 7{\cdot}29 \times 10^{-5}\ \mathrm{s}^{-1}$
$R_E = 6{\cdot}37 \times 10^{6}\ \mathrm{m}$

The value of m for any particular wave may be determined from (4.3.1) as

$$m^2 = \frac{(2\Omega_E R_E \sin^3 \theta)}{(U_{0e} - V_{pe})} \tag{4.3.6}$$

which will yield only certain characteristic values of the Doppler-shifted wave velocity. For the stationary wave

$$m^2 = \frac{(2\Omega_E R_E \sin^3 \theta)}{U_{0e}}. \tag{4.3.7}$$

Now we are already aware that Rossby waves are weak at low latitudes because of the absence of strong eastward winds at the equator. Equation (4.3.6) indicates that Rossby waves will also not exist at high latitudes if

$$\sin^3 \theta < \frac{(U_{0e} - V_{pe})}{2\Omega_E R_E}.$$

Basically, Rossby waves are a mid-latitude phenomenon.

The imposition of the boundary condition (4.3.5) enables us to determine the high frequency cut-off for Rossby waves. Using (4.3.1), (4.3.3) and (4.3.5) gives the period of Rossby waves as $12m/\sin^2 \theta$ hours, resulting in a high frequency cut-off corresponding to 12 hours.

The main pressure centres are in the temperate zones. That is, the semi-permanent system of highs and lows is related to the Rossby wave pattern. The Rossby wave pattern is in turn influenced by the mean zonal winds which are the result of the mean meridional pressure gradients, and so these pressure gradients are an important index of motion of the main centres of meteorological action. Fig. 4.2 depicts a 'trough and ridge' diagram which plots the pressure variations at a height of 3 km, which corresponds to a mean pressure of 700 mb. The diagram plots the pressure variations at latitude 50° N as a function of time (the ordinate) and longitude (the abscissa). Rossby in 1945 showed that these pressure changes are controlled by the group velocity of the Rossby waves, which also controls the distributions of energy and wavelength.

The zonal (east-west) group velocity V_{ge} of a packet of Rossby waves is

$$V_{ge} = U_{0e} + \frac{\beta\lambda^2}{4\pi^2} \tag{4.3.8}$$

and the broken line in Fig. 4.2, which connects the troughs and ridges, indicates that the energy moves at a nearly uniform rate of 30° longitude per day.

Fig. 4.2 is roughly divided into four cells by the 698 mb isobar (labelled as 98 on the figure) which indicates that the Rossby waves

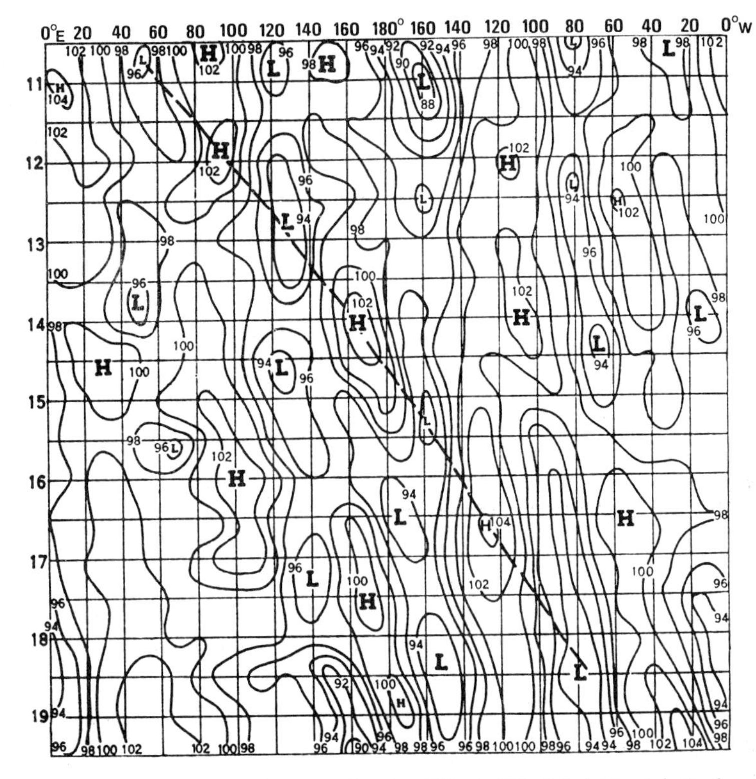

4.2 Trough and ridge diagram showing 700 mb (3 km) heights at latitude 50°N as a function of longitude (abscissa) and time (ordinate) between 11–19 October, 1948 (after Platzman, 1968)

apparent during the plotting period had $m = 4$. Furthermore the 698 mb dividing lines run almost parallel to the time axis, and it would appear that the dominant Rossby wave in the energy packet was a stationary wave. This then provides a measure of the average background zonal wind speed which, from (4.3.1) and (4.3.8), is given by

$$U_{0e} = \frac{V_{ge}}{2}$$

$$= 12{\cdot}4 \text{ m/s.}$$

Reference to Table 4.1 will show that the stationary wavelength associated with this wind speed is equivalent to a longitude of 90°—i.e. $m = 4$.

Planetary waves are geostrophic Rossby waves on the surface of a sphere. Just as Rossby waves have constant absolute vorticity on the β plane, so idealized planetary waves have constant absolute vorticity on the surface of the sphere (Longuet–Higgins, 1964).

If ζ is the vorticity measured relative to the rotating sphere, then the absolute vorticity ζ_a has a vertical component

$$\zeta_a = \zeta_z + 2\Omega_E \cos\theta \tag{4.3.9}$$

where θ is the colatitude. As the absolute vorticity remains constant in a Lagrangian formulation,

$$\frac{D\zeta_a}{Dt} = 0. \tag{4.3.10}$$

For small relative motions, this becomes

$$\frac{\partial\zeta_z}{\partial t} - \frac{2\Omega_E U_\theta \sin\theta}{R_E} = 0 \tag{4.3.11}$$

where the velocity in the θ coordinate, U_θ, is the rate of change of θ following a particle multiplied by the earth's radius.

At this stage, it becomes convenient to express the velocity in terms of a stream function, traditionally represented by ψ, where the components of the velocity in polar coordinates are

$$U_\theta = \left(\frac{1}{R_E \sin\theta}\right)\frac{\partial\psi}{\partial\phi} \tag{4.3.12}$$

and

$$U_\phi = -\frac{1}{R_E}\left(\frac{\partial\psi}{\partial\theta}\right)\sin\theta \tag{4.3.13}$$

so that the relative vorticity satisfies the equation

$$\nabla^2\psi+\zeta_{0z} = 0 \tag{4.3.14}$$

where ∇^2 denotes the two-dimensional Laplacian which, in polar coordinates, is

$$\nabla^2 = \frac{1}{R_E^2}\left[\frac{1}{\sin\theta}\frac{\partial}{\partial\theta}\left(\sin\theta\frac{\partial}{\partial\theta}\right)+\frac{1}{\sin^2\theta}\frac{\partial^2}{\partial\phi^2}\right]. \tag{4.3.15}$$

Substituting into (4.3.11), and noting that $R_E\dot{\theta} = U_\theta$, we obtain a differential equation for ψ, the stream function,

$$\frac{\partial(\nabla^2\psi)}{\partial t}+\frac{2\Omega_E}{R_E^2}\frac{\partial\psi}{\partial\phi} = 0. \tag{4.3.16}$$

The boundary condition at a fixed boundary is that ψ must remain constant everywhere along the boundary.

It is also possible to express the beta-plane equation for Rossby waves (2.2.7) in terms of the stream function ψ. It then becomes

$$\frac{\partial(\nabla^2\psi)}{\partial t}+\beta\frac{\partial\psi}{\partial x} = 0 \tag{4.3.17}$$

which has an appropriate solution (e.g. plane waves, cylindrical waves, etc.) subject to the relevant boundary conditions. The correspondence of equation (4.3.16) to the Rossby wave equation (2.2.7) arises because $\nabla^2\psi$ is the vorticity, and the induction term $(2\Omega_E/R_E^2)\times \partial\psi/\partial\phi$ is the same as $\beta^1 U_{1y}$. Platzman (1968) expressed this emphatically when he wrote:

'Rossby's beta-plane approximation of the planetary envelope is, I believe, characteristic of his style of striving to mould an awkward mathematical formalism to fit his physical intuition and I can well imagine that for him there may have been an element of mischievous delight in hitting upon such an invulnerable way to insinuate his irreverence for mathematical rigour! It would be a mistake, however, to think that the beta-plane is merely a shabby substitute for spherical coordinates—a temporary expedient that must give way as soon as the path to mathematical precision is opened up.

On the contrary, the reason his approximation has been so remarkably fruitful is that it establishes the unique frame of reference in which the Rossby wave emerges as a homogeneous plane harmonic wave unencumbered by boundaries or by the quantum restrictions of a finite domain such as a sphere.'

In order to examine the solutions of (4.3.16) it will be necessary to invoke certain functions of mathematical physics that arise from the solution of Laplace's equation, $\mathbf{\Delta}\psi = 0$ in spherical coordinates, where $\mathbf{\Delta}$ is the three-dimensional Laplacian. Laplace's equation is solved by separation of variables and it leads to a differential equation for the θ dependence of the form

$$(1-t^2)\frac{d^2\Theta}{dt^2} - 2t\frac{d\Theta}{dt} + \left[l(l+1) - \frac{m^2}{1-t^2}\right]\Theta = 0 \qquad (4.3.18)$$

where $t = \cos\theta$. One solution to this equation is the associated Legendre polynomial

$$\Theta = P_l^m(\cos\theta)$$
$$= \sin^{m/2}\theta\frac{d^m P_l(\cos\theta)}{d(\cos\theta)^m}.$$

Associated Legendre Polynomials

There are three different types of associated Legendre polynomials that differ only by their normalization terms. These are the Neumann form, the Schmidt form, often called the semi-normalized form, and finally the fully normalized form.

The function $P_l(\cos\theta)$ obeys equation (4.3.18) when $m = 0$. It can be represented as a power series

$$P_l(t) = \sum_{s=0}^{\infty} a_s t^s$$

where the coefficients satisfy the recurrence relation

$$a_{s+2} = \frac{s(s+1)-l(l+1)}{(s+2)(s+1)}a_s.$$

This gives two linearly independent solutions for a given l, one an odd function and one an even function.

As $s \to \infty$ the recurrence relation shows that $a_{s+2} \to a_s$ for any finite value of l. Both solutions of Legendre's equation (equation (4.3.18) with $m = 0$) thus seem to exhibit singularities as $t \to \pm 1$, so that neither appears to be an acceptable solution for applications to physical problems in which t can have the values ± 1. It is this property which leads to discrete eigenvalues, because physically acceptable wave functions can be found only for certain values of some parameter. In this case, the parameter is the separation constant l, which gives an acceptable function provided that l is equal to an integer. For each of these integral values, one of the functions defined by the recurrence relation is a polynomial of degree l and the other has a singularity. This latter function, which is the Legendre function of the second kind $Q_l(\cos\theta)$ may be eliminated by setting its coefficient equal to zero. On the other hand, the presence of singularities is related to the occurrence of sources and sinks of energy, so that for certain problems of geophysical interest we may wish to retain the Legendre solutions of the second kind. The three forms of the Legendre polynomials and the associated Legendre polynomials then arise from different choices of the normalization constants. The Neumann form for the Legendre polynomial, which we shall represent by N_l, is obtained by choosing a normalization factor so that $N_l(t) = 1$ for $t = 1$. The Neumann form of the associated Legendre polynomial N_l^m is then given by

$$N_l^m(t) = (1-t^2)^{m/2}\,\frac{d^m N_l(t)}{dt^m}.$$

Actually m can be positive or negative, so that $|m|$ should always be used when referring to any form of the associated Legendre polynomial. It is rather clumsy to write $|m|$ continually, so that it is customary not to do so, but to leave it understood that absolute values are to be used.

The mean value over the surface of a sphere of the square of the Neumann form of a spherical harmonic is given, for $m \neq 0$ by

$$\frac{1}{4\pi}\int_0^{2\pi}\int_0^{\pi}[N_l^m(\cos\theta)\cos(m\phi)]^2\sin\theta\,d\theta\,d\phi$$

$$= \frac{1}{2(2l+1)}\,\frac{(l+m)!}{(l-m)!}$$

For the special case $m = 0$, the corresponding mean value is $(2l+1)^{-1}$. These mean values vary greatly with m for a given value of l. For example, the ratio of this mean value for $l = 3$, $m = 1$ to that for $l = 3$, $m = 3$ is 1:60. This is an inconvenience in trying to deduce the spherical harmonics from geophysical data, because the coefficients of the terms resulting from such an analysis differ greatly because of this factor alone; differences due to the inherent importance of the terms in the distribution are obscured. Because of this, Schmidt's polynomials $\mathscr{P}_l^m(\cos\theta)$ are often used in geophysics.

The relation between the Schmidt form and the Neumann form is, for $m = 0$

$$N_l^m = \mathscr{P}_l^m = N_l = \mathscr{P}_l$$

and for $m \neq 0$

$$\mathscr{P}_l^m = \left[2\frac{(l-m)!}{(l+m)!}\right]^{\frac{1}{2}} N_l^m.$$

In order to obtain the fully normalized Legendre polynomials, P_l^m, where

$$\int_{-1}^{1} [P_l^m(t)]^2\, dt = 1$$

one multiplies by further constants. The fully normalized and semi-normalized functions are related for $m = 0$ by

$$\frac{P_l^m}{\mathscr{P}_l^m} = (l+\tfrac{1}{2})^{\frac{1}{2}}$$

and for $m \neq 0$ by

$$\frac{P_l^m}{\mathscr{P}_l^m} = (2l+1)^{\frac{1}{2}}/2.$$

Table 4.2 lists the first few fully normalized associated Legendre polynomials. Each expression has two parts. The first part in front of the multiplicative dot is the Neumann form, whereas the total expression is the fully normalized polynomial. Throughout this book the fully normalized $P_l^m(\cos\theta)$ is used exclusively.

Table 4.2 Fully normalized associated Legendre polynomials

$P_0(t) = 1{\cdot}2^{-\frac{1}{2}}$		
$P_1(t) = t\,.\,\sqrt{6}/2$	$P_2(t) = \frac{1}{2}(3t^2-1)\,.\,\sqrt{10}/2$	$P_3(t) = \frac{1}{2}(5t^3-3t)\,.\,\sqrt{14}/2$
$P_1^1(t) = (1-t^2)^{\frac{1}{2}}\,.\,\sqrt{3}/2$	$P_2^1(t) = 3t(1-t^2)^{\frac{1}{2}}\,.\,\sqrt{15}/6$	$P_3^1(t) = \frac{3}{2}(5t^2-1)(1-t^2)^{\frac{1}{2}}\,.\,\sqrt{42}/12$
	$P_2^2(t) = 3(1-t^2)\,.\,\sqrt{15}/12$	$P_3^2(t) = 15t(1-t^2)\,.\,\sqrt{105}/60$
		$P_3^3(t) = 15(1-t^2)^{\frac{3}{2}}\,.\,\sqrt{70}/120$

$$P_l(t) = \frac{1}{2^l l!}\frac{d^l}{dt^l}(t^2-1)^l\,.\,(l+\tfrac{1}{2})^{\frac{1}{2}}$$

The Legendre function of the second kind can be defined in terms of the Legendre polynomial. Using $P_l(t)$ this is

$$Q_l(t) = \frac{1}{2}P_l(t)\ln\frac{1+t}{1-t} - Z_l$$

where

$$Z_l = \frac{2l-1}{1\,.\,l}P_{l-1} + \frac{2l-5}{3(l-1)}P_{l-3} + \ldots$$

so that

$$Q_0(t) = \left(\frac{2^{-\frac{1}{2}}}{2}\right)\ln\frac{1+t}{1-t}$$

$$Q_1(t) = \left(\frac{\sqrt{6}}{4}\right)t\ln\frac{1+t}{1-t} - \left(\frac{\sqrt{2}}{2}\right)$$

$$Q_2(t) = \left(\frac{\sqrt{10}}{8}\right)(3t^2-1)\ln\frac{1+t}{1-t} - \left(\frac{3\sqrt{6}}{4}\right)t$$

$$Q_3(t) = \left(\frac{\sqrt{14}}{8}\right)(5t^3-3t)\ln\frac{1+t}{1-t} - \left(\frac{5\sqrt{10}}{4}\right)t^2 + \frac{(5\sqrt{10}-\sqrt{2})}{12}.$$

The associated Legendre function of the second kind $Q_l^m(\cos\theta)$ is defined by

$$Q_l^m(\cos\theta) = \sin^{m/2}\theta\frac{d^m Q_l(\cos\theta)}{d(\cos\theta)^m}$$

The Legendre function of the second kind is less often dealt with in mathematical physics. Whittaker and Watson (1962, Chapter 15) deal extensively with both Legendre polynomials and they point out that both solutions have the same integral representation, namely

$$\int (t^2-1)^l(t-\cos\theta)^{-l-1}\,dt$$

but it is evaluated around different contours for the two functions. They both therefore satisfy the same recurrence relations.

The general solution to the differential equation (4.3.18) is then a superposition of the two associated Legendre polynomials. The full solution to Laplace's equation includes the radial and azimuthal dependence and may be written as

$$\begin{aligned}\psi &= \sum_{m=0}^{l} (a_1 r^l + a_2 r^{-l-1})[P_l^m(\cos\theta) \pm Q_l^m(\cos\theta)] \exp(\pm im\phi) \\ &= S_l(r, \theta, \phi)\end{aligned} \quad (4.3.19)$$

where S_l is called a solid harmonic, or a spherical harmonic of degree l. In our case, there is no radial dependence, so we are interested in the surface harmonics which are represented by S_l $(r = 1, \theta, \phi)$. The curves on this unit sphere on which $P_l(\cos\theta)$ vanishes are l parallels of latitude which divide the surface of the sphere into zones, and so $P_l(\cos\theta)$ is called a zonal harmonic; the curves on which $\exp(\pm im\phi)\, P_l^m(\cos\theta)$ vanishes are $l-m$ parallels of latitude and $2m$ meridians which divide the surface of the sphere into quadrangles whose angles are right angles, and these functions are called tesseral harmonics. Spherical, zonal and tesseral harmonics are dealt with in greater detail in Chapter 5 of Lamb (1945).

Let us now try to find time-dependent solutions of the planetary wave equation (4.3.16). It is possible to have solutions in the form

$$\psi = f(\theta, \phi - [U_\phi t / R_E \sin\theta]) \quad (4.3.20)$$

where

$$U_\phi = R_E \sin\theta \left(\frac{\partial\phi}{\partial t}\right)$$

provided that

$$\frac{\partial}{\partial t}\left(\nabla^2\psi - \frac{2\Omega_E \sin\theta\psi}{R_E U_\phi}\right) = 0. \quad (4.3.21)$$

Now because ∇^2 has been used for the two-dimensional Laplacian and $\boldsymbol{\Delta}$ for the three-dimensional Laplacian, any surface harmonic satisfies the relation

$$\boldsymbol{\Delta} S_l = \nabla^2 S_l + \frac{l(l+1)}{R_E^2} S_l = 0 \quad (4.3.22)$$

and consequently solutions to the planetary wave equation exist in the form

$$\psi = S_l\left(\theta, \phi - \frac{U_\phi t}{R_E \sin\theta}\right)$$

provided that

$$\frac{-(2\Omega_E \sin\theta)}{R_E U_\phi} = \frac{l(l+1)}{R_E^2}$$

or

$$U_\phi = \frac{-2R_E\Omega_E \sin\theta}{l(l+1)} \tag{4.3.23}$$

Therefore planetary waves always drift westward in belts of constant latitude around the geographic pole with an angular velocity of $2\Omega_E/l(l+1)$. When $l = 1$, the angular velocity of the wave system matches the rotational velocity of the planet, so that the system is stationary relative to fixed axes. On the other hand, if viewed by an observer on the rotating globe, the wave system tends to follow the sun. For $l > 1$ the stationary observer viewing from fixed axes will see the wave system being carried around with the rotating sphere. To the rotating observer however the movement is still westward, but slower.

As an example, consider the tesseral harmonic

$$S_l = P_l^m(\cos\theta')\exp(im\phi') \tag{4.3.24}$$

where θ' and ϕ' are the zenith and azimuthal angles measured relative to an arbitrary pole which need not coincide with the geographic pole. In Fig. 4.3, this arbitrary harmonic pole has been represented by the point P and some tesseral harmonics around it are depicted. Even when the harmonic pole does not coincide with the geographic

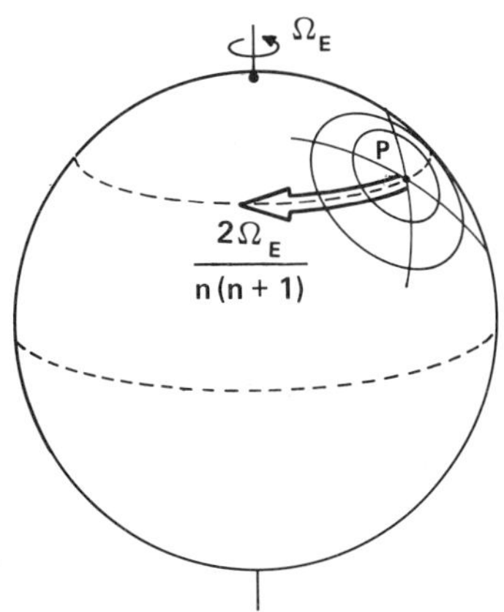

4.3 Westward drift of a global planetary wave. Some tesseral harmonics around the arbitrary pole P are shown

pole, the wave system still moves around the geographic pole, with a constant angular phase velocity. It is possible to express the eccentric tesseral harmonic of (4.3.24) in terms of a sum of tesseral harmonic that have a pole coincident with the geographic pole. For global planetary waves in a prototype atmosphere, the theory of the $2\Omega_E/l(l+1)$ drift was given by Haurwitz (1940) for centred tesseral harmonic and by Barrett (1958) for eccentric harmonics.

The planetary wave has a westward zonal angular phase velocity of $2\Omega_E/l(l+1)$ yet the Rossby wave has a westward angular phase velocity $2\Omega_E \sin^2 \theta/m$. Obviously for very large values of l the two velocities become identical but there appears to be a problem for lower values of l. The problem arises because surface harmonics have a nodal configuration similar to a waveguide mode. For example, Longuet-Higgins (1964) has shown that, for tesseral harmonics, the two constituent waves are given by

$$\frac{1}{2}\left[P_l^m(\cos\theta') \pm i\left(\frac{\pi}{2}\right)^{-1} Q_l^m(\cos\theta')\right] \exp im\phi'.$$

They are illustrated in Fig. 4.4. Each constituent wave is the analogue of a homogeneous plane wave. The poles of the constituent waves are singularities, one of which is an energy source and the other

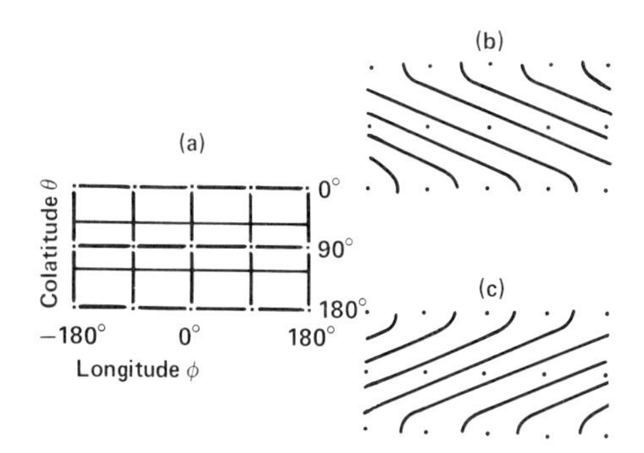

4.4 Tesseral harmonic nodes of $P_5^2(\cos\theta)$ sin 2ϕ and its decomposition into $\frac{1}{2}P_5^2(\cos\theta)$ sin $2\phi \pm \pi^{-1}Q_5^2(\cos\theta)$ cos 2ϕ. The constituent with the plus sign is shown in (b), with minus sign in (c). The north harmonic pole is an energy sink in (b) an energy source in (c) and vice versa for the south harmonic pole (after Platzman, 1968)

an energy sink. The presence of these singularities is a consequence of the finiteness of the spherical surface and an equivalent pair of singularities exists in the case of the ordinary plane wave, but the singularities in this case are located at infinity.

By attempting to synthesize the spherical harmonic functions present in the global pressure variations, it is possible to test whether the observed long waves do indeed fit the theoretically expected planetary wave picture. Deland (1965) found that the observed amplitude and phase of certain large-scale geographically centred tesseral harmonics can be interpreted as coming from two components. One component is stationary and, superimposed on it, there is a travelling wave that conforms well to the global planetary wave theory.

The solution of (4.3.16) for a stationary wave can be readily found by setting $\partial/\partial t = 0$, whence $\psi = f(\theta)$ only, and so all the oscillations of the particle velocity must be in the azimuthal (ϕ) direction along lines of constant latitude. Any barrier in the north–south direction, such as a mountain range, will inhibit the stationary wave unless there is some external energy source that is generating forced Rossby waves for which the meridional barrier is acting as an energy sink. The most common forced stationary planetary waves are the atmospheric tides. Eliasen and Machenhauer (1965) isolated the transient parts of tesseral harmonics $(m, l) = (1, 2), (2, 3)$ and $(3, 4)$ as shown in Fig. 4.5, and found westward drifts of 70, 40 and 20 degrees longitude per day. The corresponding numbers computed from the theory for a prototype atmosphere are 115, 53 and 28.

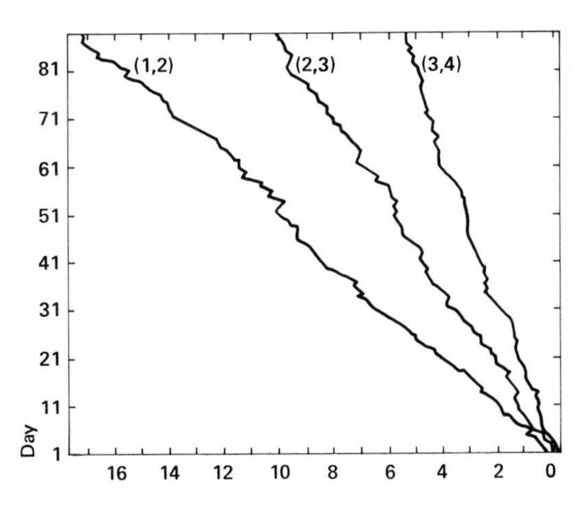

4.5 Successive daily values of the phase angle of the field of 24-hr. change of tesseral harmonic $(m, l) = (1, 2), (2, 3)$ and $(3, 4)$ of the 500 mb (6 km) stream function during the 90 day interval from December 1, 1956. The abscissa denotes the number of westward circuits around the earth (after Platzman, 1968)

The agreement between theory and experiment can be greatly improved by allowing for departures from geostrophic balance. This is equivalent to adding the terms of higher order onto the vorticity equation and the procedure is outlined in Chapter 2 of Lorenz's (1967) book.

Equivalent Depth

We have now seen how a Rossby or planetary wave is a two-dimensional movement of pressure disturbances in an incompressible constant-density fluid. But the only way that one can produce variations like this in such a fluid is to have wavelike movements of the fluid mass at the surface. In other words, a Rossby wave in an isothermal atmosphere is inextricably associated with an evanescent wave in the atmosphere, characterized in an incompressible atmosphere by a surface wave of phase velocity

$$\frac{\omega^2}{k^2} = \frac{[g \tanh(kh)]}{k}.$$

By substituting the known phase velocity for a particular evanescent wave motion in this equation, one can find the equivalent depth.

For Lamb waves, the phase velocity is the speed of sound, $\sqrt{\gamma g H}$. Thus, provided $k \ll 1/H$,

$$h = \gamma H.$$

In our intuitive discussion in the previous section, we pointed out that the equivalent depth should equal the scale height. This is true only if the changes of state in an isothermal atmosphere are also isothermal, so that $\gamma = 1$. For acoustic and gravity waves in general, $\gamma = 1{\cdot}4$.

Lamb (1910) showed that for a constant-density atmosphere with adiabatic changes of state, but with a temperature distribution following the dry adiabatic lapse rate, h is equal to H_g, the scale height at ground level. If the density also decreases adiabatically, then we showed in § 1.2 that

$$h = \frac{\gamma H_g}{(\gamma - 1)}.$$

For Rossby waves, however, there will no longer be a unique value of h. Restricting ourselves to a shallow isothermal atmosphere, we find that

$$h = \frac{V_p^2}{g} = \frac{(\beta/k^2)^2}{g} \qquad (4.3.25)$$

on a beta-plane. For centred tesseral harmonics

$$V_p = \frac{-2\Omega_E R_E \sin\theta}{l(l+1)}$$

so that

$$h \approx \frac{100 \sin^2\theta}{l^2(l+1)^2}\,\text{km.}$$

which, for all values of l, justifies the original choice of a shallow atmosphere.

At this stage, it may be pertinent to discuss the possibility of the planetary waves propagating into the upper atmosphere. This could be interpreted as a simple boundary problem from equation (4.3.19) by postulating the requirement of finite solutions as $r \to \infty$. Then $a_1 = 0$, and for all values of l, which has been assumed positive throughout this work, the Rossby wave will decrease in amplitude with height. This simple-minded approach agrees with current theoretical views on this topic but suffers from the following weaknesses:

(i) It does not apply to stationary waves, which are not necessarily described by a solution of the form (4.3.19).

(ii) A finite solution as $r \to \infty$ may not be a physically realistic boundary condition. The problem has already been mentioned in the discussion on the radiation condition and it hinges on the possibility of an amplified wave propagating to a large height and then being reflected.

(iii) The atmosphere we have been considering is highly idealized. Any departure from a two-dimensional, isothermal inviscid atmosphere will introduce extra terms into the vorticity equation (4.3.10) so that the absolute vorticity need no longer be conserved. The vorticity equation for a compressible viscous atmosphere becomes

$$\frac{D\zeta}{Dt} = (\zeta\,.\,\nabla)\mathbf{U} - \zeta(\nabla\,.\,\mathbf{U}) - \nabla\frac{1}{\rho}\times\nabla p + \eta\nabla^2\zeta$$

which is the curl of the Navier–Stokes equation. The gradient term is sometimes written in terms of the potential temperature or it may be expressed by using the Vaisala–Brunt frequency.

Charney and Drazin (1961) showed that for forced disturbances to propagate upward, it is necessary to have a mean wind speed that is west to east relative to the phase velocity of the disturbance but, when these eastward winds exist and are stronger than a few tens of meters per second, then stationary planetary waves are trapped in the troposphere. When they took the observed mean winds into account, Charney and Drazin concluded that only during equinoctial months, when the stratospheric winds are eastward but very weak, can a large amount of the planetary wave energy escape from the troposphere into the lower thermosphere. Dickinson (1969a) then included the atmosphere's cooling effects and found vertical propagation of the planetary waves to be unlikely during the equinoxes. Deland (1970) claimed that Rossby waves of very small wavenumber would have exponentially growing amplitudes and he predicted that these waves would be important in the upper atmosphere. However, these particular waves are evanescent. They grow in amplitude but they have most of the total wave energy concentrated near the ground.

On the other hand, experimental data by Deland and Friedman (1972) and by Muller (1972) report evidence for long period waves that can be interpreted as planetary waves which are in phase with tropospheric planetary waves. Paulin's (1970) results, however, suggest several days' phase lag between the tropospheric and upper atmospheric planetary waves.

Because the theoretical consensus is that planetary waves in the troposphere do not normally penetrate to the upper atmosphere, alternative explanations for the experimental data have been offered. None of these are wholly satisfactory and so far no definite answer can be given to the question: do planetary waves propagate into the upper atmosphere?

4.4 Tropical Wave Disturbances

The meteorology of the tropics is dominated by the Inter-Tropical Convergence Zone (ITCZ). The ITCZ is a region of concentrated rising air streams that stretches around the whole earth. It is especially evident in satellite photographs which clearly show the bands of cumulonimbus clouds which characterize it. The ITCZ is the location where most of the enormous quantity of latent heat evaporated from the tropical oceans into the trade winds is converted into useful heat. The ITCZ is situated slightly away from the equator and is often as far away as 10°. It advances furthest from the equator in the summer hemisphere and there are many cases of two distinct ITCZ's, one in either hemisphere.

Many tropical disturbances form directly on the ITCZ and there is a certain body of opinion which believes that if the earth was completely covered with oceans then all tropical disturbances would be associated with the ITCZ. It should be appreciated that the tropical wavelike disturbances which are going to be mentioned in this section are observed primarily over the oceans. This may be due to the paucity of observations over Asia, but in the case of Africa it represents a real situation. The Atlantic tropical disturbances are found to originate in the lower troposphere over North Africa and grow as they move westward. Though at one stage it was believed that the airflow over the Ethiopian mountains was generating these waves (Frank, 1970), more recent research indicates that the source is in the mid-tropospheric westward jet in the baroclinic zone south of the Sahara (Burpee, 1972).

Recent data analyses (Wallace, 1971) confirm the existence of two different types of waves, known jointly as *A* scale waves both of which have periods of the order of 4 to 5 days. The first type has wavelengths of a few thousand kilometres and phase speeds of 6–8 degrees longitude per day (i.e. approximately 9 m/s), westward. The wave amplitudes of this type of wave are a maximum at the ITCZ and the wave itself is confined to the troposphere. These waves appear to be the easterly waves first mentioned by Riehl (1954) in connection with disturbances in the Caribbean. There appears to be a strong amplification of these waves as they move westwards that results in typhoon disturbances in the Asiatic region (Fig. 4.6). They are dealt with in detail by Yanai (1971).

A typical sequence of an easterly wave passage begins with uniform westward winds and a fair sky. As the trough advances the winds above the moist surface layer swing southward. As the trough passes, it carries the characteristic ITCZ cumulonimbus clouds with it and showers and scattered thunderstorms develop. As the trough continues the weather becomes fair again.

The second type of wave has a maximum amplitude of the wind speed at the equator and it has a large proportion of its energy in

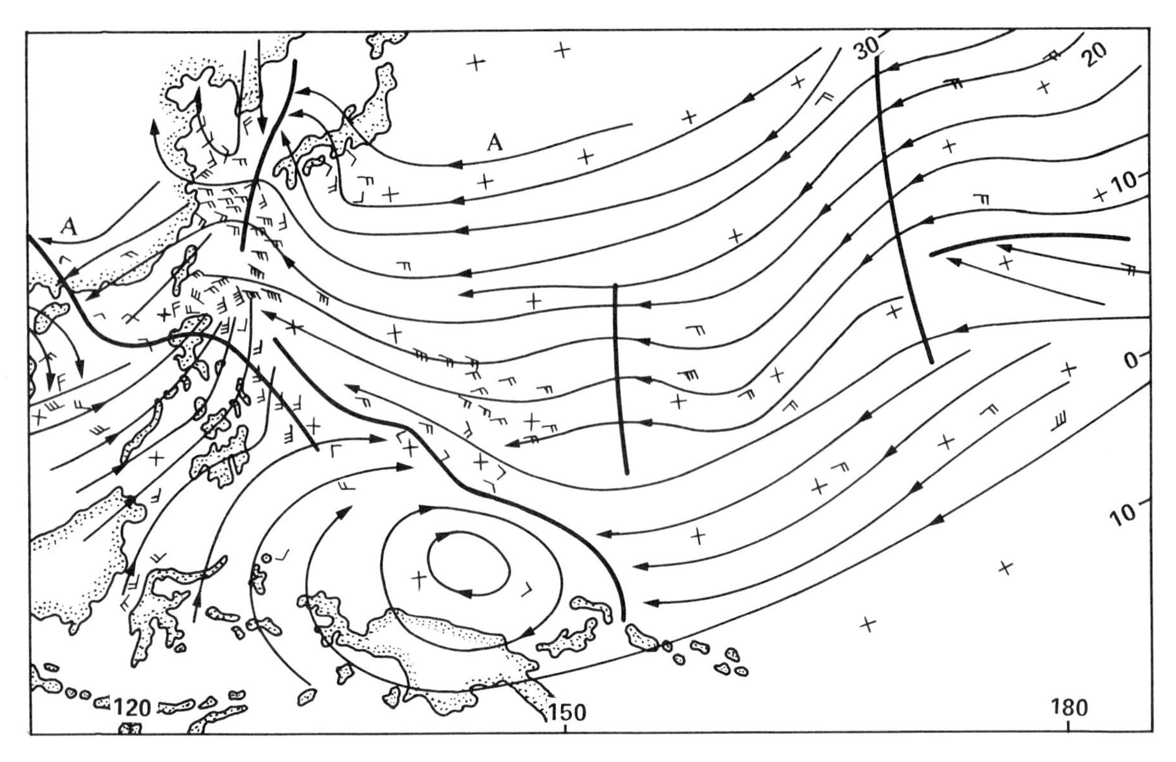

4.6 Easterly waves depicted in the winds at 1000 ft. above the Pacific (after Pettersen, 1969)

the lower stratosphere. The wave has a wavelength of about 10 000 km. This wave has been termed an equatorial wave, though it is rather doubtful whether the original equatorial wave model of Palmer (1952) is applicable any longer.

Waves in the tropical stratosphere have been reviewed by Wallace (1969), Holton and Wallace (1971) and Yanai (1971). All observed tropical stratospheric waves exhibit downward phase propagation (which implies upward energy propagation) and the particular waves discussed in this paragraph show upward phase propagation in the lower troposphere. This appears to place the energy source at the tropopause. The direction of the phase propagation also suggests that the disturbances are forced waves. Holton (1970) has shown that these waves may be interpreted as forced equatorial Rossby waves with the forcing being provided by latent heat release in the precipitation zones embedded within the waves.

The existence of equatorial Rossby waves may at first seem surprising when we recall that $f = 0$ at the equator. On the other hand, the β parameter $\partial f/\partial n$ is a maximum at the equator and it is this term that controls the Rossby wave formation. The nature of the Rossby waves is intimately tied to the strength and direction of the background wind. At ground level at the equator we have the doldrums and the wave structure of tropical disturbances is exceedingly weak at the ground. However, the wave structure in the middle and upper troposphere is far stronger because of the strong jet streams at these heights. Because there are strong currents flowing in the equatorial oceans, the oceanographers have used equatorial Rossby waves far more extensively than have the atmospheric physicists.

The solution for free atmospheric Rossby waves on a mid-latitude β plane are expressible in terms of spherical harmonics, as we have already indicated. The forced Rossby waves, which are closely related to the atmospheric tides, are expressible in terms of the so-called Hough functions, which are dealt with in the next chapter. The Rossby waves on the equatorial β plane, whether free or forced satisfy the frequency equation

$$\left(\frac{k}{\omega}\beta - k^2 + \frac{\omega^2}{gh}\right)\frac{\sqrt{gh}}{\beta} = 2N+1$$

where k is the zonal wavenumber and h the equivalent depth. The solutions can be described in terms of Hermite polynomials, $H_N(n)$, of order N where n is a normalized distance from the equator and N is equal to the number of times that the zonal particle velocity (U_{1e}) changes sign between $n = \pm\infty$. The Hermite polynomials are defined by

$$H_N(n) = (-1)^N \exp(n^2)\frac{d^N[\exp(-n^2)]}{dn^N}.$$

The identity in terms of Hermite polynomials arises because the associated Legendre polynomial $P_l^m(\cos\theta)$ can be expressed in terms of the Hermite polynomials as

$$P_l^m(\cos\theta) \propto \exp(-\tfrac{1}{2}l\mu^2)H_{(l-m)}(\mu\sqrt{2l}) \qquad (4.4.1)$$

provided l is large, m is finite and $l\mu^2$ is also finite and of the order of unity. $\mu = \pi/2-\theta$, $N = l-m$ and $\mu\sqrt{2l}$ represents the term n in the definition of the Hermite polynomial. The solution will be sinusoidal in $m\phi$ and so $m = 2\pi R_E/\lambda$ where λ is the wavelength of the observed waves. Longuet–Higgins (1964) has shown that provided $m \neq 0$, then, if one includes the latitude variations of β, all

Rossby wave solutions are trapped within a belt of latitudes around the equator that is of thickness $2l^{-\frac{1}{2}}$ radians.

Matsuno (1966) further showed that the mode $N = -1$ is also a permissible solution, albeit not a Rossby wave solution. The $N = -1$ mode represents a wave for which $U_{1n} = 0$ at all latitudes. This wave was first discovered by Kelvin when examining the waves in a bounded canal of finite depth h and is known as a Kelvin wave. It has particle velocities

$$U_{1e} = A\frac{g}{c}\exp\left(-\frac{f}{c}n\right)\cos(\omega t - kx)$$

$$U_{1n} = 0$$

where A, the wave amplitude, is a function of the canal width. The wave velocity $V_p = -\sqrt{gh}$ is not influenced by the earth's rotation though the wave itself is. Kelvin waves are essentially long surface waves trapped between vertical boundaries by the Coriolis effects but otherwise uninfluenced by them.

Rhines (1970a, b) mentions that only trapped waves can escape the low-frequency boundary $\omega = f = 2\Omega_E \sin\theta$ that acts to contain gravity waves between the frequencies f and ω_B. This may explain the meteorological importance of Kelvin waves (trapped by rotation) in the equatorial dynamics and of Lamb waves (acoustic waves trapped by density gradients) which carry much of the energy after atmospheric explosions in the short period fundamental acoustic mode (Garrett, 1969).

The observed tropical wave disturbances fall predominantly into the $N = -1$ and $N = 0$ modes. The $N = 0$ mode represents a gravity wave at very short wavelengths and a Rossby wave at the relevant tropical disturbance wavelength and because of this is often called the mixed Rossby-gravity mode. The $N = -1$ and $N = 0$ modes are illustrated in Fig. 4.7.

4.7 Pressure and steamlines for (a) the Kelvin wave and (b) the mixed Rossby-gravity wave (after Wallace, 1971)

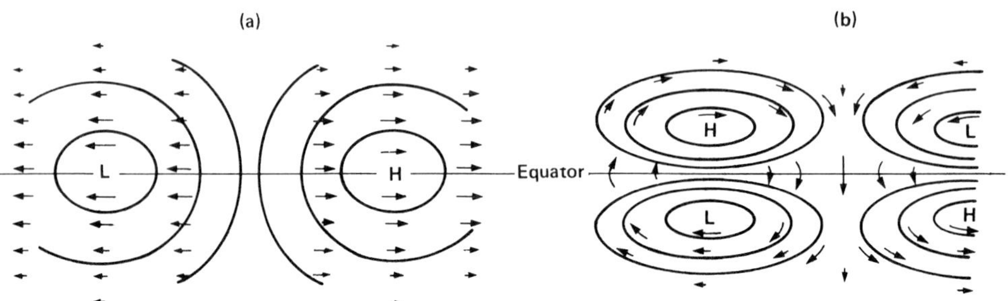

Lindzen (1967b) showed that, sufficiently close to the equator, disturbances in any given frequency band can propagate vertically away from a level of excitation. For a specified mode the vertical wavelength of a forced disturbance can be found from the zonal wavelength and frequency of the forcing excitation. This relationship between vertical wavelength, zonal wavelength and frequency, each of which can be estimated independently by various methods of analysis, provided a useful check on the identity of the normal modes (i.e. $N = -1$ and $N = 0$ in general).

Finally we should mention that there is some evidence accumulating for the existence of tropical disturbances in the period range of 5 to 10 days as well as for waves with periods longer than 10 days. Observation of these waves has been very scanty and our theoretical and empirical knowledge about them are both slight. Wallace (1971) has reviewed the available literature.

4.5 Barotropic and Baroclinic Instability

Barotropic Instability

Provided that a wave has an energy source upon which it can feed, then it is possible for it to become unstable. Mathematically this occurs for a plane wave when the $\exp i(\omega t - kx)$ term has an imaginary part for the angular frequency ω that is negative, and the rate of growth of the wave is $|\text{Im}(\omega)|^{-1}$ expressed in the appropriate units of time. This represents the time that it takes the wave to grow to $e = 2{\cdot}7\ldots$ times its initial value. The waves in both barotropic flows and baroclinic flows can become unstable. In the barotropic case the waves increase their kinetic energy as a result of the transport of vorticity, whereas the baroclinic waves grow as a result of the conversion of the potential energy associated with the baroclinicity into kinetic energy of motion.

The energy source for the waves with which we are dealing is the shear in the background wind. Sheared flow in a fluid is always unstable, though the stabilizing effects produced by the beta effect, the density gradients and the fluids compressibility can reduce, or completely eliminate the range of unstable wavelengths.

In barotropic flow there can be no wind shear in the vertical direction, and the meteorologically most relevant shear is then a zonal flow with a meridional shear. In a baroclinic atmosphere vertical shears of the zonal or meridional winds can exist. Between the mesopause and the Ekman boundary layer, these shears are well represented by the thermal wind equations.

This section and the next two sections deal with unstable and stable waves in sheared flow. This section deals with long waves, the next with mountain waves, and §4.7 with gravity waves.

The equation for the prototype Rossby wave in a constant background flow was

$$\frac{D(\zeta_z + f)}{Dt} = \frac{\partial \zeta_z}{\partial t} + \beta^1 U_{1y} = 0.$$

If the speed of the zonal flow varies, then the contribution of the $\mathbf{U}.\nabla\zeta_z$ term needs to be included, so that the equation for the vertical component of the relative vorticity in a β plane becomes

$$\left(\frac{\partial}{\partial t} + U\frac{\partial}{\partial x}\right)\nabla^2\psi + (\beta^1 - U'')\frac{\partial \psi}{\partial x} = 0 \qquad (4.5.1)$$

where ψ is the perturbation stream function and the primes represent a differentiation with respect to y. Then waves whose x component of the wavenumber is k will have

$$\psi = \Psi(y)\exp i(\omega t - kx)$$

where

$$(U - V_p)(\Psi'' - k^2\Psi) + (\beta^1 - U'')\Psi = 0$$

and $V_p = \omega/k$, which can be rewritten as

$$\Psi'' - \left(k^2 + \frac{\beta^1}{V_p - U} + \frac{U''}{U - V_p}\right)\Psi = 0. \qquad (4.5.2)$$

This equation has the same form as (3.9.1) which represents gravity waves in a sheared flow. In (4.5.2) the compressibility term ω_B^2 has been replaced by the β-effect term. Both these terms act to stabilize the shear flow that would normally be unstable. If these stabilizing terms are absent, then the resulting differential equation is known as the inviscid Orr–Sommerfeld equation.

There are some mathematical advantages to be gained by including the viscosity term (Kuo, 1949). It removes the singularity when

$U = V_p$. The equation becomes

$$(U-V_p)(\Psi''-k^2\Psi)+(\beta^1-U'')\Psi$$
$$= i\eta k^{-1}\left(\frac{\partial^4\Psi}{\partial y^4}-2k^2\Psi''+k^4\Psi\right). \qquad (4.5.3)$$

Now if there exists a critical point y_k where $\beta^1 - U'' = 0$, then the only stable waves that can exists will have a phase velocity $V_p = U(y)$ and a wavenumber $k = k(y)$ where U and k are evaluated at the critical point y_k. Waves whose phase velocities lie between the minimum and maximum wind velocities are all unstable, except for the stable wave whose phase velocity is $U(y_k)$. If the phase velocity is greater than the maximum wind speed, or below the minimum wind speed then the waves are stable. In practice, this means that wavelengths below about 300 km are stable (Fig. 1.2).

The instability depends on the strength and concentration of the wind and on the strength of the β parameter. For a wind of the form $U = U_0 \tanh y/L$ there will be no unstable waves if

$$\frac{U_0}{\beta L} < \frac{4}{3\sqrt{3}}.$$

Baroclinic Instability

There are two ways of tackling the problem of baroclinic instability. One way is to examine the waves at a surface of discontinuity between two barotropic surfaces. This concentrates all the baroclinicity into the boundary. Analyses of the waves and disturbances that appear on frontal surfaces follow this approach, though in principle the same possibility for the conversion of potential energy into kinetic energy exists in a flow with a continuously distributed baroclinicity. It appears that the waves on the frontal surfaces are an amalgam of barotropic waves and baroclinic waves. Let us now attempt to tackle the rather simple baroclinic instability problem by assuming a continuous vertical baroclinicity described by the thermal wind equation, but limiting ourselves to constant zonal winds and only meridional temperature changes.

If we now express our equations in terms of T, the temperature, we know that, if there is no heat source or sink, then the zero-order equation is

$$\frac{DT}{Dt}+(\gamma-1)T\nabla \,.\, \mathbf{U}_0 = 0$$

so that if $\nabla \,.\, \mathbf{U}_0 = 0$ then

$$\frac{\partial T}{\partial t}+U_{0e}\frac{\partial T}{\partial e}+U_{0n}\frac{\partial T}{\partial n} = 0. \qquad (4.5.4)$$

But the geostrophic wind equation tells us that

$$\mathbf{U}_g = \left(-\frac{1}{f\rho}\frac{\partial p}{\partial n}, \frac{1}{f\rho}\frac{\partial p}{\partial e}, 0\right)$$

so that for an incompressible constant density fluid

$$\mathbf{U}_g = \frac{R}{Mf}\left(-\frac{\partial T}{\partial n}, \frac{\partial T}{\partial e}, 0\right)$$

and (4.5.4) becomes

$$\frac{\partial T}{\partial t}+\frac{Mf}{R}U_{0e}U_{gn}-\frac{Mf}{R}U_{0n}U_{ge} = 0. \qquad (4.5.5)$$

If (4.5.5) is differentiated with respect to e then

$$\frac{\partial U_{gn}}{\partial t}+U_{0e}\frac{\partial U_{gn}}{\partial e}-U_{ge}\frac{\partial U_{0n}}{\partial e} = 0$$

where both zonal winds are assumed to be constant. If the meridional winds are assumed to have periodic variation

$$U_{0n} = \mathscr{U}_{0n}\exp i(\omega_1 t-k_1 e)$$
$$U_{gn} = \mathscr{U}_{gn}\exp i(\omega_2 t-k_2 e)$$

where the amplitude $\mathscr{U}_{0n}$ is constant and $\mathscr{U}_{gn}$ is a function of height then

$$\frac{\mathscr{U}_{gn}}{U_{ge}} = \frac{k_1\mathscr{U}_{0n}\exp\{i[(\omega_1-\omega_2)t-(k_1-k_2)e]\}}{k_2U_{0e}-\omega_2}. \qquad (4.5.6)$$

However the ratio of the meridional geostrophic amplitude to the constant geostrophic zonal wind must be independent of both position and time so that

$$\omega_1 = \omega_2 = \omega \quad \text{and} \quad k_1 = k_2 = k$$

so that the pressure wave described by U_{0n} and the temperature wave described by U_{gn} will have the same wavelength and period but may differ in phase. (4.5.6) becomes

$$\frac{U_{gn}}{U_{ge}} = \frac{U_{0n}}{U_{0e} - V_{\mathrm{p}}}$$

where $V_{\mathrm{p}} = \omega/k$ is the phase velocity. If V_{p} is broken into its real and imaginary parts; $V_{\mathrm{p}} = V_{\mathrm{p}r} + iV_{\mathrm{p}i}$ then

$$U_{gn} = U_{ge}U_{0n}([U_{0e} - V_{\mathrm{p}r}]^2 + V_{\mathrm{p}i}^2)^{-\frac{1}{2}} \exp(-i\theta)$$

where

$$\tan\theta = \frac{V_{\mathrm{p}i}}{(U_{0e} - V_{\mathrm{p}r})}.$$

For instability $V_{\mathrm{p}i} < 0$ and when this occurs then $0 < \theta < \pi$ depending on the sign of $U_{0e} - V_{\mathrm{p}r}$. In other words, for unstable waves, the troughs and peaks of the temperature variations lag behind the troughs and peaks of the pressure variations. Similarly, for decaying waves, the temperature variations precede the pressure variations.

For stable waves $V_{\mathrm{p}i} = 0$ so that $\theta = 0$ or π. When $\theta = 0$ then U_{gn} and U_{0n} are in phase, $U_{0e} - V_{\mathrm{p}r} > 0$ and low pressure areas coincide with temperature minima. When U_{gn} and U_{0n} are in anti-phase, then the low-pressure troughs coincide with temperature maxima and $U_{0e} - V_{\mathrm{p}r} < 0$. In summary then, for stable waves, cold troughs travel slower than the surface zonal flow whilst warm troughs travel faster.

To further examine the baroclinic instability of the thermal wind, we must return to the z component of the vorticity equation

$$\begin{aligned}\frac{D}{Dt}\zeta_a &= \frac{D(\zeta_z + f)}{Dt} \\ &= \frac{\partial \zeta_z}{\partial t} + U_e\frac{\partial \zeta_z}{\partial e} + U_n\frac{\partial f}{\partial n} \\ &= -(\zeta_z + f)\nabla . \mathbf{U}\end{aligned}$$

where the total wind components

$$U_e = U_{0e} + U_{ge}\,;\; U_n = U_{0n} + U_{gn}$$

so that

$$\frac{U_{0n}}{U_{0e} - V_p} = \frac{U_{gn}}{U_{ge}} = \frac{U_n}{U_e - V_p}.$$

Since only the meridional components have variations

$$\zeta_z = \frac{\partial U_n}{\partial e} = \frac{U_e - V_p}{U_{0e} - V_p}\frac{\partial U_{0n}}{\partial e}$$

$$= -ikU_{0n}\frac{(U_e - V_p)}{(U_{0e} - V_p)}$$

$$\frac{\partial \zeta_z}{\partial t} = k^2 V_p U_n$$

and

$$\frac{\delta \zeta_z}{\partial e} = -k^2 U_n$$

so that the vorticity equation is

$$k^2 V_p U_n - U_e k^2 U_n + \beta U_n$$

$$= \frac{U_e - V_p}{U_{0e} - V_p} U_{0n} k^2 \left(V_p - U_e + \frac{\beta}{k^2} \right) = -f \frac{\partial U_z}{\partial z} \qquad (4.5.7)$$

since $f \gg \zeta_z$ and U_n is independent of n. The term $U_e - \beta/k^2$ is the phase velocity of a Rossby wave measured relative to the ground, equation (4.3.1), so that we shall represent β/k^2, the phase velocity of the Rossby wave, by V_R. The vertical wind velocity vanishes at the lower boundary, and by postulating a rigid upper boundary it also vanishes there, so that if we integrate (4.5.7) with respect to the height of the equivalent atmosphere h, then

$$\int_0^h \left(\frac{\partial U_z}{\partial z} \right) dz = 0$$

so that

$$\overline{(U_e - V_p)(V_p - U_e + V_R)} = 0 \qquad (4.5.8)$$

where the bar represents the mean value of the quantities with respect to height. Equation (4.5.8) is quadratic in V_p and has a solution

$$V_p = \overline{U}_e - \frac{V_R \pm (V_R^2 - 4\overline{U_e^2} + 4\overline{U}_e^2)^{\frac{1}{2}}}{2}. \qquad (4.5.9)$$

This may be compared with the barotropic case in which $U_z = 0$ and $\overline{U_e^2} = \overline{U_e^2}$ so that $V_p = U_e - V_R$: the phase velocity of the prototype Rossby wave.

In terms of the thermal wind velocities

$$V_p - \overline{U}_{0e} = \overline{U}_{ge} - \frac{V_R \pm (V_R^2 - 4\overline{U_{ge}^2} + 4\overline{U}_{ge}^2)^{\frac{1}{2}}}{2}$$

where

$$\frac{\partial(U_{ge}/T)}{\partial z} = -\frac{g}{f T^2}\frac{\partial T}{\partial n}.$$

If we assume that $\partial T/\partial z = 0$ and that $\partial T/\partial n$ is constant then the thermal wind will increase linearly upwards. If we recall that the mean is defined by

$$\overline{U}_{ge} = \frac{1}{h}\int_0^h U_{ge}\, dz$$

then it may be shown that $\overline{U_{ge}^2} = \frac{4}{3}\overline{U}_{ge}^2$ so that

$$V_p - \overline{U}_{0e} = \overline{U}_{ge} - \frac{V_R \pm (V_R^2 - \overline{U_{ge}^2})^{\frac{1}{2}}}{2}.$$

In this case, the wave becomes unstable when the root mean square thermal wind exceeds the Rossby wave phase velocity. For a given thermal wind, the longer waves (greater V_R) are stable whilst shorter waves are unstable.

For our vertically linear zonal wind model, the growth rate, Im (ω), will be given by

$$\mathrm{Im}\,(\omega) = \frac{k(\overline{U_{ge}^2} - V_R^2)^{\frac{1}{2}}}{2}$$

$$\approx \frac{\pi}{\lambda}(\overline{U_{ge}^2})^{\frac{1}{2}} \approx \frac{3{\cdot}5\overline{U}_{ge}}{\lambda}$$

for the shorter waves. If λ, the zonal wavelength, is about 5000 km and $\overline{U}_{ge}$ is 14 m/s, then the growth rate is about 0·8 day^{-1}. The most unstable baroclinic waves are a good deal shorter than the most unstable barotropic waves. It is not realistic to assume that, as the wavelength tends towards zero, the growth rate increases indefinitely. This difficulty is removed when the effect of the vertical motions upon the meridional temperature gradients is taken into account. When this is done, then the maximum growth rate occurs for waves of lengths from 2500 km to 3000 km and the perturbations formed by the baroclinic instability will have the bulk of their energies in this wavelength range.

While there is enough evidence for the importance both of baroclinic and barotropic effects, there must be some limit to which phenomena can be explained purely barotropically or purely baroclinically. In general, one may expect some interaction between the two systems. On the other hand, it is possible to explain many phenomena without considering any interaction. The intense low-pressure cyclones are explicable in terms of a barotropic instability. Similarly the baroclinic instability is believed to account for the jet streams of fast flowing air at the tropopause. Jet streams have been created in the laboratory in a rotating annulus filled with a liquid such as water, heated at the outside and cooled in the middle. Dye tracings show a concentration of the disturbance into a narrow jet (Rogers, 1959), exactly as in the tropopause disturbance of the atmosphere itself. It has been found (Fultz and Long, 1951) that a westward flow of the fluid relative to an obstacle will produce a concentrated jet on the leeward side of the obstacle, in contrast with eastward flow which sets up Rossby waves.

The laboratory experiments with rotating annuli have proved most useful in simulating the atmospheric circulation and in separating the essential aspects of the general circulation from the minor ones (Hide, 1970). For example, while condensation of water vapour may play an essential role in the tropics, it is not an essential effect at mid-latitudes because the hydrodynamical phenomena found in the atmosphere, including even cyclones, jet streams and fronts, occur in a rotating annulus where there is no analogue of the condensation process.

In addition to these effects, rotating annulus experiments revealed a phenomenon which Hide called vacillation. In this case the Rossby type waves in the annulus are not steady in the rotating coordinate system, but they alter their shape and speed of progression in a regular periodic manner, returning to their original configuration at the completion of a vacillation cycle. This phenomenon seems to be completely analogous to fluctuations in the Rossby wave zonal wavenumber and attempts to describe it in terms of a theory of non-linear baroclinic instability seem to provide a useful insight into atmospheric processes (Drazin, 1970).

Baroclinic instability was first studied by Charney (1947), Eady (1949), and Fjortoft (1951). In contrast to the barotropic stability problem, we have seen that the coefficients in the linear equations become constants when the vertical wind shear is sufficiently simple, in which case the normal modes are easily determined. Despite the differing models used by the investigators, all agree that instability is favoured by a large Coriolis parameter, large vertical wind shear and low hydrostatic stability. Work on the theory of both baroclinic and barotropic instability continues. Some recent contributions to the theory of barotropic instability are those of Lorenz (1972) and Hoskins (1973), whilst the baroclinic instability was studied by Phillips (1964), McIntyre (1965, 1970) and Drazin (1970).

4.6 Mountain Waves

When an airflow approaches an obstacle such as a mountain there will be two dominant effects. Firstly, the change in height of the earth's surface will introduce a vorticity component into the air flow which may be balanced by the β effect if the mountain is of sufficient meridional extent. This will generate Rossby waves. Secondly, the air packets are being lifted up and then dropped again, which will induce gravity-wave oscillations of much smaller wave-length.

If the Rossby number is less than about 0·5, then other strange things may happen. At these low Rossby numbers, the column of air above the mountain is stagnant. This is known as a Taylor column and this effect has been postulated as an explanation for sunspots and for the great red spot visible on the planet Jupiter. For sheared air motion at low Rossby numbers, unstable Taylor vortices can be set up. On the earth, however, sufficiently small values of the

Rossby number are going to occur only for very small values of $U < h\Omega_E/\pi$, which are unlikely to have any meteorological interest.

The gravitational mountain waves on the other hand are of exceptional interest both to meteorologists and to aviators. The World Meteorological Organization has prepared an excellent technical note entitled *The Airflow over Mountains* (Alaka, 1960) which provides the most extensive compendium to date on the observational and theoretical aspects of the non-rotational mountain waves. More recent references may be found in the bibliography of an article by Vergeiner (1971).

Rotational Effects

For Rossby numbers of the order of unity, the presence of a region of high ground of finite area in the horizontal plane (an isolated mountain) or of infinite area (a mountain range) is to insert into the flow a patch of negative vorticity in the northern hemisphere or positive vorticity in the antipodes. The value of the inserted vorticity ζ_{1z} is given by

$$\frac{f+\zeta_{1z}}{h-h'} = \frac{f}{h}$$

provided the flow is of limited meridional extent and the Coriolis parameter stays constant. h' is the height of the mountain and is a function of e, the eastward co-ordinate.

On the other hand, if the meridional extent of the flow field (L) is such that $\zeta_{1z}/L \approx f/R_E$ and $L \ll R_E$, then the vorticity equation on the β plane gives

$$\frac{f+\beta n+\zeta_{1z}}{h-h'} = \frac{f+\beta n_0}{h}$$

where n_0 is the n coordinate of an element of flow as it approaches the ridge. To simplify the problem let us assume no meridional wind and that h' remains constant for $e > 0$ and is zero for $e < 0$. In this case the equation of mass conservation gives the new velocity as $U_{e>0} = U_{e<0}[h/(h-h')]$. The wind speed increases as it blows over a mountain. Taking $n_0 = n$, the relative vorticity changes from zero to $-(f+\beta n)h'/h$.

At $e = 0$ we have $\psi = -U_{e>0}n$ so that the vorticity equation yields an equation for the stream function:

$$\nabla^2\psi = \zeta_{1z} = -\frac{h'}{h}f - \beta n - q^2\psi \qquad (4.6.1)$$

where $q^2 = \beta(h-h')/U_{e>0}h$.

Equation (4.6.1) has solutions given by

$$\psi = (a-n)F(e) - \frac{(fh'/h+\beta n)}{q^2}$$

where a is a constant and $F(e)$ satisfies the equation

$$\frac{d^2F}{de^2} + q^2F = 0. \qquad (4.6.2)$$

The boundary conditions at $e = 0$ give $\partial\psi/\partial n = -U_{e>0}$ and $\partial\psi/\partial e = 0$ so that if $\psi = 0$ at $e = 0$, $n = 0$ then

$$\psi = U_{e>0}h'(f+\beta n)\frac{(\cos qe - 1)}{h\beta} - U_{e>0}n$$

which gives a set of sinusoidal streamlines which differ in shape only by an amplification in the n direction arising from the variation of the Coriolis parameter with n. The wavelength of the resulting disturbance is $2\pi\sqrt{U_{e>0}h/\beta(h-h')}$ which is about 1600 km at latitude 45° when $U_{e>0}$ is 1 m/s. The amplitude of the wave in the n direction is also quite large, being about 1000 km for a step of $0{\cdot}1h$ at latitude 45°. The disturbance due to any subsequent obstructions will depend on the air velocity at the obstruction and so will depend on the distance between the mountains.

Equation (4.6.2) further indicates that wavelike disturbances only occur for eastward flowing currents when $U_e > 0$. If the wind has a westward flow then the streamlines will have a cosh qe dependence that leads to jet stream formations. The above treatment, however, becomes invalid because cosh $qe > 1$ for all values of e so that the incoming stream cannot have a constant wind speed which, however, was assumed in the derivation.

Rossby waves will be generated by any disturbance in the eastward zonal flow. Even an isolated point mountain will set up a train of Rossby waves in the lee of the flow. A point mountain is one whose dimensions are very much smaller than $2\pi\sqrt{U/\beta}$ and it acts as a source of vorticity which sends out a stream of wave packets in all directions and at all speeds. For example, consider a wave packet whose phase velocity has magnitude $\beta \cos \alpha/k^2$ and makes an angle of $\pi+\alpha$ with the eastward direction. In the absence of the zonal flow, this packet would move out from the source at an angle of 2α with a speed β/k^2 (see Fig. 2.1), but the zonal flow causes it to be convected eastward with a speed U. Its resultant velocity can have an angle anywhere between 0 and 2α and a speed anywhere between β/k^2

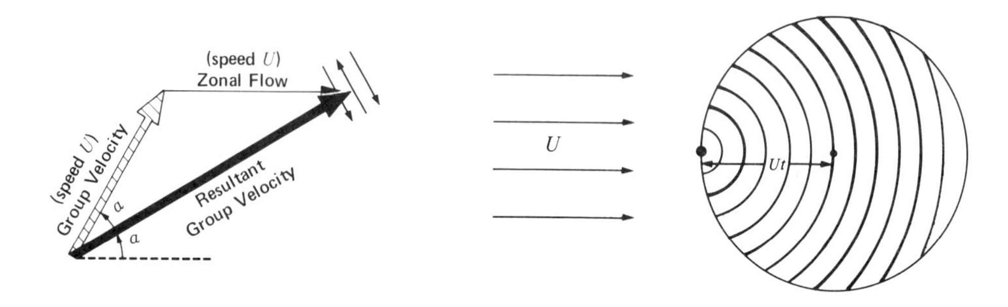

4.8 Pattern of constructive interference of group velocity packets generated by an eastward flow U over a point mountain (after Platzman, 1968)

and $U+\beta/k^2$ depending on the relative sizes of U and β/k^2. However the only packet that can make a constructive contribution to the interference pattern is the one with $\beta/k^2 = U$, because the resultant group velocity of such a packet makes an angle α with the zonal flow which is the same angle as the phase velocity. The interference of all other packets is destructive. The point mountain thus produces a lee pattern of stationary cylindrical waves of a wavelength $2\pi\sqrt{U/\beta}$ which is the same as the wavelength from the extended mountain. The cylindrical lee waves after a time t will be confined to an expanding circle of radius Ut centred on a point Ut downstream of the obstacle (Fig. 4.8). This is because the resultant group velocity is $2U \cos \alpha$ and the curve $r = 2U \cos \alpha$ in plane polar coordinates represents a circle of radius U centred on the point $r = U, \alpha = 0$.

Vertical Motions

Provided that the zonal wind flow is independent of time then the solution for the lee flow is also independent of time. This solution represents the perturbations that would be observed after a sufficiently long length of time and it represents the steady-state, or forced, perturbations due to the mountain.

The actual wind of course is never steady, but this approximation makes a convenient starting point. Then if we neglect friction, condensation, heat conduction or radiation and linearize the basic equations describing the atmosphere for a horizontally constant wind $U_{0x}(z)$ blowing perpendicular to the mountain and in the steady-state case ($\omega = 0$) a straightforward, but rather lengthy, elimination process leads to the following second-order partial differential equation for the vertical perturbation velocity $U_{1z}(x, z)$

$$U''_{1z} - \left(\mathscr{s} + \frac{m'}{m}\right) U'_{1z} + \left[\left(\frac{\omega_B^2}{g} + \frac{m'}{m}\right)\frac{g}{U_{0x}^2} + \left(\mathscr{s} + \frac{m'}{m}\right)\frac{U'_{0x}}{U_{0x}} - \frac{U''_{0x}}{U_{0x}}\right] U_{1z} + m\frac{\partial^2 U_{1z}}{\partial x^2} = 0. \tag{4.6.3}$$

The primes denote partial differentiation with respect to z, which is equivalent to total differentiation for m and U_{0x} where we have used the abbreviations

$$m = 1 - \frac{U_{0x}^2}{c^2}$$

$$\mathscr{s} = -\frac{d(\ln \rho)}{dz} = \frac{\omega_B^2}{g} + \frac{g}{c^2}.$$

c is the speed of sound and U_{0x}/c is the Mach number of the flow. In the normal atmospheric case the Mach number is very small and $m \approx 1$. Equation (4.6.3) can be written more compactly by defining

$$\mathscr{S} = \frac{d}{dz}\ln\left(\frac{m}{\rho}\right) = \mathscr{s} + \frac{m'}{m}$$

and

$$\mathscr{M} = \frac{d}{dz}\ln(m\theta) = \frac{\omega_B^2}{g} + \frac{m'}{m}$$

where θ is the potential temperature, so that

$$U''_{1z} - \mathcal{S} U'_{1z} + \left(\frac{\mathcal{M} g}{U_{0x}^2} + \frac{\mathcal{S} U'_{0x}}{U_{0x}} - \frac{U''_{0x}}{U_{0x}} \right) U_{1z} + m \frac{\partial^2 U_{1z}}{\partial x^2} = 0 \qquad (4.6.4)$$

which is a far more general form of equation (2.5.13) in the steady state case. If we then use the standard transformation of § 2.5 and set

$$U_{1z} = \sqrt{\frac{m}{m_g} \cdot \frac{\rho_g}{\rho}} \, \mathcal{U}_{1z}$$

where the subscripts g refer to the values of m and ρ at the ground then

$$m \frac{\partial^2 \mathcal{U}_{1z}}{\partial x^2} + \mathcal{U}''_{1z} + \mathcal{F}(z) \mathcal{U}_{1z} = 0 \qquad (4.6.5)$$

where

$$\mathcal{F}(z) = \frac{\mathcal{M} g}{U_{0x}^2} + \frac{\mathcal{S} U'_{0x}}{U_{0x}} - \frac{1}{4} \mathcal{S}^2 + \frac{1}{2} \mathcal{S}' - \frac{U''_{0x}}{U_{0x}} \qquad (4.6.6)$$

is known as the Scorer parameter. Now in a typical atmosphere $m = 1$ and the equation reduces to the wave equation

$$\nabla^2 \mathcal{U}_{1z} + \mathcal{F}(z) \mathcal{U}_{1z} = 0. \qquad (4.6.7)$$

Furthermore, provided U_{0x} is not exceptionally large then

$$U'_{0x} U_{0x} \ll g.$$

Also if the temperature lapse rate is not near the adiabatic lapse rate then ω_B is sufficiently large that $\omega_B^2 / U_{0x}^2 \gg (2\mathcal{S}' - \mathcal{S}^2)/4$. In that case if we assume an exponential dependence in x with a horizontal wavenumber k_x then we obtain Scorer's equation for the height variation of the transformed vertical wind velocity

$$\mathcal{U}''_{1z} + \left(\frac{\omega_B^2}{U_{0x}^2} - \frac{U''_{0x}}{U_{0x}} - k_x^2 \right) \mathcal{U}_{1z} = 0. \qquad (4.6.8)$$

For an incompressible atmosphere without density gradients $\omega_B^2 = 0$ and equation (4.6.8) reduces to the inviscid Orr–Somerfeld equation of hydrodynamics.

In a constant lapse rate atmosphere with a uniform airstream,

$\mathscr{F}(z)$ is a constant and it represents the square of the total wave-number $k_x^2 + k_z^2$ of the flow. For a troposphere with lapse rate of 6° K/km and average temperature of 250°K this corresponds to a wavelength of 10 km for a wind speed of 20 m/s or 1 km for a wind speed of 2 m/s. These waves are obviously very much shorter than the rotational waves induced by mountainous terrain.

In general the solution of equation (4.6.7) does not give a unique solution, though the problem may be overcome by invoking the radiation condition or its horizontal analogy which demands that all waves should vanish at large distances upstream from the region of disturbance. Lyra (1940, 1943) obtained results for the step function plateau mountain and a square mountain. His results were in terms of Bessel functions of the first kind

$$J_l(r\sqrt{\mathscr{F}(z)}) \qquad \text{where } r = \sqrt{x^2 + z^2}.$$

Once U_{1z} has been found then it is also possible to find U_{1x}. The steady-state form of equation (2.5.9) gives the wind divergence

$$\chi = \frac{\partial U_{1x}}{\partial x} + \frac{\partial U_{1z}}{\partial z} = \frac{g}{c^2} U_{1z}$$

and then equations (2.6.6) to (2.6.9) give the other atmospheric parameters.

Various models for $\mathscr{F}(z)$ have been chosen, but in general it is necessary to approximate the atmosphere by a number of layers. Scorer (1949) very successfully examined the airflow over a symmetrical mountain in a two layer model with a constant value of $\mathscr{F}(z)$ in each layer. He found that the occurrence of lee waves was controlled far more by the wind distribution than by the temperature distribution and the strongest lee waves are found in regions where the Scorer parameter is decreasing; thus lee waves are strongest in regions where there is a strong increase in the wind speed. On the other hand waves in a sheared flow can become unstable, so that the airstream in the lee of a mountain is strongly predisposed to turbulence. This is further discussed in the next section.

The amplitude of the waves depends on the width of the mountain and it has been found that this is a more important factor than the mountain height. The greatest lee wave amplitude occurs for a mountain whose width is about 1 km at an altitude that is one half the maximum mountain height. There may then be a further resonance of these lee waves if there is another mountain placed at a multiple of $2\pi/\sqrt{\mathscr{F}(z)}$ downwind. This situation actually can occur in California where the Owens Valley is bordered by the Sierra mountains on the west and the Inyo mountains on the east.

Wave Clouds

The most spectacular manifestation of the mountain waves occurs when the air becomes sufficiently saturated to produce clouds. These clouds are stationary and are continually reforming at the upwind edge and dissolving at the downwind edge. The mountain wave clouds may be divided into three broad groups; cap clouds, rotor clouds and lenticular clouds though there is a rich variety within the groups.

Cap clouds are examples of the saturation of an air mass that is raised to a high level in a forced ascent, and in the form of hill fog are common over high ground affected by maritime air masses. When the mountain shape is simple and regular, as in the case of an isolated conical mountain, then the cap cloud may take the form of a symmetrical lenticular (lens shaped) cloud or a stack of shallow lenticulars resting on the mountain top or occurring above it (Fig. 4.9a). Symmetrical cap clouds can only occur in stable, non-turbulent flow.

On the other hand if the air becomes saturated over a considerable portion of its forced ascent then a cloud sheet will occur generally above and to the windward of the mountain top. There may be long fibrous streamers extended into the leeward side where they are dissolved by adiabatic warming. This cloud arrangement can bear some resemblance to a waterfall and is sometimes termed a cloud fall. There are also various regional names for it, such as the Helm in the Crossfell range of the English Pennines but the name foehn wall has gained general acceptance. The foehn wall often produces prolonged rain on the windward side of the mountain.

A further type of cap cloud is the banner cloud photograph in the frontispiece of this book. This cloud gives the impression that the mountain is smoking and when it occurs over Mont Blanc the locals refer to 'Mont Blanc fumant sa pipe.' Banner clouds are a feature of steep sided isolated mountain peaks and occur as a pennant of cloud to the lee of the peak. Their formation is due to

4.9 (*a*) *Stacked lenticular cap clouds over Mt. Dom in the Swiss Alps (Neville Goodman)*

4.9 (*b*) *Turbulent rotor cloud in the first wave to the lee of Mt. Meru, Tanzania (by permission of* Weather *and H. H. Coutts)*

the pressure reduction associated with the horizontal deformation of the airflow around the peak.

If the streaming air on the downwind side of the mountain attains a sufficient velocity then it may become turbulent and exhibit rotor clouds. These clouds, also known as roll clouds, appear as large stationary rolls having the appearance of a line of cumulus or stratocumulus parallel to and downwind of the mountain crests. If the mountain does not form an extended ridge then the line of the rotor cloud will be considerably shortened (Fig. 4.9b). The base of the rotor cloud is near the level of the crest while the top may be several thousand feet higher. There may be a number of parallel lines of rotor clouds if the amplitude of the lee wave is sufficiently strong to produce a number of strong oscillations in the lee wave. Normally however only one rotor cloud develops in the first antinode of the lee wave. In the Crossfells this is called the Helm bar.

One finds the lenticular lee clouds at heights well above the mountain peak. Their most definite characteristic lens shape occurs when the relative humidity is low (30–60 per cent) because this confines the cloud to the regions of maximum ascent in the wave crests and the lenticulars are then seen in isolation. The smoothness of their outlines is an indication of the laminar airflow.

Lenticular clouds can form above simultaneous rotor clouds (Fig. 4.9c). At other times wave clouds may show little or no obvious wave

4.9 (c) Lenticular clouds overlaying rotor clouds at Glen Gairn, Scotland (by permission of Weather *and M. G. Pearson)*

4.9 (d) Unstructured wave clouds (by permission of Professor R. S. Scorer)

structure because of the complexity of the terrain that is responsible for their formation. As the lee waves arising from particular terrain features are superimposed on those from other features and more particularly on the disturbances immediately above other hills downstream, the distribution of wave clouds may show little or no obvious relation to the ground below. An example of this is shown in Fig. 4.9d.

The wind shear may produce a small scale instability in the lee waves which results in billows. In their simplest form billow clouds appear as bands, or striations of cloud which move through the stationary wave cloud. The wavelength of the billows is much shorter than the main lee wavelength. A well developed set of billows is shown in Fig. 4.10.

The lenticular clouds may occur at heights well into the troposphere and may appear as stratocumulus, altocumulus, or as cirrus. It also appears that lee waves can propagate into the stratosphere

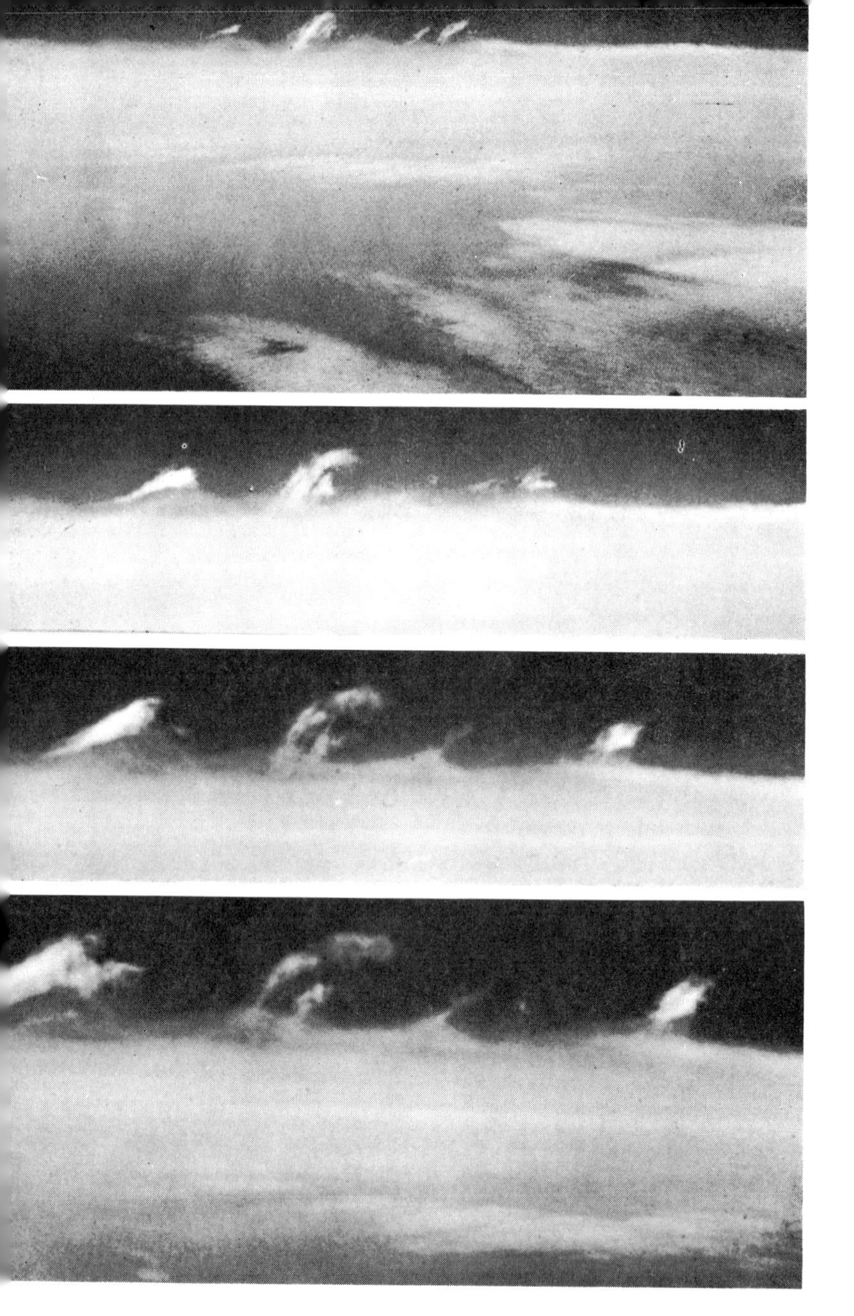

4.10 Billow clouds formed by breaking gravity waves in the jet stream (by permission of Weather *and J. R. C. Young)*

(Berkshire and Warren, 1970) where they can become visible as nacreous (mother of pearl) clouds at heights of 20 to 30 km. These clouds are mainly seen at high latitudes, but lee wave penetration can occur at other latitudes provided that there is a strong deep airstream extending to heights of 30 km or more over a large chain of mountains. These conditions are normally only provided in the strong eastward currents to the south of an intense high latitude depression during the northern hemispheric winter.

Prediction

The stability and wind requirements for the occurrence of the lee waves are:

(i) A statically stable layer, such as a temperature inversion, near the mountain peak. The stability must decrease above the stable layer. Mountain waves will occur when a frontal zone passes over a mountain but can also occur in the absence of fronts.

(ii) It is also necessary to have winds within 30 degrees of being perpendicular to the mountain. The wind speed must increase upwards through a substantial part of the troposphere. Observations indicate that a minimum speed of about 10 m/s is required at the peak in order to generate waves.

(iii) If both of these conditions are not satisfied then it is necessary to recourse to the Scorer parameter

$$\mathscr{F}(z) \approx \frac{\omega_B^2}{U_{0x}^2}$$

which must have a maximum in the lower or middle troposphere with lesser values above.

The Foehn

Locations in the lee of mountain ranges can experience a warm dry wind. Some of the local names for this wind were mentioned in the introductory Chapter and I will continue to use the term foehn for it. The warmth and dryness of the foehn arises because, as the air rises on the windward slopes it cools, moisture is condensed and removed from it and the latent heat liberated during this process causes the temperature of this air to be higher, altitude for altitude, in the lee of the mountain range than on its windward side. Furthermore the winds in the lee of the mountain occur lower than on the windward side so there will be an additional adiabatic heating due to this height difference. There appear to be both anticyclonic foehns and cyclonic foehns. In the cyclonic foehn there is a ridge of high pressure on the windward side and a trough in the lee, with a steep pressure gradient across the mountain. This type of foehn is often accompanied by the full gamut of lee wave clouds (cap, lenticular and rotor), though the bad weather associated with the foehn wall may be absent. Synoptic weather maps of this condition show a bulge, or bulges, in the isobars known as the foehn nose.

In the anticyclonic case there is no foehn nose which suggests that when there is a high pressure system in the lee of the mountain then the foehn is due to dry subsiding air. This dryness precludes the presence of wave clouds, and there is some evidence (Vergeiner, 1971) that lee waves are not a necessary part of foehn winds though large amplitude lee waves may be responsible for the generation of some of the foehn winds. This problem is further compounded by the uncertainty in the use of the name foehn. Meteorologists and laymen have somewhat different criteria for its occurrence (Brinkmann, 1971).

4.7 Clear Air Turbulence (CAT)

CAT, as its name implies, is a form of atmospheric turbulence that can occur without any visible sign of its presence. It occurs in the troposphere and it presents an extreme menace to aircraft which may suffer extensive damage from encounters with moderate or severe CAT. These dangers have endowed the study and elucidation of CAT with a high degree of importance, but despite these factors and the need to understand CAT's role in the atmospheric energy cycle, a full theoretical explanation that is capable of predicting CAT occurrence has not yet been achieved, though the use of ultra-sensitive radars and of lidars (radar systems using laser light instead of radio waves) has greatly improved the ability to detect CAT.

CAT is a particularly difficult phenomenon to study for two reasons. Firstly, the typical scale of the motions is much smaller than the normal scale of free atmospheric motions. Secondly, it is a phenomenon that occurs at the end of a long cycle of events in which motion is induced by thermodynamic causes and the kinetic energy is eventually lost by dissipation.

The cycle begins as the atmosphere is forced into motion by differential heating, and, in the process of attempting to find an equilibrium distribution, these motions create narrow layers in which both the wind and temperature variations are concentrated. When the gradients become sufficiently strong then small-scale turbulence occurs. The turbulent motions cause mixing and tend to smooth the variations of wind and temperature and to reduce the gradients. In the process, a considerable amount of heat and momentum may be transported from one region to another, and with all turbulence there is conversion of kinetic energy to thermal energy. The energetics of CAT have been summarized by Dutton (1971).

The small scale turbulence we are dealing with occurs primarily in four distinct regions: in the mountain lee waves that occur in the Ekman boundary layer (the lowest 2 km of the atmosphere); at the fronts associated with the tropospheric weather systems; at the turbulent tops of large thunder clouds which can extend into the stratosphere; and in the vicinity of the jet stream—a belt of strong winds near the turbopause.

At present the theoretical explanations for the occurrence of CAT, like those for the occurrence of billow clouds, invoke the Kelvin–Helmholtz instability which is a form of hydrodynamic instability that occurs at the interface between two fluids of different density when there is a relative horizontal velocity between them. The formal theory of the Kelvin–Helmholtz instability (KHI) is treated by Chandrasekhar (1961). At a fluid interface, with the heavier fluid on the bottom, the shearing vorticity will be stabilized by the solenoids concentrated in the interface. As a result waves longer than a critical wavelength will be stable. In the absence of surface tension, waves proportional to $\exp i(\omega t - kx)$ will grow if

$$k > \frac{2g(\delta\rho/\rho)}{(\delta U)^2} \tag{4.7.1}$$

and the growth rate is given by the imaginary part of ω, namely

$$\mathrm{Im}\,(\omega) = \frac{1}{2}k\left[(\delta U)^2 - 2\left(\frac{g}{k}\right)\frac{\delta\rho}{\rho}\right]^{\frac{1}{2}}. \tag{4.7.2}$$

This theory has been applied to billow clouds (Ludlam, 1967) and the wavelengths calculated theoretically are in good agreement with the observed spacing of the clouds.

On the other hand it may be possible that there are types of CAT which are not manifestations of the Kelvin–Helmholtz instability—there is insufficient solid data on CAT to be sure. The analysis of wind and temperature records from instrumented research aircraft that do happen to pass through a patch of CAT generally yield values for the Richardson number close to $\frac{1}{4}$, which supports the view that a smooth laminar flow has become unstable. Further support comes from the measurements of the Richardson number in warm fronts by means of radar (Browning *et al.* 1970) and from a study of billow clouds (Ludlam, 1967). The aircraft measurements show that the initial instability typically has a wavelength of a few kilometres, and the resulting billows have a height of 100 to 500 metres. Young (1971) describes the passage of an aircraft through CAT associated with the jet-stream and has produced a sequence of photographs of the billows that marked the breaking of the waves in the CAT region. (Fig. 4.10). Rather more technical descriptions of all aspects of the theory and detection of small-scale atmospheric turbulence may be found in the December 1969 issue of the journal *Radio Science* which contains the proceedings of a conference on the spectra of meteorological variables.

The motion of fluid vortices is described by the vorticity equation which, for an inviscid compressible atmosphere, may be written in terms of the potential temperature as

$$\frac{D\zeta}{Dt} = (\zeta \,.\, \nabla)\mathbf{U} - \theta^{-1}\nabla\theta \times \left[\mathbf{g} - \frac{D\mathbf{U}}{Dt}\right] \qquad (4.7.3)$$

where ∇p has been expanded out by using the equation of motion neglecting the Coriolis term.

Now let us consider one case in which it is possible to obtain vertical wind shear—meridional temperature gradients. Then by the thermal wind equations we have approximately that

$$\frac{\partial U_{ge}}{\partial z} = -\frac{g}{f}\frac{\partial \ln T}{\partial n} \qquad (4.7.4)$$

where U_g is the geostrophic wind. If we now consider the meridional temperature distribution along a constant pressure surface, which can be related to the constant height surfaces through $\partial p/\partial z = -\rho g$ then

$$\frac{\partial U_{ge}}{\partial z} = -\frac{g}{f}\left(\frac{\partial \ln \theta}{\partial n}\right)_p \qquad (4.7.5)$$

and the slope of an isentropic surface along which the potential temperature remains constant with respect to an isobaric surface is approximately

$$\tan\psi = \left(\frac{\partial z}{\partial n}\right)_\theta = -\frac{\partial \ln\theta/\partial n}{\partial(\ln\theta)/\partial z} \qquad (4.7.6)$$

and at the same time the geostrophic Richardson number is approximately

$$\mathrm{R}_{ig} = \frac{\omega_B^2}{(\partial U_{ge}/\partial z)^2} = \frac{f^2}{\psi^2[g\partial(\ln\theta)/\partial z]}$$

$$= \frac{f^2}{\psi^2\omega_B^2}. \qquad (4.7.7)$$

To return to (4.7.3), if we are examining the growth of unstable vortices then we find that the fluid acceleration is small within the time scales we are dealing with. Also the vorticity cannot be adequately concentrated into thin layers by the advection mechanism represented by the first term on the right. Consequently if we rewrite the equation in two dimensions it effectively becomes

$$\frac{D\zeta}{Dt} = \omega_B^2 \sin\psi \doteqdot -\frac{g}{\theta}\left(\frac{\partial\theta}{\partial n}\right)_p \qquad (4.7.8)$$

where ζ is the component of the vorticity normal to the flow.

Equations (4.7.7) and (4.7.8) indicate that where ω_B—a measure of the static stability—is greatest, the Richardson number will be the least and the vorticity changes will be the greatest. At the same time the passage of a gravity wave will cause variations in ψ with the largest wave amplitudes causing the greatest variation. The thickness of the shearing layer in which the instability takes place δz is related to R_i through (4.7.1) and is given by

$$k\,\delta z \geqslant 7\mathrm{R}_i$$

or if we equate R_i with $\frac{1}{4}$ we find that $\delta z = \lambda/\pi$.

Eventually the wave breaks and spreads horizontally into a turbulent layer whose thickness h is related to the wavelength by (Scorer, 1951)

$$\lambda = 2{\cdot}7h$$

and we thus find that the thickness of the turbulent layer created by the instability can be expected to be about 1·5 times as thick as the originally stable layer which, on becoming tilted, became the dynamically unstable one.

The turbulence in the thick layer dies as the Richardson number approaches unity but there is weak turbulence at the edges of the layer, which slowly thins the layer back to its original thickness. The process is illustrated in Fig. 4.11.

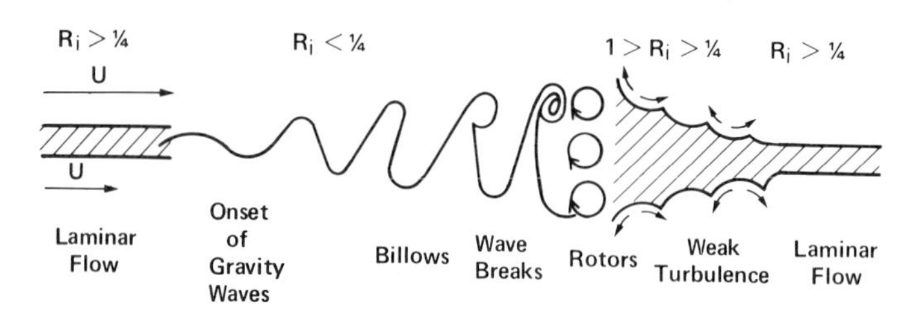

4.11 Cycle of the Kelvin–Helmholtz instability on gravity waves in a sheared flow

4. Background Reading

Platzman, G. (1968) The Rossby wave, *Quarterly Journal of the Royal Meteorological Society*, **94**, 225.

The airflow over mountains, *World Meteorological Organization Technical Note No. 34*, WMO, Geneva (1960).

Spectra of meteorological variables, *Radio Science*, **4**, No. 12 (1969).

Roach, W. T. (1972) *Clear Air Turbulence*, Merrow Publ. Co., Watford.

Holton, J. R. (1972) *An Introduction to Dynamic Meteorology*, Academic Press, New York.

Monin, A. S. (1969) Hydrodynamic theory of short-range weather forecasting, *Soviet Physics Uspekhi*, **11**, 746 (English translation of *Usp. Fiz. Nauk*, **96**, 327).

5

Atmospheric Tides

5.1 Introduction

Both the sun and the moon exert periodic external forces upon the earth's atmosphere. In the case of the moon these forces are wholly gravitational, except for the minute heating effect from the reflected radiation at the full moon. The sun, however, exerts a strong thermal effect as well as a much weaker gravitational effect. The earth's atmosphere will respond to these forces in a manner analogous to forced mechanical vibrations. As it is possible to analyse the forcing term into harmonic components, the steady-state responses of the atmosphere to these forces are known as atmospheric tides. They will have periods that are submultiples of the solar or lunar day.

The semi-diurnal period, which is one half the period of the earth's rotation, occupies a central position in atmospheric wave theory because it acts as a cutoff frequency. For waves in a deep body of a homogeneous incompressible fluid in rotation (inertia waves), at the poles the semidiurnal frequency $2\Omega_E$ acts as a high-frequency cut-off. This is because small changes in vorticity satisfy

$$\frac{\partial}{\partial t}(\nabla \times \mathbf{U}) = 2\Omega_E \frac{\partial \mathbf{U}}{\partial z} \tag{5.1.1}$$

if the Coriolis parameter stays constant, so that taking the curl again

$$-\frac{\partial^2}{\partial t^2}(\nabla^2 \mathbf{U}) = 4\Omega_E^2 \frac{\partial^2 \mathbf{U}}{\partial z^2} \tag{5.1.2}$$

since $\nabla . \mathbf{U} = 0$ for an incompressible homogeneous fluid. Equation (5.1.2) has plane wave solutions only if

$$\omega^2(k_x^2 + k_y^2 + k_z^2) = 4\Omega_E^2 k_z^2 \tag{5.1.3}$$

so that the period of the oscillation must be longer than one half the earth's rotation period. Planetary waves are inertial waves with a meridionally variable Coriolis parameter and we have seen in the previous chapter that they also must have periods longer than the semidiurnal period.

On the other hand gravity waves have a low-frequency cutoff at the semidiurnal frequency. For a homogeneous fluid with a free surface at $z = h$ and a bottom at $z = 0$, equation (5.1.3) still specifies the waves that will occur for $\omega < 2\Omega_E$ with k_x, k_y and k_z real. The free surface however, makes possible, in addition, solutions with k_z wholly imaginary so that for $\omega > 2\Omega_E$, if $\Gamma = \text{Im}(K_z)$

$$\Gamma^2(\omega^2 - 4\Omega_E^2) = \omega^2(k_x^2 + k_y^2). \tag{5.1.4}$$

These solutions have a dispersion relation that we have already found, namely

$$\omega^2 = g\Gamma \tanh [\Gamma \, . \, h]. \tag{5.1.5}$$

Equations (5.1.4) and (5.1.5) show that for $\omega > 2\Omega_E$, when inertial waves propagating internally are impossible, then surface waves become possible with their dispersion relation modified by rotation. These waves have a low-frequency cutoff at the semi-diurnal frequency. On the other hand Rossby waves with large l values may be thought of as surface waves with $\omega < 2\Omega_E$.

The varying depth induced by the surface waves gives a disturbed value of the vertical vorticity which is f times the ratio of the actual depth to the undisturbed depth. The presence of this disturbed vorticity makes possible solutions in which yet another component of wave number is purely imaginary, say $K_y = \text{Im}(K_y)$. In this case the boundary conditions can be satisfied by the vorticity falling off exponentially as $\exp[-\text{Im}(K_y)y]$ with distance y from the boundary. The vorticity propagates one dimensionally along the boundary by rises and falls in depth. This is the Kelvin wave.

Internal gravity waves in a rotating fluid must be analysed by starting with equation (5.1.1) and adding a term that represents the rate of production of horizontal vorticity due to potential temperature gradients. The z component of (5.1.2) is then

$$-\frac{\partial^2}{\partial t^2}\nabla^2 U_z = 4\Omega_E^2 \frac{\partial^2 U_z}{\partial z^2} + \omega_B^2\left(\frac{\partial^2 U_z}{\partial x^2} + \frac{\partial^2 U_z}{\partial y^2}\right) \tag{5.1.6}$$

so that plane wave solutions of (5.1.6) satisfy

$$\omega^2(K_x^2+K_y^2+K_z^2) = 4\Omega_E^2 K_z^2 + \omega_B^2(K_x^2+K_y^2) \tag{5.1.7}$$

Since K_x, K_y and K_z all have real components, plane wave solutions only exist when ω lies between $2\Omega_E$ and ω_B. In the unusual case when $\omega_B < 2\Omega_E$ then gyroscopic waves with $\omega_B < \omega < 2\Omega_E$ would be the only ones extant. Thus we see that internal gravity waves of period longer than the semi-diurnal period cannot exist. This applies to all latitudes, though the $4\Omega_E^2 K_z^2$ term is replaced by $4\Omega_E^2 K_z^2 \cos^2\theta$ producing a low-frequency cutoff at $2\Omega_E \cos\theta$, the Coriolis parameter, where θ is the colatitude.

This shows that except for the diurnal component all the tides are gravity wave modes. Between latitudes $\pm 30°$ the diurnal tide may also be a gravity wave but if it exists at higher latitudes then it must be a Rossby mode. To a first approximation the tidal modes at low altitudes will be evanescent. This arises because the wavelength for large period evanescent waves satisfies

$$\lambda > \frac{0{\cdot}9(2\pi)c}{\omega} \tag{5.1.8}$$

(see § 2.9). Wavelengths that do not satisfy this inequality are internal gravity waves. For a tidal mode the wavelength is given by $2\pi R_E/m$ and the frequency is $\omega = 2\pi m/24\,\text{hours}^{-1}$ so that the inequality in (5.1.8) is satisfied provided that the atmospheric temperature, which affects the sound speed c, is less than about 750°K. In the upper atmosphere this will not be true and thus above about 150 km altitude one may expect to find a vertical phase structure in the tidal modes. Therefore we would intuitively expect to find no vertical phase variation from the ground up to an altitude equal to the equivalent depth. Similarly we would expect to find a vertical phase variation in the upper atmosphere. The following sections will attempt to examine how good an approximation (5.1.8) is.

5.2 Theory

Just as a full discussion of planetary waves requires the use of spherical coordinates, so it proves necessary to recast the equations of tidal motion into spherical form. This neglects the earth's ellipticity which is unlikely to be a very serious fault. The equations of motion on a rotating earth are

$$\frac{DU_\phi}{Dt} = F_\phi - \left(2\Omega_E + \frac{U_\phi}{r \sin\theta}\right)(U_\theta \cos\theta + U_r \sin\theta)$$

$$\frac{DU_\theta}{Dt} = F_\theta + \left(2\Omega_E + \frac{U_\phi}{r \sin\theta}\right) U_\phi \cos\theta - \frac{U_r U_\theta}{r}$$

$$\frac{DU_r}{Dt} = F_r - g + \left(2\Omega_E + \frac{U_\phi}{r \sin\theta}\right) U_\phi \sin\theta + \frac{U_\theta^2}{r}$$

where the centifugal acceleration is incorporated into g, and F is the acceleration due to the external pressure and frictional force. The velocity components are defined by

$$U_r = \frac{Dr}{Dt}; \qquad U_\theta = r\frac{D\theta}{Dt}; \qquad U_\phi = r \sin\theta \frac{D\phi}{Dt}.$$

Now for tidal motions the atmosphere is shallow so that we take

$$r = R_E$$

and replace $\partial/\partial r$ by $\partial/\partial z$ whenever it occurs. The coordinate is zero when $r = R_E$. In this case it is also necessary and reasonable to assume $U_{0r} = 0$ (Phillips, 1966, 1968). Then the perturbed equations of motion may be written as

$$\frac{\partial U_{1\theta}}{\partial t} - 2\Omega_E U_{1\phi} \cos\theta = -\frac{1}{R_E}\frac{\partial}{\partial\theta}\left(\frac{p_1}{\rho_0} + \Phi\right)$$

$$\frac{\partial U_{1\phi}}{\partial t} + 2\Omega_E U_{1\theta} \cos\theta = -\frac{1}{R_E \sin\theta}\frac{\partial}{\partial\phi}\left(\frac{p_1}{\rho_0} + \Phi\right)$$

and

$$\frac{\partial p_1}{\partial z} = -g\rho_1 - \rho_0 \frac{\partial \Phi}{\partial z},$$

since

$$\frac{D}{Dt} = \frac{\partial}{\partial t} + U_r \frac{\partial}{\partial z} + \frac{U_\theta}{R_E}\frac{\partial}{\partial\theta} + \frac{U_\phi}{R_E \sin\theta}\frac{\partial}{\partial\phi}.$$

We have simplified by problem by neglecting any background prevailing winds $\mathbf{U}_0$. The Φ term is a scalar potential that describes the gravitational tide-producing forces (Lamb, 1945, p. 359). The dissipative effects of viscosity, thermal conduction, etc., have been ignored.

The other equations necessary to describe the dynamics of the motion are the equation of continuity

$$\frac{D\rho}{Dt} = \frac{\partial \rho_1}{\partial t} + U_{1z}\frac{\partial \rho_0}{\partial z} = -\rho_0 \chi$$

and the adiabatic equation

$$\frac{R}{M(\gamma-1)}\frac{DT}{Dt} = \frac{R}{M(\gamma-1)}\left(\frac{\partial T_1}{\partial t} + U_{1z}\frac{\partial T_0}{\partial z}\right)$$

$$= \frac{gH}{\rho_0}\frac{D\rho}{Dt} + J$$

where the velocity divergence is

$$\chi = \nabla . \mathbf{U} = \frac{1}{R_E \sin\theta}\frac{\partial}{\partial \theta}(U_{1\theta}\sin\theta) + \frac{1}{R_E \sin\theta}\frac{\partial U_{1\phi}}{\partial \phi} + \frac{\partial U_{1z}}{\partial z}.$$

J is the thermotidal heating per unit mass per unit time and it represents the periodic driving force acting on the free atmospheric oscillations.

It is now most convenient to seek periodic solutions of these equations for the quantity $-(\gamma p_0)^{-1} Dp/Dt$. For the free oscillations, in the absence of the driving force J, this term corresponds to the wind divergence χ. However, when J is included it no longer equals χ and we shall denote it by

$$-\frac{1}{\gamma p_0}\frac{Dp}{Dt} = G(\theta, z)\exp[i(\omega t + m\phi)]$$

where $2\pi/\omega$ represents either a solar or lunar day or some suitable fraction thereof and

$$m = 0, \pm 1, \pm 2, \ldots.$$

All the fields have the same exponential dependence so that all the $\partial/\partial t$ and $\partial/\partial\phi$ terms may be eliminated. This leads directly to expressions for the velocity functions $U_{1\theta}(\theta, z)$ and $U_{1\phi}(\theta, z)$ in terms of the pressure perturbation $p_1(\theta, z)$. The details of the derivation are given by Siebert (1961) and Chapman and Lindzen (1970, p. 109) and provided that the term $\partial^2\Phi/\partial z^2$ can be ignored then the equation for $G(\theta, z)$ is

$$H\frac{\partial^2 G}{\partial z^2}+\left(\frac{dH}{dz}-1\right)\frac{\partial G}{\partial z} = \frac{g}{4R_E^2\Omega_E^2}\hat{F}\left(\left(\frac{dH}{dz}+\frac{(\gamma-1)}{\gamma}\right)G-\frac{(\gamma-1)J(\theta,z)}{\gamma^2 gH}\right) \tag{5.2.1}$$

where $\hat{F}$ is an operator

$$\hat{F} \equiv \frac{1}{\sin\theta}\frac{\partial}{\partial\theta}\left(\frac{\sin\theta}{f^2-\cos^2\theta}\frac{\partial}{\partial\theta}\right) - \frac{1}{f^2-\cos^2\theta}\left(\frac{m}{f}\frac{f^2+\cos^2\theta}{f^2-\cos^2\theta}+\frac{m^2}{\sin^2\theta}\right) \tag{5.2.2}$$

and

$$f = \frac{\omega}{2\Omega_E}$$

$$J = J(\theta, z)\exp\left[i(\omega t+m\phi)\right].$$

Equation (5.2.1) may be solved by the method of separation of variables. If we assume that

$$G(\theta, z) = \sum_n L_n(z)\Theta_n(\theta)$$

then J may be expanded in terms of the function Θ_n, which is generally called the Hough function, as

$$J(\theta, z) = \sum_n J_n(z)\Theta_n(\theta)$$

so that equation (5.2.1) breaks up into two equations: Laplace's tidal equation and the vertical structure equation.

Laplace's Tidal Equation

The method of separation of variables breaks the partial differential equation into two or more parts, each of which is a function of only one variable. As the two parts are equal in this case, each part of (5.2.1) must equal a constant. If we denote this constant by h_n then

$$\hat{F}(\Theta_n) = -\frac{4R_E^2\Omega_E^2}{gh_n}\Theta_n \tag{5.2.3}$$

The set of values of h_n are the set of eigenvalues of Laplace's tidal equation and by analogy with the analysis of oceanic tidal theory, h_n is called the equivalent depth of the atmosphere, which may be rewritten as

$$\frac{d}{d\mu}\left(\frac{1-\mu^2}{f^2-\mu^2}\frac{d\Theta_n}{d\mu}\right) - \frac{1}{f^2-\mu^2}\left(\frac{m}{f}\frac{(f^2+\mu^2)}{(f^2-\mu^2)}+\frac{m^2}{1-\mu^2}\right)\Theta_n + \frac{4R_E^2\Omega_E^2}{gh_n}\Theta_n = 0 \tag{5.2.4}$$

where $\mu = \cos\theta$.

This equation has simple solutions in only three rather special cases; for a non-rotating earth, for an infinite equivalent depth and for the semi-diurnal tide with $m = 0$.

In the last case $f = 1$ and $m = 0$ so that

$$\frac{d^2\Theta_n}{d\mu^2} + \frac{4R_E^2\Omega_E^2}{gh_n}\Theta_n = 0. \tag{5.2.5}$$

The equivalent depths may be found either by imposing mathematical conditions; that Θ be a uniquely defined, single valued, monotonic function with continuous derivative; or by imposing physical conditions that $U_{1\theta}$ should be zero at the north and south poles. These conditions can only be satisfied if

$$\frac{2R_E\Omega_E}{\sqrt{gh_n}} = \frac{n\pi}{2}$$

so that the equivalent depths in this case are

$$h_n = \frac{16R_E^2\Omega_E^2}{n^2\pi^2 g}$$

$$= \infty, 35{\cdot}04, 8{\cdot}76, 3{\cdot}89, \ldots \quad \text{km}$$

for $n = 0, 1, 2, 3, \ldots$.

In general, however, it is possible to obtain third order recursion relations by expressing Hough's functions as a sum of associated

Legendre Polynomials:

$$\Theta_n^{\omega,m} = \sum_{l=m}^{\infty} C_{n,l}^{\omega,m} P_l^m(\cos\theta).$$

though it is equally possible to express the Legendre functions as a sum of Hough functions. For example,

$$P_2^2(\cos\theta) = 0{\cdot}97\Theta_2^{S2,2}(\theta) + 0{\cdot}22\Theta_4^{S2,2}(\theta) + \ldots.$$

The three most important tides in the atmosphere are the solar semidiurnal tide (S_2^2) with $m = 2$ and $f = 1$,* the solar diurnal tide (S_1^1) with $m = 1$ and $f = 0{\cdot}5$ and the lunar semidiurnal tide (L_2^2) with $m = 2$ and $f = 0{\cdot}965$. The general solar tide is represented by S_s^m and the general lunar tide by L_s^m where $\omega = 2\pi s/24$ sidereal or lunar hours^{-1} respectively. Tables 5.1, 5.2 and 5.3 give the Legendre function expansions of the Hough functions for the principal modes of the

Table 5.1 Diurnal Tide Hough function expansions (from Chapman and Lindzen, 1970)

	Equivalent depths (km)
$\Theta_{-1}^{S1,1} = P_2^1$	803
$\Theta_1^{S1,1} = 0{\cdot}3P_1^1 - 0{\cdot}6P_3^1 + 0{\cdot}6P_5^1 - 0{\cdot}3P_7^1 + 0{\cdot}1P_9^1$	0·69
$\Theta_{-2}^{S1,1} = 0{\cdot}9P_1^1 + 0{\cdot}4P_3^1$	−12·3
$\Theta_2^{S1,1} = 0{\cdot}3P_4^1 - 0{\cdot}6P_6^1 + 0{\cdot}6P_8^1 - 0{\cdot}4P_{10}^1 + 0{\cdot}2P_{12}^1$	0·24
$\Theta_{-3}^{S1,1} = 0{\cdot}8P_4^1 + 0{\cdot}6P_6^1 + 0{\cdot}2P_8^1$	−1·81
$\Theta_3^{S1,1} = -0{\cdot}1P_1^1 + 0{\cdot}2P_3^1 - 0{\cdot}1P_5^1 - 0{\cdot}2P_7^1 + 0{\cdot}5P_9^1 - 0{\cdot}6P_{11}^1$ $+ 0{\cdot}5P_{13}^1 - 0{\cdot}3P_{15}^1 + 0{\cdot}1P_{17}^1 - 0{\cdot}1P_{19}^1$	0·12

Table 5.2 Semi- and terdiurnal Hough function expansions (S_2 from Chapman and Lindzen (1970); S_3 from Siebert (1961))

	Equivalent depths (km)
$\Theta_2^{S2,2} = P_2^2 - 0{\cdot}2P_4^2$	7·85
$\Theta_3^{S2,2} = 0{\cdot}9P_3^2 - 0{\cdot}4P_5^2$	3·67
$\Theta_4^{S2,2} = 0{\cdot}2P_2^2 + 0{\cdot}8P_4^2 - 0{\cdot}5P_6^2 + 0{\cdot}1P_8^2$	2·11
$\Theta_5^{S2,2} = 0{\cdot}3P_3^2 + 0{\cdot}6P_5^2 - 0{\cdot}6P_7^2 + 0{\cdot}2P_9^2$	1·37
$\Theta_6^{S2,2} = 0{\cdot}1P_2^2 + 0{\cdot}4P_4^2 + 0{\cdot}5P_6^2 - 0{\cdot}7P_8^2 + 0{\cdot}3P_{10}^2 - 0{\cdot}1P_{12}^2$	0·96
$\Theta_3^{S3,3} = P_3^3 - 0{\cdot}1P_5^3$	12·9
$\Theta_4^{S3,3} = P_4^3 - 0{\cdot}1P_6^3$	7·66

* Actually $f = 0{\cdot}99727$ because the S_2^2 mode has a period of 12 sidereal hours whereas Ω_E has a period of 12 solar hours. A similar comment applies to all the solar tides.

Table 5.3 Lunar tidal Hough function expansions (from Jones, 1970)

	Equivalent depths (km)
$\Theta_{-2}^{L1,1} = -0{\cdot}9P_1^1 - 0{\cdot}4P_3^1$	$-61{\cdot}9$
$\Theta_{-1}^{L1,1} = -P_2^1 + 0{\cdot}1P_4^1$	123
$\Theta_{-3}^{L1,1} = -0{\cdot}1P_2^1 - 0{\cdot}9P_4^1 - 0{\cdot}5P_6^1 - 0{\cdot}1P_8^1$	$-9{\cdot}55$
$\Theta_{1}^{L1,1} = -0{\cdot}2P_1^1 + 0{\cdot}6P_3^1 - 0{\cdot}6P_5^1 + 0{\cdot}4P_7^1 - 0{\cdot}1P_9^1 + 0{\cdot}1P_{11}^1$	2·36
$\Theta_{2}^{L2,2} = P_2^2 - 0{\cdot}3P_4^2$	7·60
$\Theta_{3}^{L2,2} = 0{\cdot}9P_3^2 - 0{\cdot}4P_5^2 + 0{\cdot}1P_7^2$	3·49
$\Theta_{4}^{L2,2} = 0{\cdot}2P_2^2 + 0{\cdot}8P_4^2 - 0{\cdot}6P_6^2 + 0{\cdot}2P_8^2$	1·96

solar diurnal, semidiurnal and terdiurnal tides and for the lunar semidiurnal and diurnal tides. The coefficients are given only to the first decimal place. Both the Hough functions and the Legendre functions are completely normalized so that

$$\int_{-1}^{1} \Theta_n(\theta)\Theta_{n'}(\theta)\, d(\cos\theta) = \delta_{nn'}$$

and

$$\int_{-1}^{1} P_l^n(\mu)P_l^{n'}(\mu)\, d\mu = \delta_{nn'}$$

where

$$\delta_{nn'} = 0 \qquad \text{if } n \neq n'$$

$$\delta_{nn'} = 1 \qquad \text{if } n = n'.$$

The latitude at which the Hough function drops to zero are the nodal parallels of latitude.

Except for the diurnal mode when $m = 1$, there are $n - m$ nodal latitudes, excluding the nodes at the pole. On the other hand for the diurnal mode the number of nodal latitudes cannot be expressed by one single formula. Because of this Lindzen suggested a renaming of the diurnal modes so that they should all have $|n| - m$ nodal latitudes. Certain authors, e.g. Lindzen (1967a) Hines (1967), use the Lindzen notation. Kato (1966) uses the standard notation for $n > 0$ and the Lindzen notation when $n < 0$. Other authors, principally Chapman and Lindzen (1970) use the standard notation employed herein (see Table 5.4).

Table 5.4 Notation for oscillations of the second class

Standard notation	*Lindzen notation*
$\Theta^{\omega,1}_{-2}$	$\Theta^{\omega,1}_{-1}$
$\Theta^{\omega,1}_{-4}$	$\Theta^{\omega,1}_{-3}$
$\Theta^{\omega,1}_{1}$	$\Theta^{\omega,1}_{3}$
$\Theta^{\omega,1}_{3}$	$\Theta^{\omega,1}_{5}$
$\Theta^{\omega,1}_{5}$	$\Theta^{\omega,1}_{7}$
Antisymmetric Modes	
$\Theta^{\omega,1}_{-1}$	$\Theta^{\omega,1}_{-2}$
$\Theta^{\omega,1}_{-3}$	$\Theta^{\omega,1}_{-4}$
$\Theta^{\omega,1}_{2}$	$\Theta^{\omega,1}_{4}$
$\Theta^{\omega,1}_{4}$	$\Theta^{\omega,1}_{6,1}$
$\Theta^{\omega,1}_{6}$	$\Theta^{\omega,1}_{8}$

This confusion has arisen so recently because it is only in the past decade that an extensive study has been made of the diurnal solar and lunar tides, which has revealed the presence of negative n values and negative equivalent depths. These negative equivalent depths are associated with the energy trapping characteristic of the evanescent tidal Rossby mode mentioned in § 5.1, so that the negative n values exist only for the diurnal tide. The tidal oscillations with $n < 0$ are known as 'oscillations of the second class' and planetary waves are included within this particular set of oscillations. The other tidal modes, and internal gravity waves fall into the set of 'oscillations of the first class.'

Vertical Structure Equation

In order to further examine the equivalent depths we need to turn to the second equation obtained by the method of separation of variables applied to equation (5.2.1). This equation is

$$H\frac{d^2L_n}{dz^2}+\left(\frac{dH}{dz}-1\right)\frac{dL_n}{dz}+\frac{1}{h_n}\left(\frac{dH}{dz}+\frac{\gamma-1}{\gamma}\right)L_n=\frac{\gamma-1}{\gamma^2 gHh_n}J_n.$$

In order to convert this equation into standard form we transform the altitude into scale height units

$$\mathrm{x}=\int_0^z\frac{dz}{H}$$

so that

$$\frac{d^2L_n}{dx^2}-\frac{dL_n}{dx}+\frac{1}{h_n}\left(\frac{dH}{dx}+\frac{(\gamma-1)H}{\gamma}\right)L_n = \frac{(\gamma-1)}{\gamma^2 gh_n}J_n$$

which may be transformed by the substitution in (2.5.15)

$$L_n = y_n \exp\left(+\frac{x}{2}\right)$$

to

$$\frac{d^2y_n}{dx^2}-\frac{1}{4}\left[1-\frac{4}{h_n}\left\{\frac{(\gamma-1)H}{\gamma}+\frac{dH}{dx}\right\}\right]y_n = \frac{(\gamma-1)J_n}{\gamma^2 gh_n}\exp\left(-\frac{x}{2}\right). \quad (5.2.6)$$

It is also possible to derive expressions for the other atmospheric parameters in the same way as in § 2.6. If we expand the tidal potential in terms of Hough functions as

$$\Phi = \sum_n \Phi_n(x)\Theta_n$$

then

$$p_1 = \sum_n \frac{p_0(0)}{H(x)}\left[-\frac{\Phi_n}{g}\exp(-x)+\frac{\gamma h_n}{i\omega}\exp\left(-\frac{x}{2}\right)\left(\frac{dy_n}{dx}-\frac{1}{2}y_n\right)\right]\Theta_n$$
$$\times \exp(i(\omega t+m\phi))$$

$$U_{1z} = \sum_n \left[\gamma h_n \exp\left(\frac{x}{2}\right)\left\{\frac{dy_n}{dx}+\left(\frac{H}{h_n}-\frac{1}{2}\right)y_n\right\}-\frac{i\omega}{g}\Phi_n\right]\Theta_n$$
$$\times \exp(i(\omega t+m\phi))$$

$$U_{1\theta} = \sum_n \frac{\gamma gh_n}{4R_E\Omega_E^2}\exp\left(\frac{x}{2}\right)\left(\frac{dy_n}{dx}-\frac{1}{2}y_n\right)\cdot\frac{1}{(f^2-\cos^2\theta)}\left(\frac{d}{d\theta}+\frac{m\cot\theta}{f}\right)\Theta_n$$
$$\times \exp(i(\omega t+m\phi)$$

$$U_{1\phi} = \sum_n \frac{i\gamma gh_n}{4R_E\Omega_E^2}\exp\left(\frac{x}{2}\right)\left(\frac{dy_n}{dx}-\frac{1}{2}y_n\right)\cdot\frac{1}{(f^2-\cos^2\theta)}$$
$$\times\left(\frac{\cos\theta}{f}\frac{d}{d\theta}+\frac{m}{\sin\theta}\right)\Theta_n \exp(i(\omega t+m\phi))$$

$$\rho_1 = \sum_n \frac{p_0(0)}{(gH)^2} \left\{ -\Phi_n \exp(-\mathrm{x}) \left(1 + \frac{1}{H}\frac{dH}{d\mathrm{x}}\right) + \frac{\gamma g h_n}{i\omega} \right.$$

$$\times \exp\left(-\frac{\mathrm{x}}{2}\right) \left[\left(1 + \frac{1}{H}\frac{dH}{d\mathrm{x}}\right) \left(\frac{dy_n}{d\mathrm{x}} - \frac{y_n}{2}\right) + \frac{H}{h_n}\left(\frac{\gamma - 1}{\gamma} + \frac{1}{H}\frac{dH}{d\mathrm{x}}\right) y_n \right]$$

$$\left. - \frac{(\gamma-1)J_n}{i\omega\gamma} \right\} \Theta_n \exp(i(\omega t + m\phi))$$

$$T_1 = \sum_n \frac{M}{R} \left\{ \frac{\Phi_n}{H} \frac{dH}{d\mathrm{x}} \exp(-\mathrm{x}) - \frac{\gamma g h_n}{i\omega} \exp\left(-\frac{\mathrm{x}}{2}\right) \left[\frac{(\gamma-1)H}{\gamma h_n} + \frac{1}{H}\frac{dH}{d\mathrm{x}} \right. \right.$$

$$\left. \left. \times \left(\frac{d}{d\mathrm{x}} + \frac{H}{h_n} - \frac{1}{2}\right) \right] y_n + \frac{(\gamma-1)J_n}{i\omega\gamma} \right\} \Theta_n \exp(i(\omega t + m\phi))$$

The boundary condition $U_{1z} = 0$ at $z = 0$ implies that

$$\frac{dy_n}{d\mathrm{x}} + \left(\frac{H_g}{h_n} - \frac{1}{2}\right) y_n = \frac{i\omega}{\gamma g h_n} \Phi_n. \tag{5.2.7}$$

It is sometimes useful when dealing with planetary waves to express this lower boundary condition in terms of the meridional velocity $U_{1\theta}$

$$U_{1\theta} = \sum_n U_{1\theta}(\mathrm{x})_n U_{1\theta}(\theta)_n$$

where

$$U_{1\theta}(\theta)_n = \frac{1}{f^2 - \cos^2\theta} \left(\frac{d}{d\theta} + \frac{m \cot\theta}{f}\right) \Theta_n$$

$$U_{1\theta}(\mathrm{x})_n = \frac{\gamma g h_n}{4 R_E \Omega_E^2} \exp\left(\frac{\mathrm{x}}{2}\right) \left(\frac{dy_n}{d\mathrm{x}} - \frac{1}{2} y_n\right)$$

$$= \frac{\gamma g h_n}{4 R_E \Omega_E^2} \left(\frac{i\omega}{\gamma g h_n} \Phi_n - \frac{H_g}{h} y_n\right). \tag{5.2.8}$$

Substitution of (5.2.8) into (5.2.7) will give the lower boundary condition in terms of $U_{1\theta}(\mathrm{x})_n$ and Φ_n.

The upper boundary condition may be obtained from the vertical structure equation (5.2.6). The general solution of this differential equation in an isothermal atmosphere is

$$y = A \exp(iqz) + B \exp(-iqz)$$

where

$$q = \frac{1}{H}\sqrt{\frac{(\gamma-1)H}{\gamma h} - \frac{1}{4}}. \qquad (5.2.9)^*$$

The standard boundary condition invokes the radiation condition and sets $B = 0$. This will be satisfactory provided that q^2 is not almost zero. In this case

$$q = \sqrt{\frac{1}{Hh}\left[\frac{\gamma-1}{\gamma} + \frac{dH}{dz}\right] - \frac{1}{4H^2}} \qquad (5.2.10)$$

acts like a vertical wavenumber. If q is real at any given height the mode is an internal wave and vertical energy propagation is possible. For imaginary q however the waves are evanescent and vertical energy propagation is inhibited.

As an exercise in the application of tidal theory the reader should solve equation (5.2.4) in the case of a non-rotating earth and use the equivalent depths of the solution to compare equations (5.2.10) and (2.4.3).

Resonance Theory

At this stage it becomes possible to try and determine whether the gravitational forcing (Φ) or the thermal forcing (J) is the dominant mechanism responsible for the tides actually observed in the earth's atmosphere. To do this, we draw upon an analogy with the theory of mechanical oscillations which informs us that a system has a set of free oscillations. If the period of the forcing matches the free period then there will be a resonant amplification of the system.

Rather than find free periods, we shall find the equivalent depths for which (5.2.6) has analytic solutions subject to the boundary condition (5.2.7) and the radiation condition. If this equivalent depth matches one of the equivalent depths obtained from the eigenvalues of Laplace's tidal equation, which is independent of the driving mechanism, then that particular mode will be amplified.

* An alternate method of deriving q which does not introduce the equivalent depth is to solve the adiabatic equation and the continuity equation under the quasi-geostrophic approximation. This method was used in Charney and Drazin's (1961) paper.

The equivalent depths that will lead to non-zero solutions for the vertical structure equation are influenced primarily by the atmospheric temperature distribution and to a much lesser extent by the background winds (Lindzen, 1968a). For the free oscillations of an isothermal atmosphere subject to the radiation condition we have

$$y_n = A \exp(iq\mathrm{x})$$

where

$$q^2 = \frac{\gamma - 1}{\gamma}\frac{H}{h} - \frac{1}{4} \tag{5.2.11}$$

so that from (5.2.7) we find that $h_n = \gamma H_g$ is determined by the scale height at the ground and has only one value. If H is proportional to x then the solution for $y(\mathrm{x})$ involves Airy functions, whereas if H varies linearly with z then the solution can be obtained in terms of Bessel functions.

It is experimentally found from measurements of the variation of the atmospheric pressure at the ground that the solar semi-diurnal tide predominates. This is surprising because both the sun and moon appear every 24 hours so that one would expect a diurnal periodicity to dominate. Historically there have been many attempts to explain the dominance of the semi-diurnal tide in terms of the resonance theory. The dominant $\Theta^{S2,2}$ mode has an equivalent depth of 7·85 km and it used to be thought that the earth's stratospheric temperature variations were such as to produce two equivalent depths at 10·4 km and 7·9 km (Wilkes, 1949). However, once more accurate upper atmospheric temperature determinations became available it was apparent that even when resonances besides the 10 km resonance exist, these will not occur at 7·9 km. The resonant equivalent depths are determined by the exact shape of the temperature distribution and by the upper boundary condition that is applied. In Fig. 5.1 we present the response curve of the CIRA (1961) reference atmosphere obtained by applying the radiation condition. The magnification is defined as the ratio of the perturbation pressures for the tidal oscillations to the perturbation pressures for free oscillations. There is no resonance near 7·9 km.

Even though various resurgences of the resonance theory appear from time to time (Covez, 1971) it is most probable that atmospheric thermal sources are able to explain the amplitude of the solar

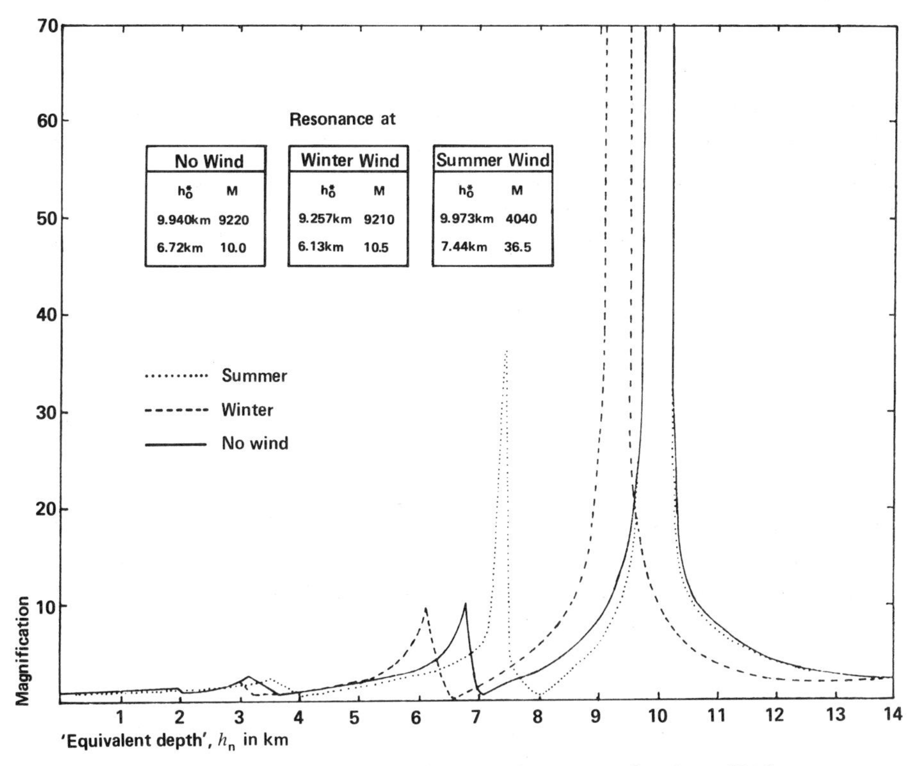

5.1 Response curve for atmospheric oscillations (after Giwa, 1969)

semidiurnal tide. According to this viewpoint we would consider the diurnal tide as being suppressed, rather than the semidiurnal tide as enhanced. Because a given energy input is less efficient for tidal excitation if it is low down in the atmosphere where the density is greater, most of the attention has focussed on the absorption of direct solar radiation by ozone and water vapour. Siebert (1961) pointed out that insolation by water vapour can account for one third the semidiurnal surface pressure variation $S_2^2(p)$; this is far more than could be accounted for by gravitational excitation or surface heating. Butler and Small (1963) showed that the ozone absorption could account for the remaining two-thirds. On the other hand the problem of $S_2^2(p)$ at the ground is not yet completely settled. There remains a discrepancy of one half-hour in phase between theory and observation.

At this point one is forced to return to the question that was put by Lord Kelvin in 1882: Why is the semi-diurnal oscillation stronger than the diurnal one? The situation has, moreover, become more complicated since then for upper atmospheric wind results reveal that at low latitudes the diurnal winds are as strong, and often stronger than the semi-diurnal winds.

Actually the theory we have outlined provides a satisfactory answer. To discuss this point we should recall once again that q, which is given by

$$q^2 = \frac{1}{Hh}\left[\frac{\gamma-1}{\gamma}+\frac{dH}{dz}\right]-\frac{1}{4H^2} \qquad (5.2.12)$$

represents a vertical wavenumber. When h is very large, or negative, as it is for all the $\Theta_{n<0}^{S1,1}$ modes, then the vertical wavenumber in a stable atmosphere becomes imaginary. These oscillations are thus completely evanescent and the associated energy is inhibited from flowing vertically away from its level of deposition. In particular, for application to ionospheric altitudes, it is apparent that these modes will have little to contribute to the tidal oscillation unless there are local diurnally varying heat sources of sufficient strength. In the ionosphere the heating of molecular oxygen is liable to be the most important heat source, but as a heat source it is liable to be rather diffuse because of the large vertical extent of the molecular oxygen. Conversely however, the $n > 0$ modes have little to contribute at high latitudes, so the $n < 0$ modes may there provide the only diurnal tidal motion, whether significant or not.

In combination these points imply that the diurnal tide at meteor and ionospheric heights can be expected to have a relatively small amplitude at latitudes of 50° or more. The detailed computations of Lindzen (1967a) show this to be the case for the heating functions, in ozone and water vapour, which he assumed.

At the ground, polewards of $\pm 30°$, the diurnal period is longer than the local cutoff-frequency, $2\pi/f$, and the diurnal tide is incapable of propagating vertically. Because of this, 80 per cent of the diurnal excitation goes into physically trapped modes near the levels of excitation which are unable to propagate to the ground. However even those modes which can propagate vertically (primarily within $\pm 30°$ latitude) have relatively short wavelengths. This arises because the equivalent depth for these modes is quite small so that q^2 will be quite large. These modes will thus be subject to destructive interference effects, but what really proves to be important is that the propagating modes receive only 20 per cent of the excitation, most of the energy going into the trapped modes.

5.3 Tides at the Earth's Surface

The detection and study of actual tidal oscillations at the ground consists of a long series of readings of recordings of meteorological data—particularly of pressure, but also of wind and temperature—made at numerous observatories widely distributed over the globe. Pressure readings are used so often in tidal analyses because they are least subject to local geographic variations but reflect the tidal nature most clearly.

Solar Semidiurnal Tide

The dominant atmospheric pressure variation is the solar semidiurnal tide. This tide is apparent from even a casual inspection of the trace of a barograph situated in the tropics. There S_2 has its greatest amplitude and is not usually obscured by the large pressure changes associated with mid-latitude planetary scale meteorological disturbances. The pressure maxima of $S_2(p)$ are observed to occur two to three hours before local noon and midnight, and the amplitude of the pressure variations near the equator is slightly over 2 millibar.

At mid-latitude stations, S_2 can only be obtained from data averaged over many months. Haurwitz (1956) used determinations from 296 stations to prepare equilines of amplitude and phase of $S_2(p)$ relative to local mean time (Fig. 5.2). These show that the phase of the tide (measured in degrees from the hours 24/1, 24/2, 24/3, . . . 24/s) is relatively constant, implying the dominance of the migrating semidiurnal tide $s = m = 2$. Other components are found as well and the most significant of these is the semidiurnal standing oscillation, $m = 0$.

If we measure local time, t, in terms of angular measure, so that there are 360° in 24 solar (or lunar) hours, then according to Haurwitz the solar semidiurnal tide is well represented by

$$S_2(p) = 1{\cdot}16 \text{ mb} \sin^3\theta \sin(2t + 158°) + 0{\cdot}054 \text{ mb}\, P_2(\theta) \sin(2t - 2\phi + 118°)$$

where θ = colatitude, ϕ = longitude measured eastward from Greenwich, $P_2(\theta) = (\sqrt{10}/4)(3\cos^2\theta - 1)$. The first term in the expression is the migrating wave $S_2^2(p)$ and the second term is the stationary $S_2^0(p)$ wave which makes a minor contribution at low latitudes but dominates at high latitudes.

The $S_2(p)$ variation is hardly affected by the seasons. There are

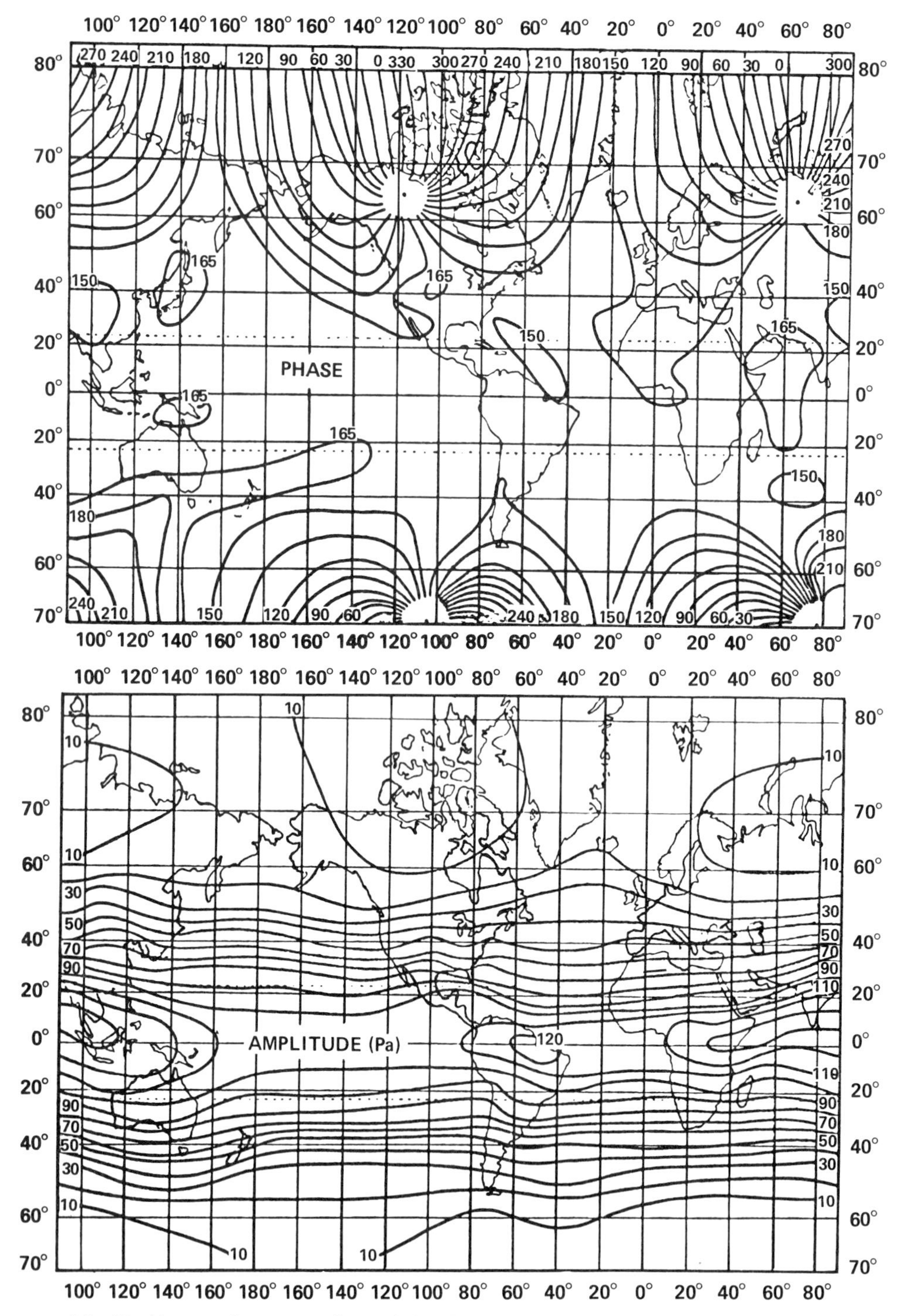

5.2 World maps showing equilines of the phase and amplitude of $S_2(p)$ relative to local time (after Chapman and Lindzen, 1970)

however local irregularities due to the earth's topography which affect the phase of $S_2(p)$. The local irregularity in phase in Fig. 5.2 over the western United States is attributed to the Rocky mountains. Similar irregularities due to the Andes and the Kenyan mountains are also apparent. Presumably more data would reveal similar irregularities over other mountainous areas such as the New Zealand Southern Alps or the Himalayas. The singular points of convergence of the phase equilines arise from the combination of the main travelling wave and the zonal standing wave.

The temperature distribution of S_2 is also found to consist of a zonal standing wave:

$$\begin{aligned} S_2^0(T) = {} & 0{\cdot}024 \sin(2t - 2\phi + 219°) \\ & + 0{\cdot}062 P_1 \sin(2t - 2\phi + 194°) \\ & + 0{\cdot}025 P_2 \sin(2t - 2\phi + 214°) \\ & + 0{\cdot}060 P_3 \sin(2t - 2\phi - 1°) \\ & + 0{\cdot}049 P_4 \sin(2t - 2\phi + 56°) \end{aligned}$$

and a migrating wave

$$\begin{aligned} S_2^2(T) = {} & 0{\cdot}27 P_2^2 \sin(2t + 65°) + 0{\cdot}28 P_3^2 \sin(2t + 65°) \\ & + 0{\cdot}075 P_4^2 \sin(2t + 68°). \end{aligned}$$

The asymmetrical terms in these expressions assume such importance because of the greater difference in land fractions on both sides of the equator.

It is possible to use the observed values of $S_2(p)$ and the equations of motion to deduce expected values of $S_2(U_\theta)$ and $S_2(U_\phi)$. In the absence of friction

$$\frac{\partial U_\theta}{\partial t} - 2\Omega_E U_\phi \cos\theta = -\frac{1}{\rho R_E}\frac{\partial p}{\partial \theta} \qquad (5.3.1)$$

$$\frac{\partial U_\phi}{\partial t} + 2\Omega_E U_\theta \cos\theta = -\frac{1}{\rho R_E \sin\theta}\frac{\partial p}{\partial \phi} \qquad (5.3.2)$$

so that if we take the main term of $S_2(p)$:

$$1{\cdot}16 mb \sin^3\theta \sin(2\Omega_E t_u + 158° + 2\phi)$$

where t_u is universal time (Greenwich mean time) then

$$\frac{\partial}{\partial t} = \Omega_E \frac{\partial}{\partial \phi}$$

for the semidiurnal tide so that

$$S_2(U_\theta) = 2{\cdot}5 C_s \cos\theta \sin(2t + 158° + 90°)$$

$$S_2(U_\phi) = C_s(1 + 1{\cdot}5\cos^2\theta)\sin(2t + 158° + 180°)$$

where $C_s = 116/\rho\Omega_E R_E = 0{\cdot}2$ m/s.

The theoretical phase for $S_2(U_\theta)$ should be 248° in northern latitudes, and 68° in southern latitudes. For $S_2(U_\phi)$ it should be 338° in all latitudes. These predictions do not agree well with the data that actually is available (Chapman and Lindzen, 1970, p. 46). The amplitudes are still more discordant. The theoretical amplitudes increase with latitude, but the observational values decrease. The inclusion of friction, which is of well-established importance in the wind systems below 2 km will modify these theoretical $S_2(\mathbf{U})$ estimates. Whether the modification is sufficient is as yet unknown. The inclusion of friction would add the term $\eta\nabla^2 U_\theta$ or $\eta\nabla^2 U_\phi$ to the right-hand side of equation (5.3.1) and (5.3.2) respectively.

Even though it thus appears that the eddy viscosity is important in determining $S_2(\mathbf{U})$, calculations by Lindzen (1970) show that eddy viscosity and eddy thermal conductivity are negligible for the thermal response of $S_2(p)$ at the surface. Furthermore Lindzen and Blake (1971) show that the atmospheric cooling effects, molecular viscosity and molecular thermal conduction are also negligible. On the other hand their theoretical results, whilst explaining the amplitude of the ground response $S_2(p)$, suffer from a discrepancy of 44 minutes in phase which, presumably, is a consequence of large daily variations in ozone heating.

Solar Diurnal Tide

The situation for $S_1(p)$ is not as simple as the semidiurnal tide. $S_1(p)$ varies with season, it has a small amplitude, and it is strongly polluted by nonmigrating diurnal oscillations. The amplitudes of $S_1^2(p)$ and $S_1^{-1}(p)$ are as large as $\frac{1}{4}$ that of the migrating $S_1^1(p)$ mode, whereas $S_2^2(p)$ was twenty times as large as its nearest competitor. Furthermore large values of m have smaller zonal wavelengths and produce larger winds for a given amplitude of pressure oscillation than $m = 1$. According to Haurwitz (1965), $S_1(p)$ is roughly representable by

$$S_1^1(p) = 593\mu b \sin^3 \theta \sin(t+12°),$$

though his data was mainly obtained from stations in mid-latitudes. Lindzen (1967a) has included the effects of the insolation by water vapour to determine $S_1(p)$. Lindzen's theoretical results, along with Haurwitz's observational results are shown in Fig. 5.3.

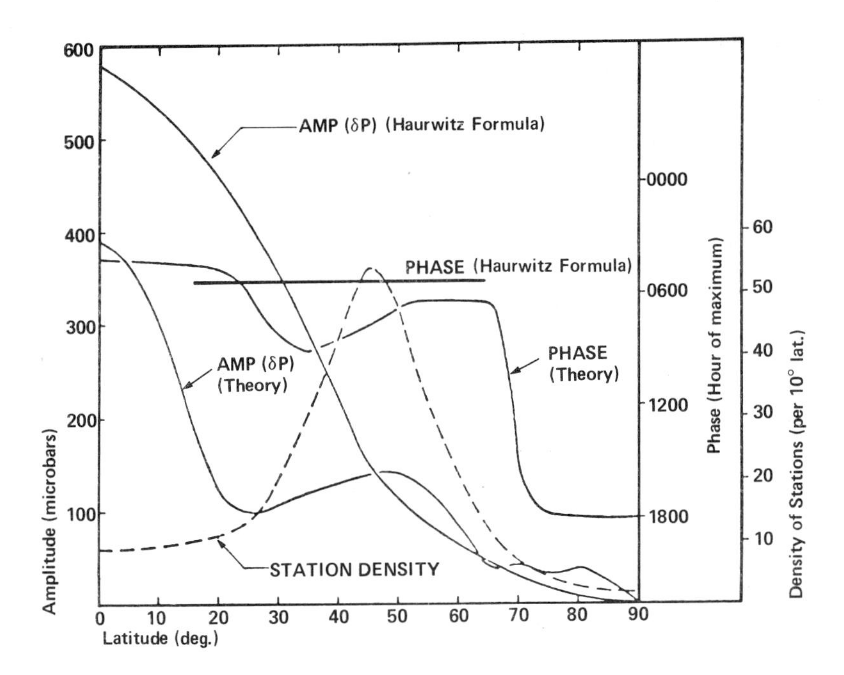

5.3 Calculated amplitude and phase of $S_1(p)$. Lindzen's theoretical results as well as Haurwitz's observed result are given along with the distribution of stations on which Haurwitz based his formula (after Chapman and Lindzen, 1970)

Lunar Semi-diurnal Tide

The lunar semi-diurnal oscillation at the tropics is also far stronger than the same oscillation at mid latitudes. The first reliable determination of the tropical lunar tide, with an amplitude of 0·06 mm of mercury, was made in 1842 yet it was not until 1918 that the existence of a 0·01 mm $L_2(p)$ tide was demonstrated in mid-latitudes. The latitude variation of $L_2(p)$ is almost the same as that of $S_2(p)$ though its amplitude is only $\frac{1}{20}$ that of $S_2(p)$.

$L_2(p)$ has a peculiar seasonal variation that occurs with both the northern and southern hemispheres in phase. One suggested reason for this is that the southern hemisphere, being primarily ocean, will have a lunar tidal lower boundary, namely the varying ocean's surface, that could interact with $L_2(p)$. There will also be a difference in the surface friction of the two hemispheres because of the land-sea dichotomy.

The heating of the atmosphere by moonlight is quite negligible, but the moon does produce a lunar semidiurnal air temperature variation, as a secondary consequence of its mechanical tidal action. The changes of air density that accompany $L_2(p)$ will then depend on whether the temperature changes are adiabatic or isothermal. This depends on whether the density variations are slow enough for the heat of compression to be conducted away, during the period of each oscillation. Chapman (1932) showed that the lunar tides will be adiabatic up to about 100 km, except over the sea (Chapman and Lindzen, 1970) where the greater thermal conductivity of the sea water annuls the adiabatic variations so that the pressure oscillations are isothermal.

In order to make quantitative predictions about L_2 from tidal theory we need to know the form of the gravitational tidal potential Φ, which is (Doodson, 1922; Chapman and Lindzen, 1970)

$$\begin{aligned}\Phi = &-(2{\cdot}3662\,\Theta_2 + {\cdot}5615\,\Theta_4 + {\cdot}2603\,\Theta_6 + \ldots) \\ &\times \cos(2[\omega t + \phi])\ \mathrm{m^2\,s^{-2}}\end{aligned} \tag{5.3.3}$$

where ω is 2π/(lunar day) and ϕ is the longitude measured eastward from Greenwich.

If we solve the vertical structure equation then (5.3.3) may be used to give the surface pressure variations for an isothermal atmosphere. In the $m = 2$ mode this is (Chapman and Lindzen, 1970).

$$(p_1)_2^{L2,2} \approx \frac{34{\cdot}17\mu\mathrm{b}\,\exp(i[2\{\omega t + \phi\} + 90^\circ])}{\left[\left(\dfrac{H}{h} - \dfrac{1}{2}\right) + i\left(\dfrac{[\gamma - 1]H}{\gamma h} - \dfrac{1}{4}\right)^{\frac{1}{2}}\right]} \tag{5.3.4}$$

For $H \approx 7$ km this gives values close to the observed ones. However Sawada (1956) computed the pressure variation for various stratospheric temperatures and found that both the amplitude and the phase were strongly influenced by very small changes in the temperature at the peak of the stratosphere. This is in direct contrast to the solar semidiurnal oscillation which is relatively insensitive to the precise choice of basic temperature distribution. The Hough functions for these two oscillations are so similar that the different behaviour can only result from the different natures of the excitations. The gravitational excitation behaves as though it were a coherent source at the ground, while the thermal excitation is distributed throughout the atmosphere. Apparently, small changes in the vertical distribution of the temperature change the effective height of levels where semidiurnal tides are partially reflected, as well as changing the reflectivities. Repeated reflections produce significant interference for coherent gravitational excitation; for distributed sources these effects tend to cancel each other.

Other Modes

Various other modes can be extracted from the ground based barometric oscillations. These are all solar modes and the principal ones that have received attention are $S_3(p)$ and $S_4(p)$. The terdiurnal mode is geographically regular with a notable seasonal reversal of phase in each hemisphere from summer to winter that makes it antisymmetric about the equator. The component $S_4(p)$ is small, with geographic variations that are greater than either $S_2(p)$ or $S_3(p)$, but it has more regularity than $S_1(p)$.

5.4 Tides above the Ground

Data above the surface are rarer and less accurate than ground based measurements. At a few stations there are frequent balloon ascents which permit tidal analyses. These indicate that in the troposphere horizontal tidal wind oscillations have amplitudes of about 0·1 m/s which grow to about 0·5 m/s in the stratosphere. (Groves, 1969a). At many stations balloon soundings are made twice a day, 12 hours apart. The soundings are made simultaneously at all stations. Thus if one subtracts the average of measurements taken at 1200 UT from the average of measurements taken at 0000 UT one should obtain a fair approximation to the diurnal component of the flow field at 0000 UT. The results (Wallace and Hartranft, 1969) show that at 700 mb (3 km) the diurnal flow field is dominated by gyres associated with relatively small orographic features. At 60 mb (20 km) these gyres remain but they are associated with large features like the Pacific Ocean. By 15 mb (30 km) orographic effects have died out and diurnal oscillations above this level will be representative of the migrating tide.

In the region between 30 and 60 km most of the data is from meteorological rocket soundings. At these heights the tidal wind makes a significant contribution to the total wind so that even infrequent soundings are able to provide useful information. The results that are obtained seem to show diurnal and semidiurnal meridional wind components with roughly equal amplitudes that increase from about 1 m/s at 30 km to 10 m/s at 60 km. The semidiurnal phase above 50 km stays constant at 270° (i.e. a maximum south wind at 0600) but below this height the phase seems to depend on the observer's latitude, remaining at 270° at high latitude stations but decreasing for low latitudes (Groves, 1969a).

In the ionosphere tidal winds are about 20 m/s averaged over the E region but they tend to be outweighed in magnitude by the irregular winds generated by the gravity waves. On the other hand the tidal winds produce strong geomagnetic effects that will be discussed in the next section. Lunar tides can be detected in the ionosphere but there appear to be no theoretical or experimental determinations of $L(p)$ between the ground and 60 km.

Lindzen (1968a, b; 1970, 1971; Lindzen and Blake, 1970, 1971) and his co-workers have launched an extensive theoretical attack on atmospheric solar tides below the ionosphere. To do this he assumed thermal heating due to ozone and water vapour to have the vertical and meridional forms shown in Fig. 5.4 during the daytime. Indications are that surface heating will not affect the tides significantly. The latitude distributions of the excitations are similar to $\Theta_2^{S2,2}$, and this mode should then receive the bulk of the excitation. Furthermore, the equivalent depth of this mode, 7·9 km is such that

$$\left[\frac{(\gamma-1)}{\gamma}+\frac{dH}{dz}\right]\Big/ hH-\frac{1}{4H^2}$$

is everywhere close to zero so that the vertical wavelength of this mode is very large—about 200 km. Thus not only does most of the

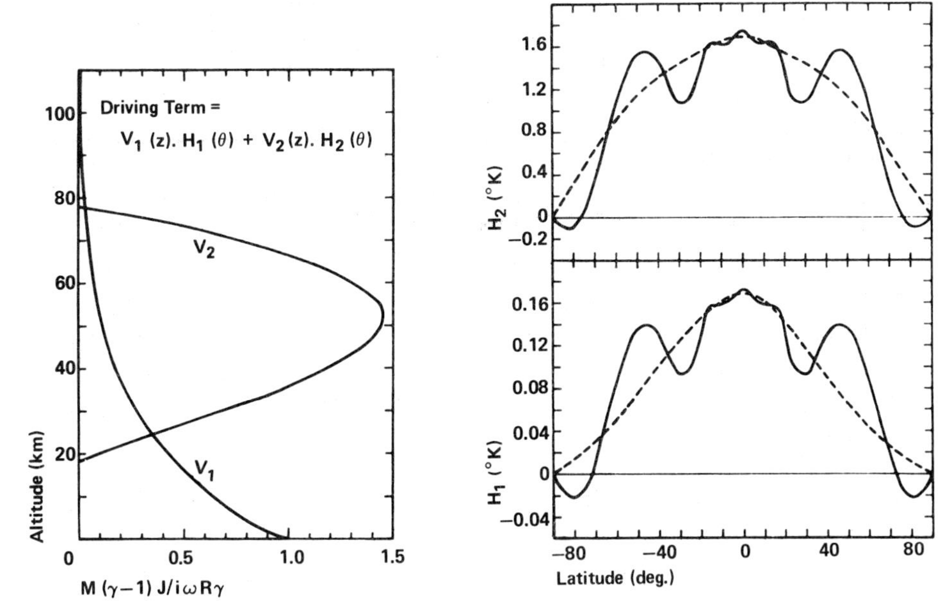

5.4 *Altitude and latitude distribution of thermal excitation due to water vapour* (V_1, H_2). *and ozone* (V_2, H_2). *The vertical distribution's abscissa is a dimensionless quantity composed of the molecular weight, M, the heating function, J, and the tidal frequency,* ω (*after Lindzen, 1968*)

excitation go into $\Theta_2^{2S,2}$, but this mode responds with particular efficiency since all excitations below 100 km act in phase. This explains the strength and regularity of S_2. The larger local variations in heating excite less efficient Hough modes, though higher order modes, which are not evanescent in the mesosphere, become increasingly important at high altitudes and latitudes.

The theoretically obtained southward wind velocities are depicted in Fig. 5.5 for the semidiurnal tide and Fig. 5.6 for the diurnal tide at various latitudes. Above the earth's boundary layer they agree well with observations up to 90 km. The eastward wind at the equator is shown in Fig. 5.7. It has a similar amplitude to the southward wind but a different phase. The vertical velocity U_{1z} ranges from 0·01 m/s at 30 km height to 0·1 m/s at 100 km altitude.

The dominant tidal modes at the equator that were obtained by these theoretical studies are listed in Table 5.5. The vertical wavelength, λ_z, was computed from equation (5.2.9). The most surprising

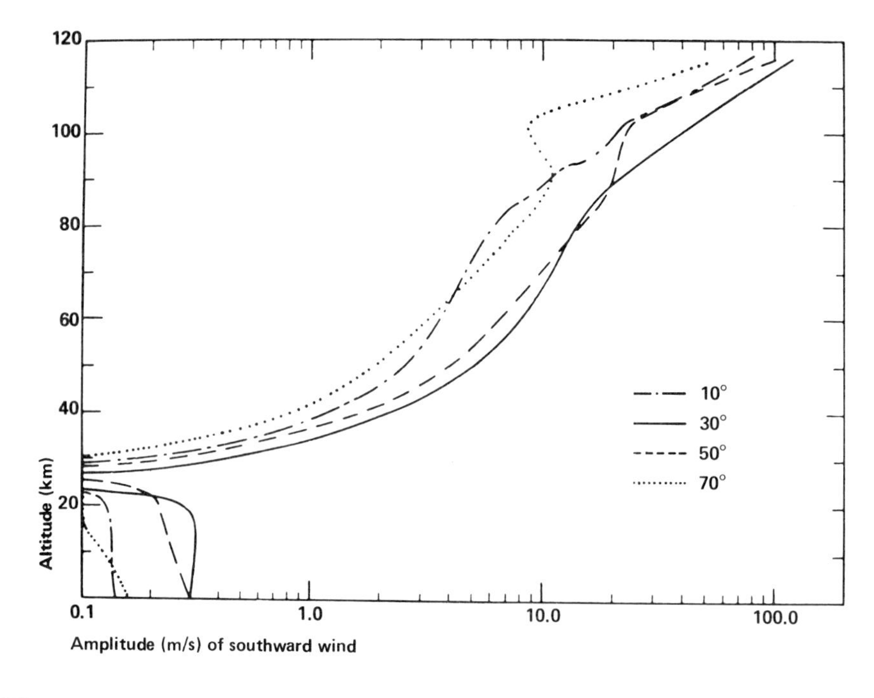

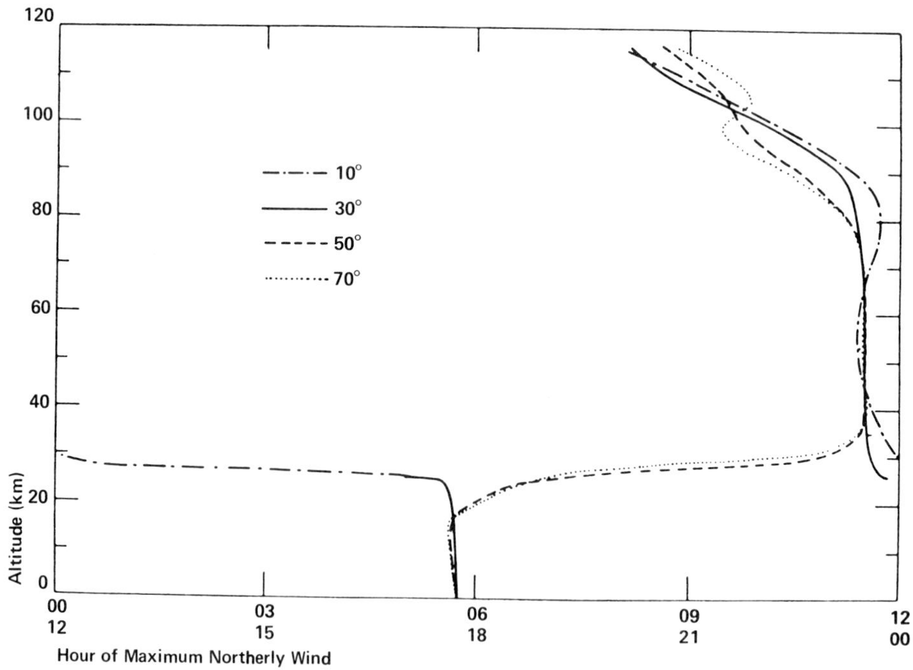

5.5 Amplitude and phase of the theoretical solar semidiurnal southward wind at various latitudes (after Chapman and Lindzen, 1970)

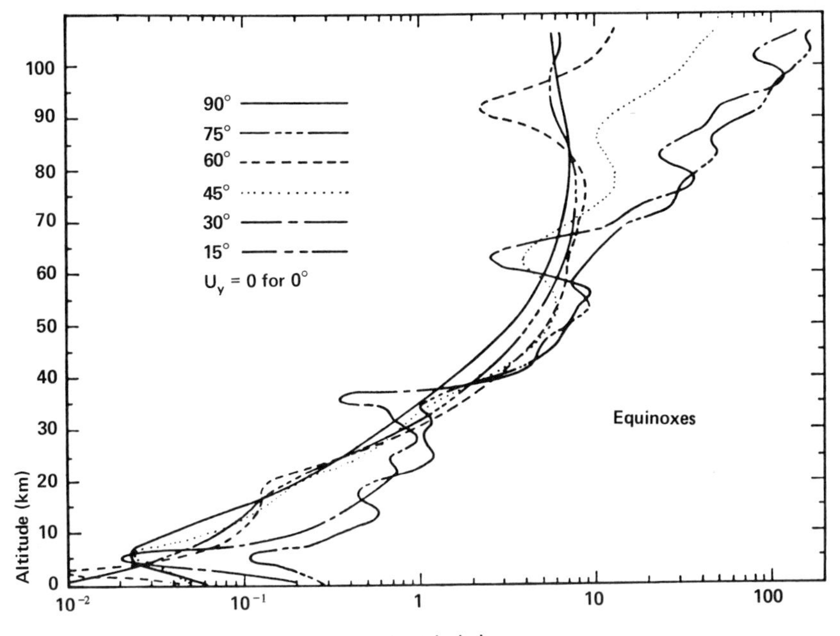

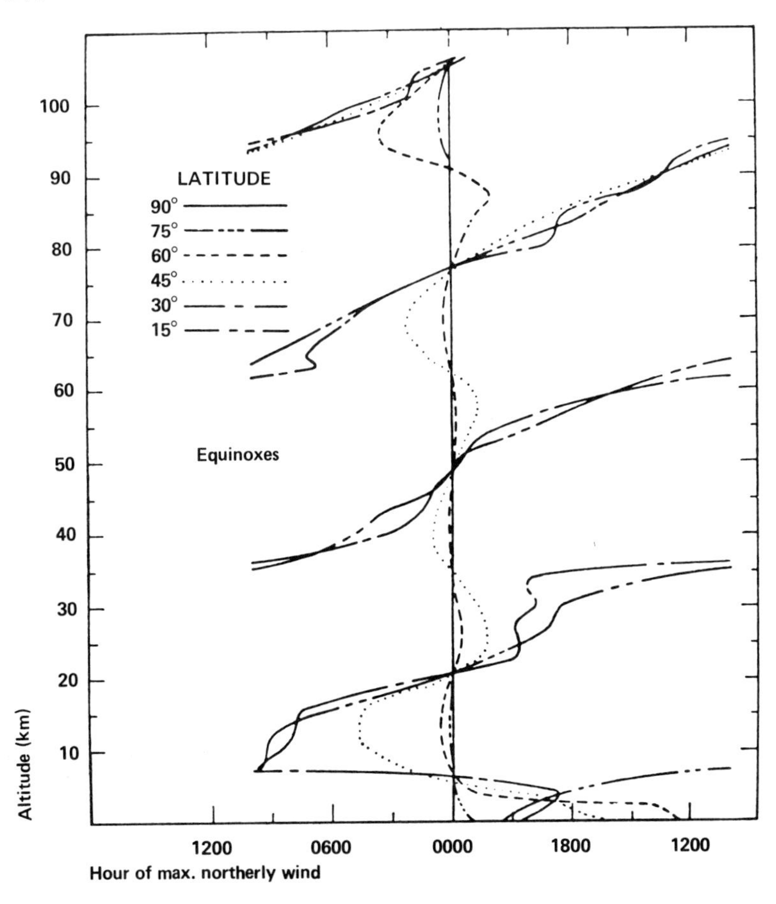

5.6 Amplitude and phase of the theoretical solar diurnal southward wind at various latitudes (after Chapman and Lindzen, 1970)

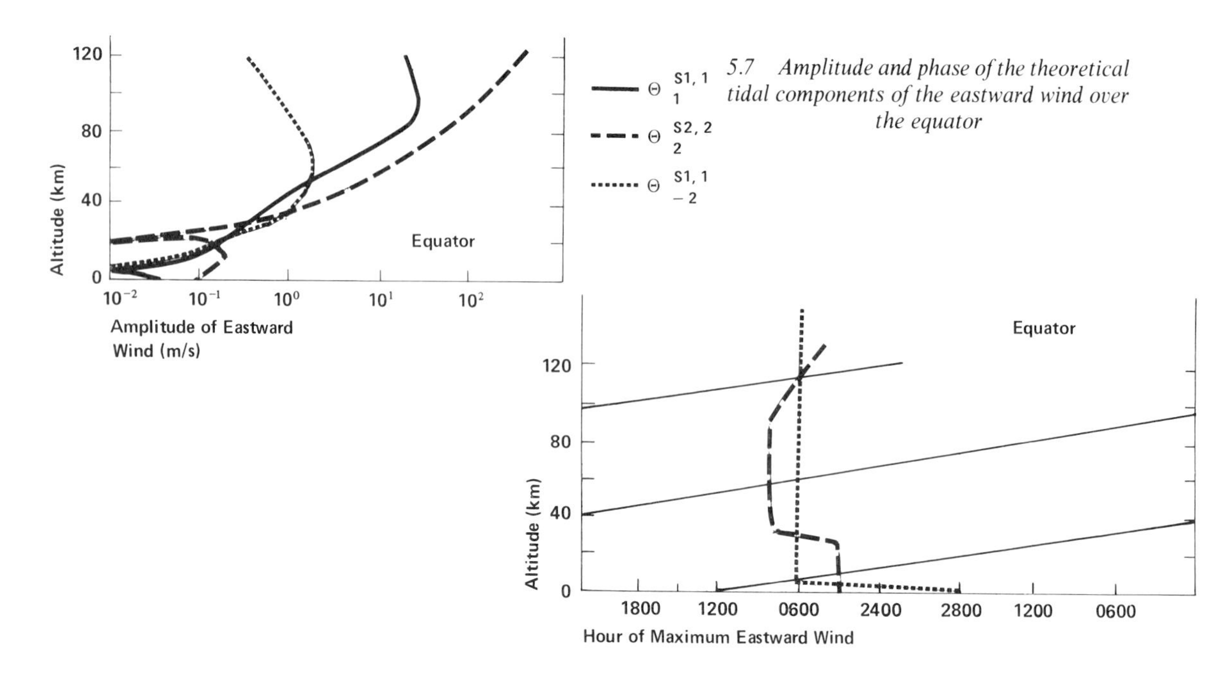

5.7 Amplitude and phase of the theoretical tidal components of the eastward wind over the equator

Table 5.5 Principal tidal modes: (0–90 km)

Mode		*Frequency* ω	m	h (*km*)	k_e (km^{-1})	k_n (km^{-1})	λ_z (*km*)
First symmetric semi-diurnal	$\Theta_2^{S2,2}$	$\pi/6\ hr^{-1}$	2	7·85	$3{\cdot}14 \times 10^{-4}$	$4{\cdot}2 \times 10^{-4}$	293
Second symmetric semi-diurnal	$\Theta_4^{S2,2}$	$\pi/6\ hr^{-1}$	2	2·11	$3{\cdot}14 \times 10^{-4}$	$0{\cdot}96 \times 10^{-3}$	53·4
First symmetric propagating diurnal	$\Theta_1^{S1,1}$	$\pi/12\ hr^{-1}$	1	0·699	$1{\cdot}57 \times 10^{-4}$	$8{\cdot}64 \times 10^{-4}$	28·3
First symmetric trapped diurnal	$\Theta_{-2}^{S1,1}$	$\pi/12\ hr^{-1}$	1	−12·25	$1{\cdot}57 \times 10^{-4}$	$2{\cdot}62 \times 10^{-4}$	∞
3-hour wave	$\Theta_{47}^{S8,47}$	$\pi/1{\cdot}5\ hr^{-1}$	47	0·64	$7{\cdot}35 \times 10^{-3}$	0	26·8

outcome of these theoretical works is the discovery that the 3-hour tide plays a role in the middle atmosphere. Whether this will be confirmed by experimental observations remains to be seen.

At upper atmospheric heights the tidal wave amplitudes have come to be quite large. At the same time the role of viscosity and thermal conduction has also increased. It appears that the dissipative processes are sufficiently strong to prevent non-linearity and a consequent instability in the semidiurnal tide. $S_2(\rho)$ has ρ_1/ρ_0

stabilize at about 0·6. This is not so for S_1, which will tend to become unstable when $\rho_1/\rho_0 > 1$. This occurs in the region from 80 km up to 108 km. At 108 km the molecular damping forces abruptly diminish the value of ρ_1/ρ_0 below unity so that any instability extant would cease above this height. This instability of the diurnal propagating mode offers a most attractive explanation for the existence of the turbopause. Richmond (1971) obtained similar results.

5.5 Tides in the Upper Atmosphere

Below 60 km the tides are difficult to measure because wind readings can only be taken by rockets or balloons, which do not supply a continuous record of wind observations. However, above 60 km the atmosphere is ionized and it is possible to use radio methods to investigate the motion of the ionization. Once this is known and an adequate theory has been constructed then it is possible to infer the motions of the neutral, unionized atmosphere from the charged particle velocities. When sufficient data has been accumulated then the tidal component and the prevailing wind can be extracted from the available data. The prevailing wind is the global scale wind driven by long-lived, though seasonally varying, pressure inequalities. Theoretically, the prevailing winds may be derived by solving the equations of motion in the same way as the thermal wind equation. In the mesosphere viscous damping is not too important and the thermal wind equation provides a useful approximation. This is no longer so in the thermosphere.

Between 60 km and 80 km there is too little data at present for tidal analysis. At these heights the neutral–charged particle collisions occur so frequently that the charged particles move with the velocity of the neutral, unionized winds. The movement of the charged particles can be measured from the partial reflections of radio waves with frequencies of about 2·5 MHz. These radio waves are partially reflected from irregularities in the ionization and a system of three receivers can measure the drift of these irregularities. This technique has only recently come to be used and hopefully it will not be too long before sufficient data has amassed to allow the extraction of tidal components. Though this technique suffers from the disadvantage that useful results can generally only be obtained during the day, it is the only ground based method available for this height range at present.

Between 80 and 105 km there is a rapid growing body of data from the observation of meteor trails by radar, made by using the Doppler shift of the received frequency to map out the path of the ionized trail, which reflects radio signals that strike it perpendicularly. Much of this data is for vertically averaged winds over the whole range of 80–105 km (Fig. 5.8) but Spizzichino (1969) has published vertical wind profiles from Garchy (47° N), France for the amplitude of the

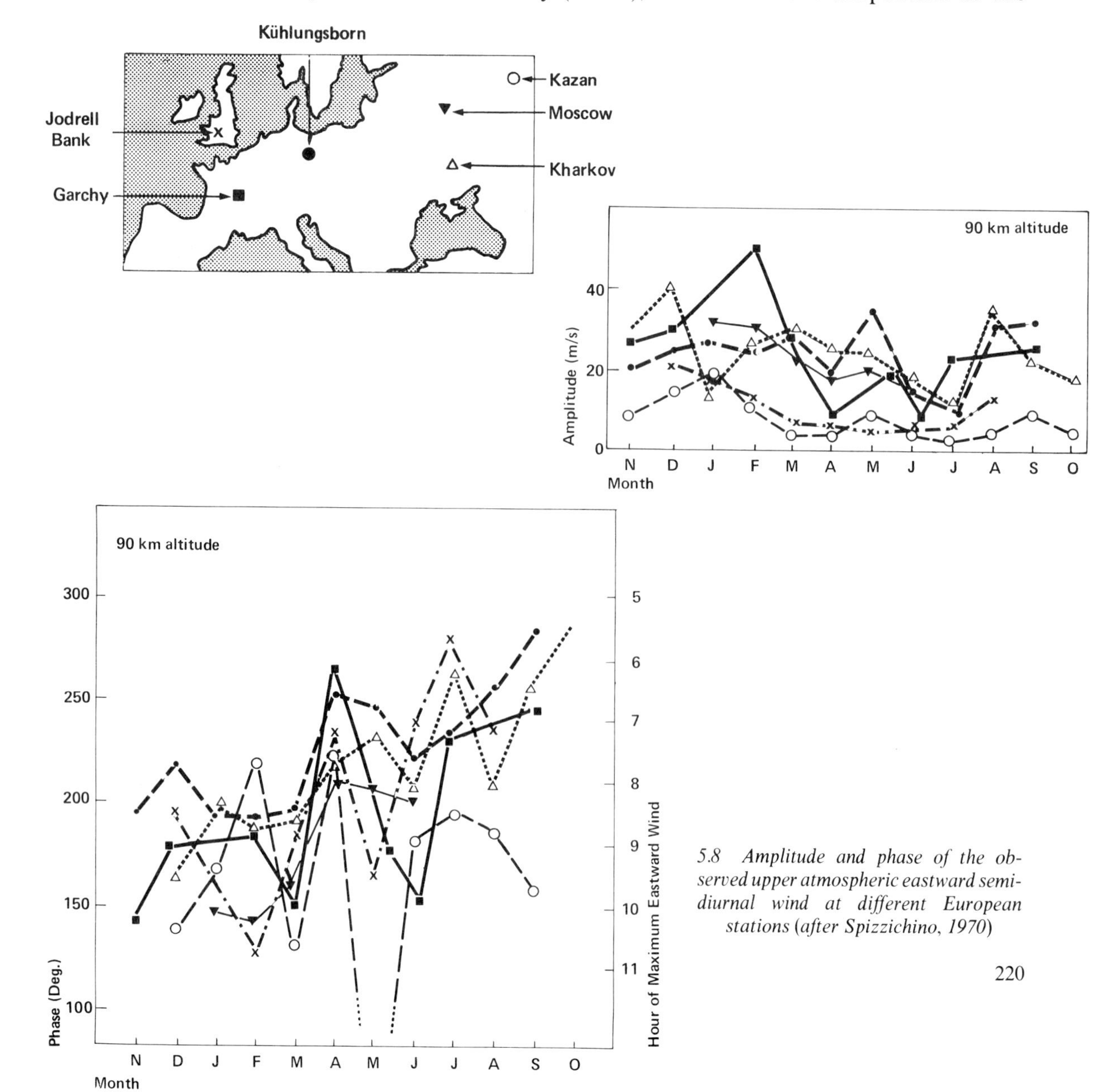

5.8 Amplitude and phase of the observed upper atmospheric eastward semi-diurnal wind at different European stations (after Spizzichino, 1970)

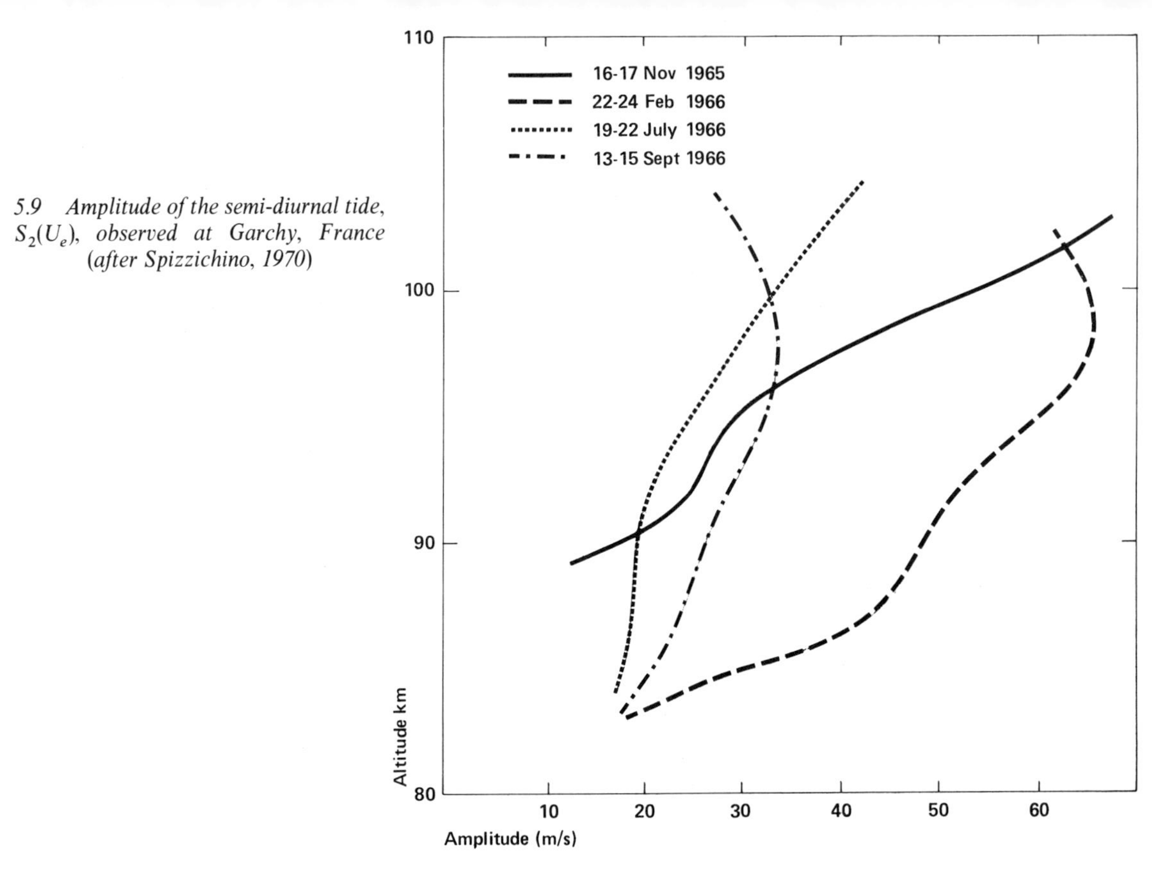

5.9 Amplitude of the semi-diurnal tide, $S_2(U_e)$, observed at Garchy, France (after Spizzichino, 1970)

observed semidiurnal wind. Two of his plots are reproduced in Figs. 5.9 and 5.10. For these experiments the tidal wind dominates over both the prevailing wind and the irregular wind component. At low latitudes the diurnal oscillations predominant whereas at high

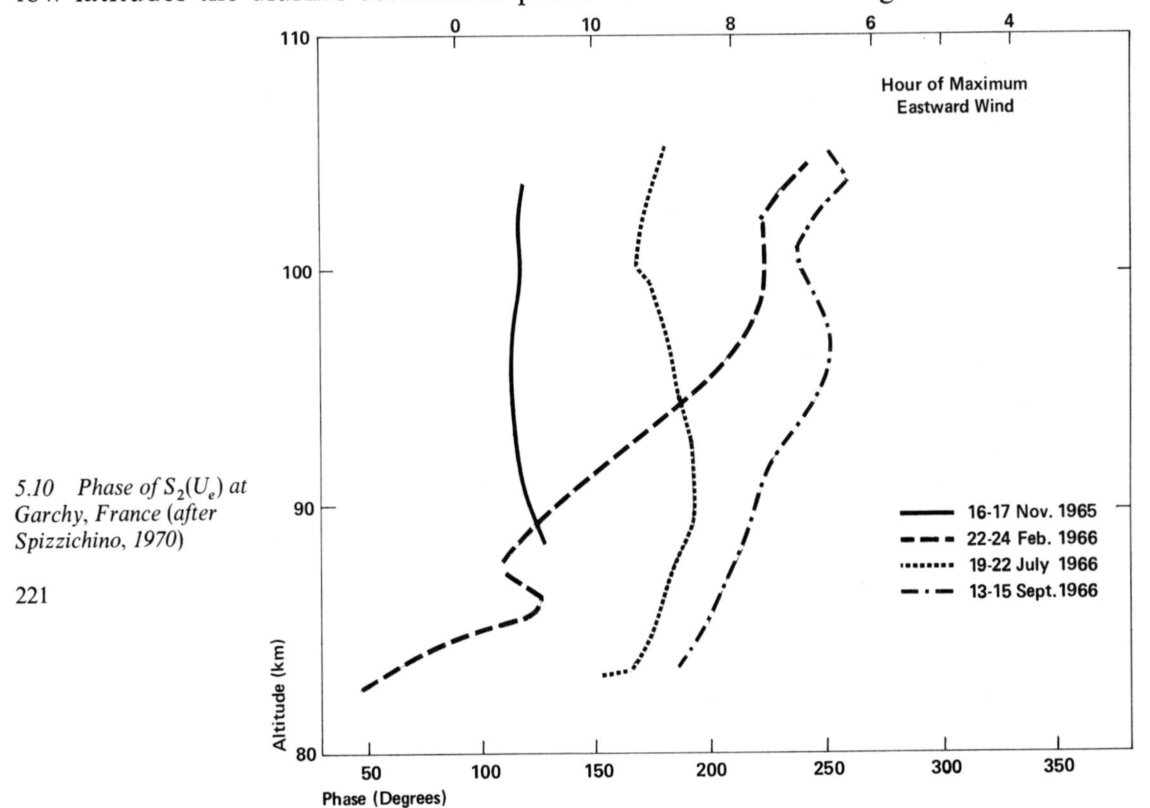

5.10 Phase of $S_2(U_e)$ at Garchy, France (after Spizzichino, 1970)

latitude stations the semidiurnal tide is strongest (Fig. 5.11). The phase of the diurnal tide at Garchy is depicted in Fig. 5.12.

The experimental results in this height range do not present any violent disagreements from the theoretically expected results, though the fine details are often unable to be matched due to the many unknown seasonal variations that will affect the tidal phases and amplitudes. On the other hand the Garchy results failed to detect any

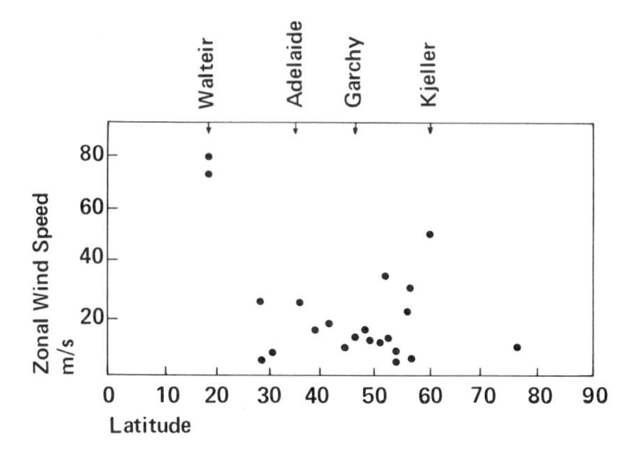

5.11 Amplitude of $S_1(U_e)$ as observed at various stations (after Spizzichino, 1970)

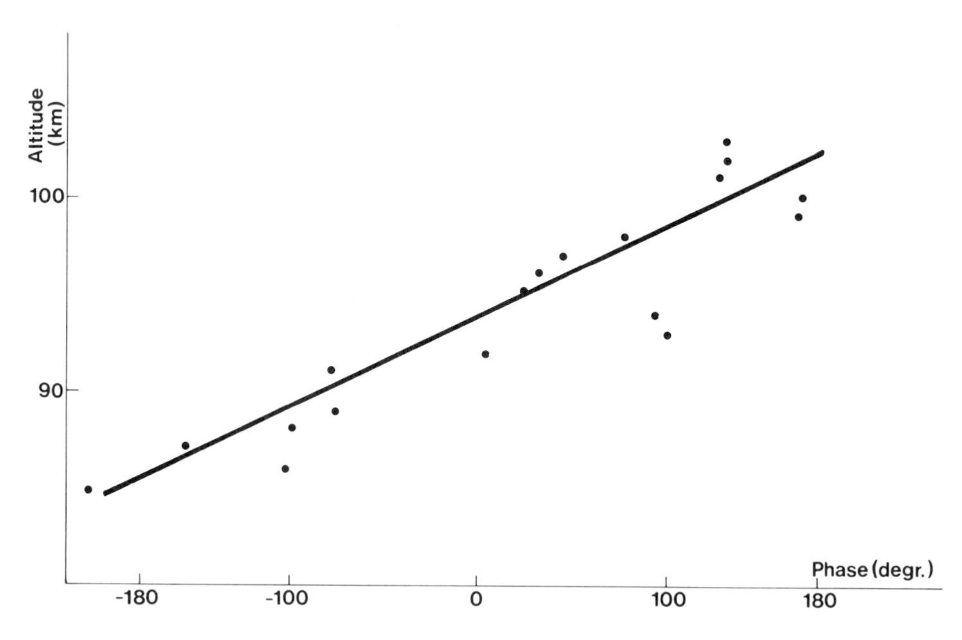

5.12 Phase of $S_1(U_e)$ observed at Garchy, 14–16 December, 1965 (after Spizzichino, 1970)

marked influence from the 3-hour tide. The dominant semidiurnal and diurnal tides (in that order) were followed in strength by an equally strong prevailing wind (which was treated as a zero frequency tide in the spectral analysis) and a 6-hour tide. Then came equally strong 4- and 3-hour tides. It still remains possible though that the 3-hour tide dominates at the equator.

Above 90 km wind data can be obtained by visually tracking luminous vapour trails emitted from rockets. In most cases this is possible only in twilight at sunrise and sundown but certain chemicals, such as trimethyl aluminium (TMA), which react with atomic oxygen in the atmosphere to produce a luminous trail have occasionally been used to make night time observations.

The rocket trail experiments at mid-latitudes show a rotation of the diurnal wind vector with height characteristics of a wave with vertical wavelength of about 20 km, presumably corresponding to the $\Theta_1^{S\,1,1}$ mode. The amplitude appears to grow with height up to 105 km and then decay. However data from above 200 km, obtained mainly by satellite drag observations show an immense daily variation of the thermospheric density. Day–night variations of almost an order of magnitude have been claimed to occur at 600 km. There are problems with this observational technique and real variations could be much larger.

Within the ionosphere, but principally at an altitude of 100 km, the motion of the air across the geomagnetic field induces electromotive forces, which drive currents. These currents in turn affect the magnitude of the earth's magnetic field which thus exhibits strong tidal oscillations. These oscillations when recorded by ground based magnetometers can then be used to infer the tidal winds from which they originated. It is primarily from these geomagnetic oscillations that our knowledge of the lunar tide in the ionosphere is derived.

5.6 Dynamo Effects

Theory

When the oscillations of the upper atmosphere extend to ionospheric heights, they involve the motion of charged as well as of neutral particles. In the presence of the geomagnetic field the mean velocities of the neutral particles, the positively charged ions and the negatively charged electrons are generally not the same, and ionospheric currents result. In a qualitative manner, these currents may be said to be induced by the motion of conducting air across the geomagnetic field, just as the current in the winding of a dynamo is induced by the motion of a conductor in a magnetic field. This analogy cannot however, be pursued too far, because in the presence of a magnetic field the conductivity of a partially ionized gas is not isotropic.

The motion of a single particle in the ionosphere may be derived from the equations of motion:
for ions

$$\begin{aligned}\frac{m^+ D\mathbf{V}^+}{Dt} &= m^+ g - \frac{1}{N^+}\nabla(N^+ k T^+) + e(\mathbf{E} + \mathbf{V}^+ \times \mathbf{B}) \\ &\quad - 2\mathbf{\Omega}_E \times \mathbf{V}^+ - m^+ v_n^+(\mathbf{V}^+ - \mathbf{U}) - m^+ v_e^+(\mathbf{V}^+ - \mathbf{V}^-)\end{aligned} \tag{5.6.1}$$

and for electrons

$$\begin{aligned}\frac{m^- D\mathbf{V}^-}{Dt} &= m^- g - \frac{1}{N^-}\nabla(N^- k T^-) - e(\mathbf{E} + \mathbf{V}^- \times \mathbf{B}) \\ &\quad - 2\mathbf{\Omega}_E \times \mathbf{V}^- - m^- v_n^-(\mathbf{V}^- - \mathbf{U}) - m^- v_i^-(\mathbf{V}^- - \mathbf{V}^+)\end{aligned} \tag{5.6.2}$$

where $\mathbf{V}^{\pm}$ = ion (plus sign) or electron (minus sign) velocities

$m^{\pm}$ = ion or electron mass

k = Boltzmann's constant

$\mathbf{E}$ = electric field

$\mathbf{B}$ = magnetic field

$T^{\pm}$ = ion or electron temperature.

Both the electrons and ions in the atmosphere originate from the same atoms and molecules. Furthermore, the ions and electrons cannot drift very far away from each other because any separation will set up an electric field which will tend to reunite the particles. Because of this, charge neutrality is very nearly preserved so that the ion concentration N^+ is the same as the electron concentration N^-.

The last two terms in (5.6.1) and (5.6.2) represent the collision between the various species of particles. The rigorous treatment of plasma collisions is very complicated (Chapman and Cowling, 1952) so we define a coefficient which is a measure of the transfer of momentum and is commonly called the collision frequency. It is also referred to as the effective collision frequency or as the frictional collision frequency. By using this ν, the force per unit volume $\mathbf{F}_a$, experienced by a particle of a type a due to collisions with a particle of type b is written

$$\mathbf{F}_{ab} = -\mathbf{F}_{ba} = m_a \nu_b^a (\mathbf{V}_b - \mathbf{V}_a) \qquad (5.6.3)$$

so that $m^+ \nu_e^+ = m^- \nu_i^-$. The parameter ν does not represent the real frequency of collisions between the two particles and some authors use a different collision parameter based on the actual kinetic theory frequency of collisions.

The collision frequencies may be evaluated from

$$\nu_n^+ = 10^{-16}[4.2\mathrm{N}(O_2) + 7{\cdot}5\mathrm{N}(N_2) + 20\mathrm{N}(O)]\ \mathrm{s}^{-1} \qquad (5.6.4)$$

$$\nu_n^- = 10^{-16}[0{\cdot}17\mathrm{N}(N_2)T + 3{\cdot}8\mathrm{N}(O_2)T^{\frac{1}{2}} + 1{\cdot}4\mathrm{N}(O)T^{\frac{1}{2}}]\ \mathrm{s}^{-1} \qquad (5.6.5)$$

$$\nu_i^- = \left[59 + 4{\cdot}18 \log_{10}\left(\frac{T^3}{N}\right)\right] N T^{-\frac{3}{2}} 10^{-6}\ \mathrm{s}^{-1} \qquad (5.6.6)$$

where T is the neutral particle's temperature, N is the charged particle concentration in m^{-3} and N(A) is the concentration, in m^{-3}, of the neutral constituent A.

Within the dynamo region, which is the highly conducting E region of the ionosphere, ν_i^- is very small. It will be ignored.

In order to study the ionospheric conductivity one can ignore gravity, the pressure gradient forces and the Coriolis forces as well. Furthermore as we wish to know only the average drift velocities over periods long compared with $1/\nu$ and $1/\omega$, we can set $D/Dt = 0$. Hence, in a Cartesian coordinate system (me, $\|$, $\perp z$) in which $\mathbf{B}$ is directed parallel to the $\|$ axis, and me is perpendicular to the magnetic field in an eastward, horizontal direction; the equation of motion may be rewritten as

$$\begin{pmatrix} V_{\mathrm{me}} \\ V_{\|} \\ V_{\perp z} \end{pmatrix}^{\pm} = \pm \begin{pmatrix} \mu_1^{\pm} & 0 & \mp\mu_2^{\pm} \\ 0 & \mu_0^{\pm} & 0 \\ \pm\mu_2^{\pm} & 0 & \mu_1^{\pm} \end{pmatrix} \begin{pmatrix} E_{\mathrm{me}} \pm \dfrac{U_{\mathrm{me}}}{\mu_0} \\ E_{\|} \pm \dfrac{U_{\|}}{\mu_0} \\ E_{\perp z} \pm \dfrac{U_{\perp z}}{\mu_0} \end{pmatrix} \qquad (5.6.7)$$

where μ represents a mobility. The Pedersen mobility, which produces drift motions parallel to the electric field is

$$\mu_1^{\pm} = \frac{(\nu^{\pm})^2 \mu_0}{(\omega_H^{\pm})^2 + (\nu^{\pm})^2}. \tag{5.6.8}$$

The Hall mobility characterizing **E × B** drifts is

$$\mu_2^{\pm} = \frac{\omega_H^{\pm} \nu^{\pm}}{(\omega_H^{\pm})^2 + (\nu^{\pm})^2} \mu_0 \tag{5.6.9}$$

and the longitudinal mobility is

$$\mu_0^{\pm} = \frac{|e|}{m^{\pm} \nu^{\pm}}. \tag{5.6.10}$$

The subscript n on $\nu_n^{\pm}$ has been dropped and $\omega_H^{\pm}$ is the gyrofrequency

$$\omega_H^{\pm} = \frac{|e\mathbf{B}|}{m^{\pm}}. \tag{5.6.11}$$

Both **V** and **E** are measured relative to the ground. The mobilities are tabulated in Table 5.6.

Table 5.6 Ionospheric transport coefficients (mks units)

HT (*km*)	ν_n^+	ν_n^-	μ_1^+	μ_1^-	μ_2^+	μ_2^-
80	1·80*E*5	2·75*E*6	1·84*E*1	5·70*E*3	1·69*E*−2	1·82*E*4
85	7·51*E*4	1·14*E*6	4·41*E*1	2·56*E*3	9·71*E*−2	1·97*E*4
90	3·11*E*4	4·73*E*5	1·07*E*2	1·07*E*3	5·68*E*−1	1·99*E*4
95	1·22*E*4	1·91*E*5	2·74*E*2	4·36*E*2	3·75	2·00*E*4
100	5·28*E*3	8·45*E*4	6·40*E*2	1·92*E*2	2·05*E*1	2·00*E*4
105	2·32*E*3	3·87*E*4	1·47*E*3	8·81*E*1	1·08*E*2	2·00*E*4
110	1·07*E*3	1·87*E*4	3·15*E*3	4·25*E*1	5·09*E*2	2·00*E*4
115	5·12*E*2	9·73*E*3	6·12*E*3	2·22*E*1	2·09*E*3	2·00*E*4
120	2·77*E*2	5·67*E*3	9·07*E*3	1·29*E*1	5·80*E*3	2·00*E*4
125	1·58*E*2	3·48*E*3	9·92*E*3	7·92*E*0	1·13*E*4	2·00*E*4
130	1·00*E*2	2·34*E*3	8·46*E*3	5·32*E*0	1·53*E*4	2·00*E*4
135	6·79*E*1	1·66*E*3	6·51*E*3	3·77	1·76*E*4	2·00*E*4
140	4·89*E*1	1·24*E*3	4·92*E*3	2·82	1·87*E*4	2·00*E*4
145	3·52*E*1	9·18*E*2	3·62*E*3	2·09	1·93*E*4	2·00*E*4
150	2·71*E*1	7·28*E*2	2·80*E*3	1·66	1·96*E*4	2·00*E*4

(*E*1 denotes $\times 10^1$)

The mobilities are related to the Longitudinal and Pedersen conductivities, σ_0, σ_1 respectively by

$$\sigma_0 = N|e|(\mu_0^+ + \mu_0^-)$$

$$\sigma_1 = N|e|(\mu_1^+ + \mu_1^-)$$

and to the Hall conductivity by

$$\sigma_2 = N|e|(\mu_2^- - \mu_2^+).$$

This then gives the current density flowing parallel to the magnetic field, $\mathbf{J}_{\parallel}$ as

$$\mathbf{J}_{\parallel} = \sigma_0 \mathbf{E}_{\parallel} \tag{5.6.12}$$

and the current normal to $\mathbf{B}$ as

$$\mathbf{J}_{\perp} = \sigma_1 \mathbf{E}_{\perp} + \frac{\sigma_2(\mathbf{B} \times \mathbf{E})}{|\mathbf{B}|}. \tag{5.6.13}$$

Fig. 5.13 tabulates the dimensionless quantities $\sigma_0 B/Ne$, $\sigma_1 B/Ne$ and $\sigma_2 B/Ne$, which are proportional to the conductivities per unit electron concentration, as functions of the height, for representative values of the collision frequencies and gyrofrequencies at a moderate latitude. Above the E region σ_0 is extremely large. This means that currents flow freely along the magnetic field lines which thus act as equipotentials. The perpendicular electric fields that are set up by the dynamo action can thus be transmitted along these highly conducting field lines into the F region. Thus tidal effects may be detected in the ionospheric F region through electrodynamic linking of it with the E region, as well as directly through the thermospheric tidal winds. This makes F region tidal analyses which are based on charged particle observations extremely difficult to interpret.

Fig. 5.13 clearly shows that the Hall conductivity, σ_2, which

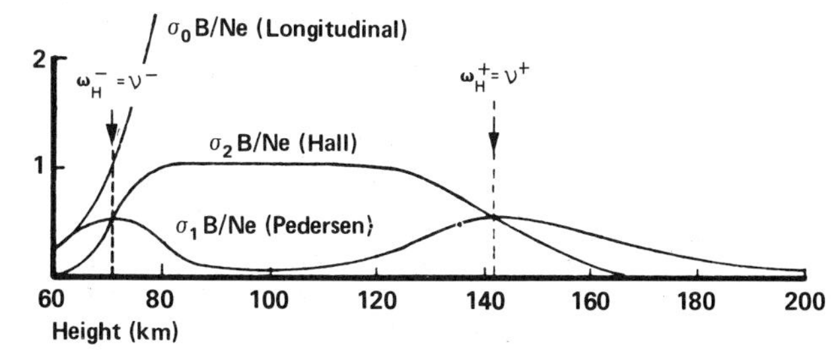

5.13 The height distribution of the three conductivities σ_0, σ_1, σ_2 at a moderate latitude, in terms of the dimensionless quantity $\sigma B/Ne$ (after Fejer, 1964)

controls the dynamo effect is largest from 70 km to 140 km. It is for this reason that we continually talk of the E region as the dynamo region.

The total electric field consists of two distinct parts. One part is due to the dynamo action of the tidal wind which sets up an induced electric field $\mathbf{U} \times \mathbf{B}$. To this must be added the polarization field that results from the separation of the positively and negatively charged particles. As we have seen, there can be no component of the polarization field parallel to the magnetic field.

In the F region and above, the current is inhibited from flowing across field lines by the low σ_1 and σ_2 values, and thus any steady-state current carried along a tube of force must be independent of position along the tube. If symmetry about the magnetic equator (the line where $\mathbf{B}$ has no vertical component) is assumed both in the wind system and in the conductivity, no current flows along the tubes of force in the F region and above. The currents are confined to a relatively thin shell and the vertical current approximately vanishes at all heights. The equation that expresses the vanishing of the vertical component of the current density can be used to eliminate the vertical component of the total electric field from the vector equations (5.6.12) and (5.6.13) so that in terms of an me, mn, z coordinate system (magnetic eastward, magnetic northward, vertically upward)* the equations

$$\mathbf{J}_{\mathrm{me}} = \sigma_{\mathrm{me\,me}}\mathbf{E}_{\mathrm{me}} + \sigma_{\mathrm{me\,mn}}\mathbf{E}_{\mathrm{mn}} \tag{5.6.14}$$

$$\mathbf{J}_{\mathrm{mn}} = -\sigma_{\mathrm{me\,mn}}\mathbf{E}_{\mathrm{me}} + \sigma_{\mathrm{mn\,mn}}\mathbf{E}_{\mathrm{mn}} \tag{5.6.15}$$

are obtained. The layer conductivities, $\sigma_{\mathrm{me\,me}}, \sigma_{\mathrm{me\,mn}}, \sigma_{\mathrm{mn\,mn}}$ are functions of the parallel and perpendicular conductivities and of the dip angle I. They are given by the equations

$$\begin{aligned} \sigma_{\mathrm{me\,me}} &= \sigma_1 + \frac{\sigma_2^2 \cos^2 I}{(\sigma_0 \sin^2 I + \sigma_1 \cos^2 I)} \\ \sigma_{\mathrm{me\,mn}} &= \frac{\sigma_0 \sigma_2 \sin I}{(\sigma_0 \sin^2 I + \sigma_1 \cos^2 I)} \\ \sigma_{\mathrm{mn\,mn}} &= \frac{\sigma_0 \sigma_1}{(\sigma_0 \sin^2 I + \sigma_1 \cos^2 I)}. \end{aligned} \tag{5.6.16}$$

* This is not the standard coordinate system used in geomagnetism but it is employed here in order to remain consistent with the rest of the book. The usual geomagnetic set of coordinates is (magnetic southward, magnetic eastward, upward).

At the magnetic equator $\sigma_{\text{mn mn}} = \sigma_0$ and $\sigma_{\text{me me}} = (\sigma_1^2 + \sigma_2^2)/\sigma_1 = \sigma_3$ where σ_3 is known as the Cowling conductivity. The Cowling conductivity is comparable in magnitude to the longitudinal conductivity σ_0. There is therefore a highly conducting zonal strip along the magnetic equator which carries a large current known as the equatorial electrojet, which is confined to the region a few degrees in width, where $\sigma_0 \sin^2 I \ll \sigma_1 \cos^2 I$.

The simple model of the dynamo region is therefore a relatively thin, horizontal stratified layer. The effects of the currents as observed at the ground can be calculated most conveniently with the use of 'integrated layer conductivities,'

$$\Sigma_1 = \int \sigma_1 \, dz; \qquad \Sigma_2 = \int \sigma_2 \, dz$$

and similar forms for $\Sigma_{\text{me me}}$, $\Sigma_{\text{me mn}}$ and $\Sigma_{\text{mn mn}}$ where the integration is made vertically through the conducting region. Since the E region controls the conductivities, the integrated layer conductivities are approximately proportional to the peak electron concentration. At night they become small because the E layer almost disappears. Typical daytime values of Σ_1, Σ_2 and Σ_3 are 10, 20, and 200 mhos respectively. $\Sigma_{\text{me me}}$, $\Sigma_{\text{me mn}}$ and $\Sigma_{\text{mn mn}}$ are depicted in Fig. 5.14.

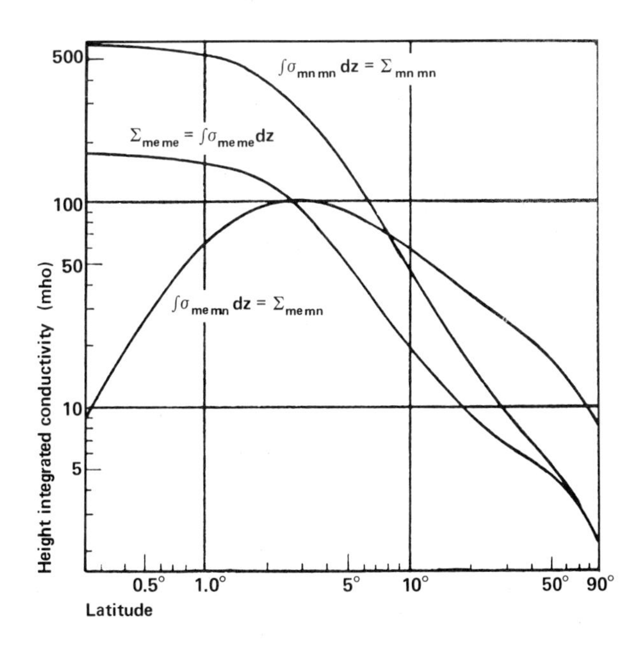

5.14 The height integrated midday layer conductivities as functions of latitude (after Fejer, 1964)

S_q and L Current Systems

Magnetic records at the ground are obtained by using magnetometers. These measure either: (a) the northward, eastward and vertical component of **B**; or (b) westward declination, horizontal intensity and vertical intensity. Both these sets of elements undergo a daily oscillation about their mean values. These disturbances are divided into two parts that are dependent on the solar activity cycle, and one part that is influenced by the gravitational attraction of the moon.

The long term averages enable one to obtain the magnetometer changes due to the quiet sun, namely the S_q variations, and those due to the moon, the L variations. Solar flares and other disturbed solar conditions produce magnetic storms that perturb the magnetometers greatly. These disturbances, though solar in origin, could not really be classified as tidal.

It is then convenient to represent the source of the regular magnetic variations by a current sheet flowing on the surface of a sphere above the earth, at an altitude of 100 km where the maximum conductivity is to be found. The height integrated current density can be expressed directly in terms of the magnetic field components, using Ampere's circuital relations

$$g'\mu_0'\mathbf{J}_{\mathrm{me}} = \mathbf{B}_{\mathrm{mn}}; \qquad -g'\mu_0'\mathbf{J}_{\mathrm{mn}} = \mathbf{B}_{\mathrm{me}} \tag{5.6.17}$$

where μ_0' is the permeability of free space. The numerical factor g' depends on the contributions of currents induced in the solid earth and in the oceans by the overhead current system. The internal S_q and L current systems induced within the earth have been reviewed by Price (1967) and Matsushita (1967) and more recent work has concentrated on the electric and magnetic field induced by the ocean tides (Larsen, 1968; Chapman and Kendall, 1970), which appears to strongly influence the L variation. If the ground were a perfect conductor, no polarization fields could be induced in it and all the magnetic variations would be due to the ionosphere. In this case g' would be unity. Chapman and Bartels (1940, pp. 692 and 694) show that the effect of the ionosphere is, on the average, 2·5 times as strong as the induced ground currents so that $g' = 0{\cdot}75$, though the value of g' varies at different localities and at the same location at different times.

By using equation (5.6.17) and data on magnetic variations collected from observatories around the world it is possible to deduce the S_q and L current systems which, if existing in a thin conducting shell at a height of 100 km, would produce the observed magnetic variations. The result of this process applied to the averaged magnetic observations during the International Geophysical Year is displayed

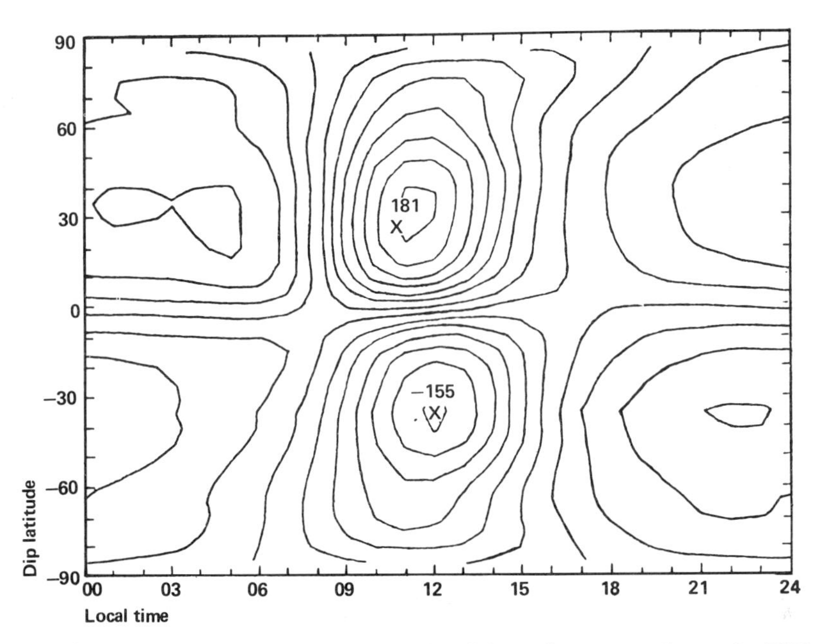

5.15 The average S_q current system deduced from observations during the IGY. The current intensity between two consecutive lines is 2.5×10^4 amp. The numbers in the figure give the total current in each vortex in thousands of amperes (after Matsushita (1968), Geophys. J. Roy. Ast. Soc., **15**, *109, figure 2d)*

in Fig. 5.15. There is however an uncertainty as to the location of the zero value of the current. In Fig. 5.15 $J = 0$ at 6 a.m. at the equator. A more usual assumption is to take $J = 0$ at midnight, though this is liable to be incorrect. Even though the conductivities do drop very sharply at night because of the drop in electron density, appreciable ionospheric currents have been observed in the night-time equatorial electrojet.

Tarpley (1970) has showed that the wind system associated with the $\Theta^{S1,1}_{-2}$ mode, (Fig. 5.16) would produce a current system that

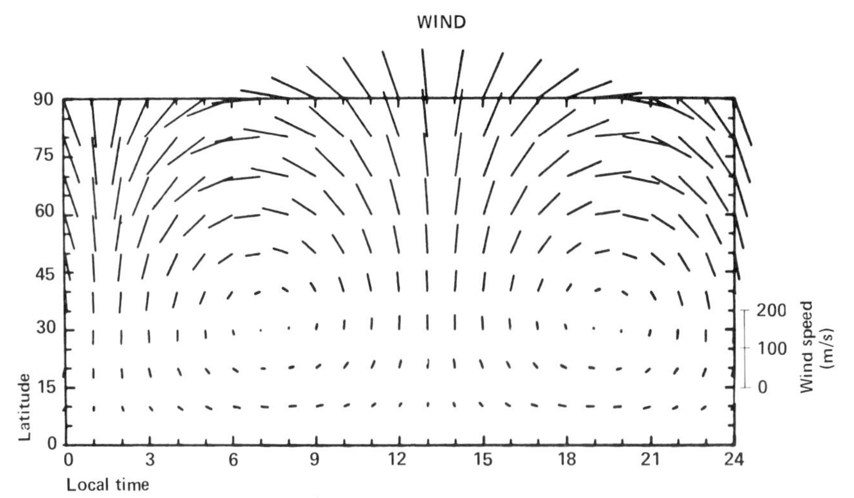

5.16 Wind vectors from the $\Theta_{-2}^{S1,1}$ mode, which give a current distribution very near to the S_q current system that is observed (after Matsushita (1971), Radio Science, **6**, *279, figure 1c)*

closely approximates the observed S_q system. A maximum wind amplitude of 130 m/s produces a total hemispherical current of $1{\cdot}22 \times 10^5$ amperes, with a focus at 36° latitude at 11·4 hours local time. Poleward of 70° latitude however, the magnetospheric currents are liable to play a significant role in the dynamo region fields and so the tidal wind system of Fig. 5.16 is unlikely to explain the S_q variations at very high latitudes, where it is likely that the current density is no longer non-divergent.

Tarpley also found that the $\Theta_2^{L2,2}$ mode winds with a maximum amplitude of 7·5 m/s would well describe the estimated lunar magnetic variations, but until the exact influence of the ocean on both the lunar semidiurnal tide and on the ground conductivity is known, this result should only be treated as a useful estimate.

As we have started to see, the dynamo calculations are subject to a large number of assumptions that cast doubt on the final results that are obtained. Slowly these barriers are being broken down as work progresses on the role of the dynamo region's finite width (Jackson, 1971), oceans, the equatorial electrojet and the magnetosphere, though a detailed analysis of the role of the height variations in **U** on the observed geomagnetic variations has not yet been done. Furthermore the interaction between the tides and the **B** field affects the tides as well. Calculating these changes in the tides is an extremely complex procedure but recently Jones (1970, 1971, 1972) has analysed the effects of Coriolis, Pedersen, Hall and viscous forces. Of exceptional interest is his conclusion that in a region where the Hall forces dominate the Coriolis forces. Hough functions cannot be used to represent the tides because the resulting equivalent depths are in error. Furthermore the Hall effect introduces negative modes, or oscillations of the second class, with their associated negative eigenfunctions. The importance of these results stems from the fact that at the equator Coriolis forces in the traditional approximation will be weak, or non-existent. Therefore any symmetric tidal mode will induce evanescent atmospheric waves of tidal periods in the ionosphere. This is likely to explain Tarpley's observation that the $\Theta^{S1,1}_{-2}$ mode dominates the S_q system. If we recall that the $\Theta^{S1,1}_{-2}$ mode is generated by local heating sources then its direct generation would be surprising. The solar insolation of atomic oxygen is not very significant in the E region though it increases in importance in the F region. It would therefore appear that in fact the dominant S_q mode is of an 'effective $\Theta^{S1,1}_{-2}$ type' generated by the interaction of the magnetic field with the atmospheric tide.

5.7 Tides in Real Atmospheres

We have already started to deal in Chapter 3 with some of the effects that will influence the simplified theoretical basis for many of our derivations. Now I wish to re-examine these for the case of atmospheric tides. Luckily, atmospheric tides are simply rotationally modified gravity waves so that much of our previous work can be directly applied. To the discerning reader, who has felt offended in this chapter at the lack of rigour, I can only apologize and point out, *a posteriori*, the amazing utility of the simplified approach that we have so far used. In general this indicates that the approximations inherent in our derivations so far are unlikely to be too seriously in error.

Neglect of Winds and Horizontal Temperature Gradients

If we include realistic background wind profiles and temperature gradients then the equations become extremely complicated, and for this reason no comprehensive study of this has yet been made. As a rule both must be studied together since the mean winds and temperatures are reasonably well equated by the thermal wind equations. Lindzen (1971b) cites references which show that the inclusion of winds and temperature gradients does not effect the resonance properties of the atmosphere. The main affect of the winds is to doppler shift the wave's frequency. The neglect of winds and horizontal temperature variations can then be justified if

(i) Latitude variations in temperature must be sufficiently small so that the variations of vertical wavenumber,

$$\frac{H}{h}\left[\frac{(\gamma-1)}{\gamma}+\frac{dH}{dz}\right]-\frac{1}{4}\Bigg\}\Bigg/H^2$$

are small compared to the vertical wavenumber itself for the mode in question. This condition is fairly well satisfied, but indications are that it is not perfectly observed. These indications arise from the observations of the seasonal variations in tides which cannot be accounted for by seasonal variations in excitation alone.

(ii) Mean zonal winds must be small compared to the zonal phase speed of a tide, V_p, which is of the order of 450 m/s. This is certainly going to be true over virtually all the earth's surface.

The mean background winds will slightly change the equivalent depth. From equation (4.3.25) we see that the equivalent depth may be written as

$$h=\frac{m\Omega_E^2 R_E^2}{g}=\frac{mV_p^2}{g}$$

where $V_p=\Omega_E R_E$ is the speed of the equatorial atmosphere relative to the earth and represents the zonal phase speed of a migrating tide. m is a number whose value depends on the horizontal variation of the oscillations and on the position on the earth. The introduction of a zonal wind U_0 will alter V_p by the same amount since the phase velocity of the wave is then convected by a speed U_0. The change δh is given by

$$\frac{\delta h}{h}=\frac{2U_0}{V_p}.$$

A more exact analysis for the free oscillations, when $h = \gamma H$, reveals (Lindzen, 1968c)

$$h = \frac{\gamma H}{1+\dfrac{U_0}{V_p - U_0}\left(2-\gamma-\dfrac{(\gamma-1)U_0}{V_p-U_0}\right)}. \tag{5.7.1}$$

In the equatorial F region above about 250 km altitude there are indications that U_0 may be a sizable fraction of V_p, so that $U_0/V_p \sim 0.5$. At these heights then, mean winds will affect the tides though viscous dissipation and ion-neutral interactions also become very important. The effect of no single one of these processes has been completely solved for tides in the F region and the solution for tides subject to all three must still be awaited.

By treating the denominator of (5.7.1) as a quadratic in $U_0/(V_p - U_0)$ one finds that the minimum positive value of the equivalent depth is $0.815\gamma H$. If we use this value of h in conjunction with the solution of the vertical structure equation (5.2.10) then we find that free oscillations in an isothermal atmosphere are always evanescent regardless of the wind profile. This is a rigorous extension of the statement given in § 5.1.

Dissipation

The role of dissipation is of great importance for it plays the major role in preventing atmospheric waves from 'breaking'. Consequently dissipation helps to justify the continued use of the perturbation approach. The atmospheric wave or tide has an amplitude that grows as $\exp(z/2H)$—or more correctly as $\exp(x/2)$—so that eventual non-linearity may be expected through the breakdown of the perturbation condition $U_1 \ll U_0$. However the dissipative effects suppress this growth at sufficiently great heights so that certain modes may remain stable at all heights.

It proves useful to consider dissipation in terms of a mechanism whose time scale τ_{diss} may be specified. If a given tidal mode has a period τ_{tide} then two distinct situations may exist

(i) If

$$\tau_{\text{diss}} < \frac{\tau_{\text{tide}}}{2\pi} \tag{5.7.2}$$

then the tides are fundamentally altered since dissipation is more important than inertia.

(ii) The presence of any dissipation will tend to reduce the $\exp(z/2H)$ growth of propagating tidal modes. If, moreover

$$\frac{\tau_{\text{tide}}}{2\pi} < \tau_{\text{diss}} < \frac{\tau_{\text{tide}}}{(\lambda_z/2H)} \tag{5.7.3}$$

where λ_z is the vertical wavelength of the tidal mode, then λ_z is relatively uneffected, but the $\exp(z/2H)$ growth above the region of excitation will be replaced by decay of amplitude.

The first condition (5.7.2) arises from molecular conductivity and viscosity whose effectiveness increases as $1/\rho$ and from hydromagnetic effects such as ion drag. For the most important tidal modes molecular effects assume importance above 100 km while hydromagnetic drag becomes important primarily above 200 km. Below 100 km infrared cooling is of moderate importance—leading to small reductions of amplitude but not to decay with height. Eddy viscosity and conductivity, which play dominant roles for acoustic gravity wave attenuation in the lowest 2 km of the atmosphere do not seem to affect the long wavelength dominant tidal modes.

Furthermore the presence of dissipative mechanisms with a ρ^{-1} dependence generates partial reflections of the atmospheric waves. The ratio of the amplitudes of the reflected and incident wave is of the order of $\exp(-2\pi^2 H/\lambda_z)$ which is unlikely to create significant effects for the dominant internal modes.

Lorentz Forces

At upper atmospheric altitudes the Lorentz force per unit mass is $(\mathbf{J}\times\mathbf{B})/\rho$ where

$$\mathbf{J} = \sigma_0\mathbf{E}_{\parallel} + \sigma_1(\mathbf{E}_{\perp} + \mathbf{V}\times\mathbf{B}) + \sigma_2\mathbf{B}\times\frac{(\mathbf{E}_{\perp} + \mathbf{V}\times\mathbf{B})}{B}.$$

The Lorentz force can be divided into two components. The ion drag force due to σ_1 opposes wind motion across magnetic field lines and dissipates tidal wind energy. It has already been mentioned. The Hall force component due to σ_2 acts to deflect the wind and partially counteracts the Coriolis force. It does not dissipate wind energy.

We have already mentioned some of the surprising results that follow from an inclusion of dominant Hall force terms. Richmond (1971) found similar results but surprisingly found that the Hall term acts differently for tides and internal gravity waves. For tides the Hall force increases the vertical wavelength, whereas for gravity waves it decreases it (Taffe, 1969). A further result is that for tides, the inclusion of Hall forces allows non-linear coupling between different tidal modes. This does not occur on a non-rotating earth so that internal gravity waves retain their linearizable perturbation forms.

Topographic Effects

Variations in surface topography do not greatly affect the dominant migrating tides. Instead they generate additional oscillations of tidal periods which do not travel with the sun (e.g. modes of the form $\Theta_n^{S2,0}$, $\Theta_n^{S2,\pm 1}$ etc.). Such modes are important for the tidal winds below 30 km, but are not of sufficient strength to affect the pressure variations.

There are also likely to be effects induced by the greater concentrations of water vapour over the sea, and of CO_2 over the land. This will affect the latitudinal distribution of insolation and consequently the solar diurnal tide. Further discussion of this may be found in Chapman and Lindzen (1970, p. 157).

5 Background Reading

Dikii, L. A. (1969) *The theory of oscillations of the earth's atmosphere*, Gidrometizdat Press, Moscow (in Russian).

Chapman, S. & Lindzen, R. S. (1970) *Atmospheric tides*, D. Reidel Publ. Co. Dordrecht, Holland.

Kato, S. (1971) Wave dynamics in the atmosphere: tidal motion, *Space Science Review*, **4**, 421.

Lindzen, R. S. (1971) Atmospheric tides, in *Mathematical problems in the Geophysical sciences*, American Mathematical Society, Providence, Rhode Island.

Siebert, M. (1961) Atmospheric tides, in *Advances in Geophysics*, **7**, Academic Press, New York.

6

Waves in the Ionosphere

6.1 Introduction

Atmospheric waves that have propagated to the upper atmosphere may have rather large amplitudes because of their exp $(z/2H)$ growth. These waves can then be manifested as appreciable variations in the vertical and horizontal neutral wind velocities and as variations in the neutral gas density, pressure and temperature. The wave types that are important in the upper atmosphere and ionosphere are the atmospheric tides that were discussed in the previous chapter and internal waves of both gravity and acoustic modes.

Originally the non-tidal irregular wind variations were ascribed to turbulence, but when it was discovered that they exist above the turbopause then the idea that they were a manifestation of internal atmospheric waves quickly gained hold. This concept has proved incredibly fruitful and has provided a quantitative explanation for many of the observations of various types of ionospheric irregularities. In fact it is common nowadays to search for the existence or generation of internal waves by seeking wavelike F region ionization irregularities. This approach should leave the purist feeling rather queasy, because the interaction between the ionized and neutral atmosphere is extremely complex and only partially understood. Yet, in justification, it does seem to yield highly useful results.

At this stage it may be appropriate to discuss the use of the words 'prevailing wind' and 'background wind'. There appears to be some confusion between the two. The prevailing wind is a seasonally changing wind that may be described by the thermal wind equations throughout a large part of the atmosphere. It is possible that it may have slight variations due to planetary wave passages and this point was discussed in § 4.3. The prevailing wind can be obtained by finding the average wind over a one or two day period, and it is found that the prevailing wind remains reasonably constant for

January

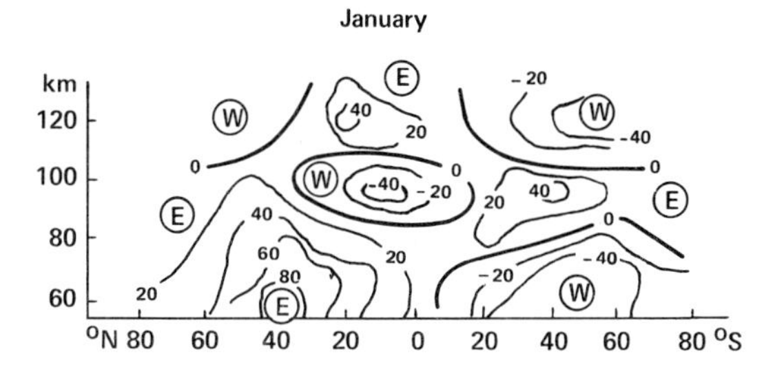

April

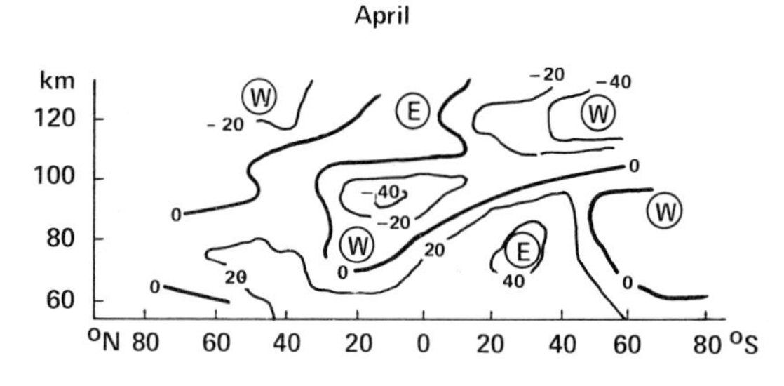

Zonal prevailing wind
(in m/s)

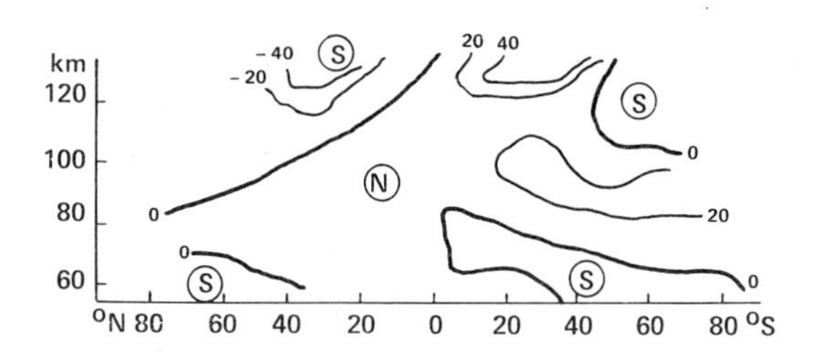

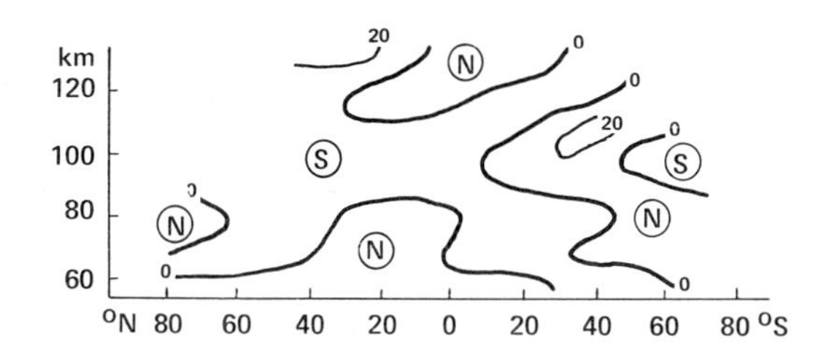

Meridional prevailing wind
(in m/s)

6.1 Prevailing wind between 60 and 130 km as a function of latitude. E denotes a region of eastward winds and N a region of northward winds

periods of about a month. Typical plots of the prevailing wind as a function of latitude were given by Murgatroyd (1957) for altitudes between 20 km and 100 km and by Groves (1969b) for heights of 60 to 130 km. These plots reveal that below the F region the prevailing wind is predominately zonal. Groves' results for January and April are reproduced in Fig. 6.1.

The background wind has so far been represented by the symbol U_0. As we have assumed that $U = U_0 + U_1$ and U_1 has been taken as a gravity wave with a fixed, defined frequency and wavenumber, the background wind is going to be different for each individual gravity component of the irregular wind. In other words U_0 is composed of the prevailing winds, the tidal winds, plus all the components of the irregular wind other than the particular gravity wave under consideration. Obviously U_0 is an extremely complicated function both of position and time and extreme simplifications are called for in order to make problems involving U_0 tractable. One method of handling these problems is to assume that the prevailing wind dominates in U_0 and to further assume that the height variation in U_0 is not very great. In this case U_0 is always constant and it is possible to ignore it in the equations of motion and to substitute the doppler shifted frequency $\Omega = \omega - \mathbf{U}_0(z) . \mathbf{k}(z)$ in the gravity wave under consideration. This approach also tends to assume that one component of the gravity wave spectrum dominates the irregular wind component.

In the E region when we wish to analyse the charged particle motions then a rather more elegant method is available. Even though we still need to assume that one gravity wave mode dominates, it is possible to separate out the prevailing and tidal winds from U_0 by pushing the tidal winds into a Lorentz force term. In this case all the tidal wind terms become equivalent to electric fields. At low and mid-latitudes these electric fields are predominantly zonal (Matsushita, 1969).

We have noted a number of times already, that in the upper atmosphere the wave amplitudes may be large but at the same time the dissipative effects are also large. It seems likely then that the success of the approach wherein only one dominant gravity wave is considered stems from the dissipative criterion. Presumably at any one height one particular gravity wave is dominant, but at higher or lower heights waves with some other frequency or wavenumber dominate.

Temperature Variations

Atmospheric wave and tides have sinusoidal variations in pressure and density, so that by using the ideal gas law one would expect temperature variations T_1 of the form

$$\frac{T_1}{T_0} = \frac{p_1}{p_0} - \frac{\rho_1}{\rho_0} = \frac{\mathrm{P}-\mathrm{R}}{\mathrm{X}} U_{1x} \tag{6.1.1}$$

in all the regions of the ionosphere. Temperature variations of this form have been found in the E region by Wand and Perkins (1968, 1970) and in the D region by Rai and Fejer (1971). Hines (1965b) has discussed the reversible adiabatic heating of the ionosphere and points out that fluctuations of $\pm 10°$K can be expected low in the E region, and of as much as 30°K at 110 km.

Hines (1965b) also considered the irreversible heating of the atmosphere caused by viscous dissipation of the waves at great heights. Since the wave spectrum is thought to be roughly white (in the sense that many waves of many wavelengths and periods are present simultaneously) and since the height at which viscous dissipation acts to remove a given wave component depends on the wave parameters, one would expect the heating stemming from this source to be rather widely distributed in height. Hines (1965b) pointed out that this heat input exceeds the radiative input at night, and it is possible that the gravity wave input exceeds radiative input when both are averaged over 24 hours. He claims that the very rapid rise in temperature that is found in the E region may be a consequence of this dissipated heat input. Whether these mechanisms can supply heat to the F region depends largely on the direction and method of energy propagation, and the role of the damping processes; consequently there is more uncertainty over gravity wave heating above 150 km than below this height.

Lindzen and Blake (1970) examined the mean heating of the thermosphere by tides. Since most tidal modes are dissipated below 130 km, only the effects of the main semidiurnal tidal mode need to be considered. The dissipation of this particular mode by molecular viscosity and conductivity appears to be able to maintain thermospheric temperatures of 600°K to 700°K. Therefore this heat source combined with the diurnally varying extreme solar ultraviolet flux (which also depends on the solar cycle) seems to explain both the phase and the amplitude of the large diurnal density oscillations that have been inferred from satellite drag analyses (Moffett, 1973; Mayr and Volland, 1973).

6.2 Waves in the D Region

The D region extends from the bottom of the ionosphere at about 50 to 60 km up to 90 km. It is almost an isothermal region within which the mesopause may be found. It is characterized by relatively high gas densities and low electron densities, whilst the gyrofrequency and collision frequency of the electrons tend to be of the same order of magnitude.

The D region is highly turbulent. Hodges (1967) pointed out that wind shears of sufficient magnitude to produce turbulence have a very slight probability of occurrence. The wind shears that are normally observed do not lower the Richardson number sufficiently. Hodges considered the density fluctuations produced by internal gravity waves and found that these would sufficiently lower the local Vaisala–Brunt frequency to create turbulence. This mechanism is attractive up to about 90 km, but above this height the atmosphere is more stable due both to its temperature rise and to the great dissipative effects at greater heights. Presumably between 90 km and the turbopause at about 110 km the observed turbulence is due to the instability of the diurnal tidal mode.

If Hodges interpretation is correct then one may assume that the fluctuations in electron density are also produced by gravity waves. These fluctuations are responsible for the partial reflections of radio signals at frequencies near 2·5 MHz. At D region heights it seems likely that a large spectrum of gravity waves still exist, so that overall one might observe an $\exp(z/2H)$ growth in the electron density fluctuations. If we represent the total electron density N, as a sum of a background and fluctuating term: $N = N_0 + N_1$, then there is no *a priori* certainty as to whether it would be N_1/N_0 or N_1 alone which would vary as $\exp(z/2H)$. In theory one should be able to solve the continuity equation for the charged particles in order to find their fluctuation with height. This is extremely difficult but seems to indicate that it is N_1/N_0 that varies as $\exp(z/2H)$. The meagre experimental data available is directly contrary to this. The data that does exist suggest that N_1 varies as $\exp(z/2H)$ up to heights of 200 km. Beer (1972a) assumed that

$$N_1 \propto \exp(pz)$$

and determined p from partial reflection data. His results are shown in Fig. 6.2 and they show a marked tendency for p to vary as $1/2H$, providing confirmation for the idea of gravity wave generation of D region electron density fluctuations.

Tchen (1970) points out that an analysis of the power spectrum of the turbulence could not differentiate between turbulence due to Hodge's variation in the buoyancy force and turbulence due to wind shears. Both would vary as k^{-3}, where k is the wavenumber of the respective fourier component of the turbulence. Experimental results at 50 km indicate that for $k < 0{\cdot}4\ \text{km}^{-1}$ the power spectrum follows a k^{-3} law, whereas for $k > 0{\cdot}4\ \text{km}^{-1}$ it follows the $k^{-\frac{5}{3}}$ law that is characteristic of homogeneous isotropic turbulence. This

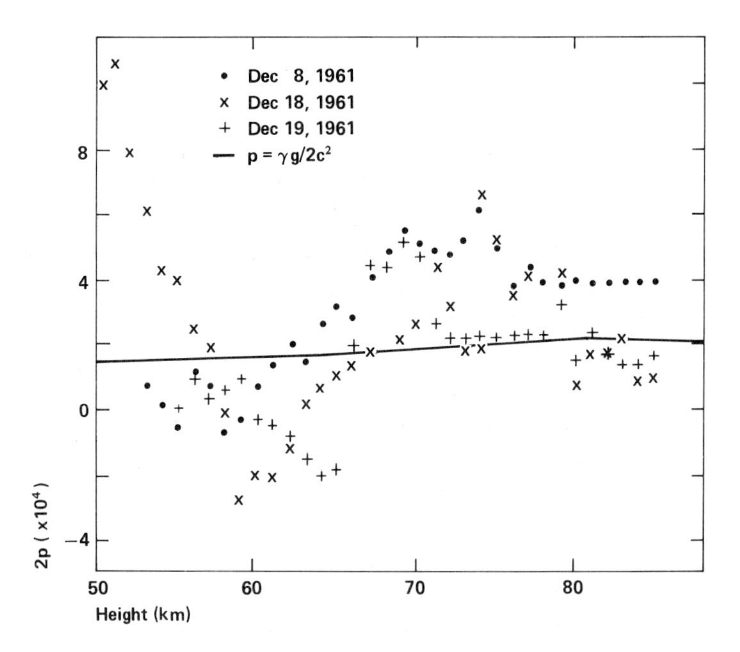

6.2 2p, the parameter representing fluctuations in electron density, as a function of height

seems to put a lower limit of 10 km onto the horizontal scale size of D region gravity waves.

Hodges (1969) has also estimated the eddy diffusion coefficients due to the unstable internal gravity waves at the mesopause. Eddy diffusion coefficients of the order of 10^3 m^2/s are readily produced by gravity waves of horizontal wavelengths of 10 km–1000 km. Since the average coefficient of 10^2 m^2/s has been deduced for the D region, on the basis of heat transport, a random but frequent occurrence of turbulence-producing gravity waves seems plausible.

6.3 Waves in the E Region

The normal E region (90–150 km) is a well behaved ionospheric layer whose major properties can be described by Chapman's theory of the formation of ionized layers. Because of the high, anisotropic conductivity there, this region plays a dominant role in the electrodynamics of the ionosphere, a role that has already been reviewed in the last chapter. Like most of the upper atmosphere the E region is permeated by a collection of tidal and gravity waves. Fig. 6.3 illustrates the power spectrum in the frequency domain of the zonal wind in the low E region. The contributions of the

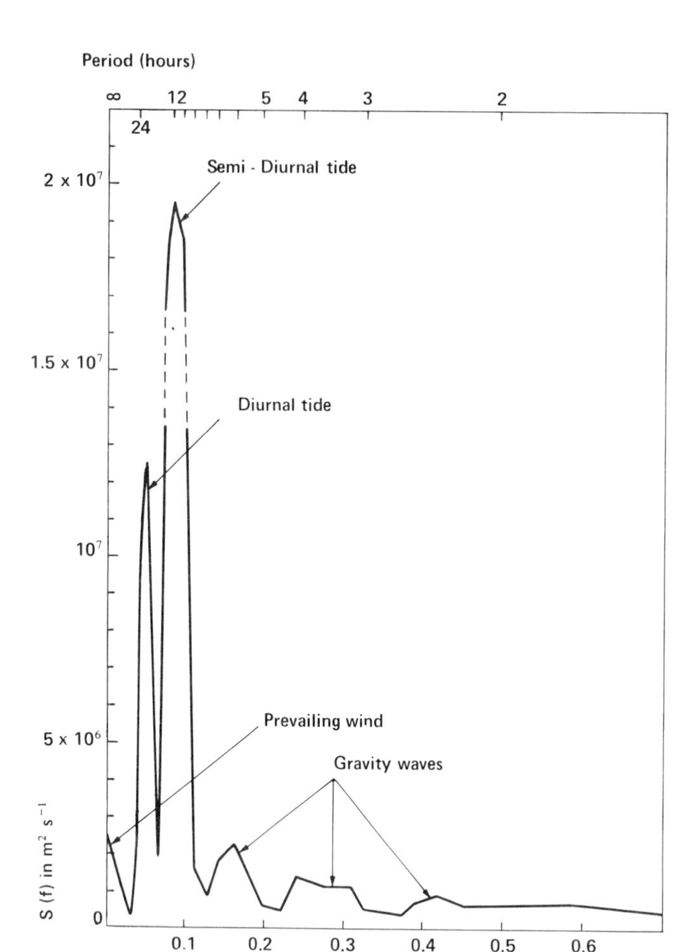

6.3 Power spectrum S(f) of the zonal wind at Garchy, 29 March–1 April 1966 (after Spizzichino, 1970)

prevailing wind (frequency equal to zero), the diurnal and semi-diurnal winds and of the gravity waves are obvious. The average power spectrum for the gravity waves averaged over a large number of occasions does not exhibit the peaks that are found on single recording runs. The average power spectrum of these oscillations decreases with increasing frequency ω proportionally to $\omega^{-\delta}$, where δ is 0·82 at 90 km and 0·47 at 100 km. The coefficient δ regularly decreases with increasing height (Spizzichino, 1969).

Observations of E region gravity waves (Revah, 1970) reveal that they are composed of both an upward propagating and a partially reflected component. The reflection coefficient varies between 0·1 and 0·7, the most frequent values being 0·3–0·6. This seems to provide experimental confirmation of Yanowitch's theory of partial reflections of gravity waves due to the effects of viscosity. For a reflection coefficient of 0·37, the vertical wavelength of the wave before partial reflection is approximately given by $20H$ where H is the scale height. In the lower E region H varies from 5 km to 8 km, indicating that the dominant gravity waves are of the order of one or two hundred kilometres vertical wavelength. This agrees with other estimates (Fig. 3.7).

When a gravity wave propagates through the ionosphere it sweeps the ionization into a wavelike distribution through collisions between the charged and neutral particles. The presence of the magnetic field complicates this process and it is no longer obvious that in the E region the ionization density will have the same form as the neutral particle density. Kato *et al.* (1970) demonstrate that in fact when a gravity wave propagates through the E region, it produces a wavelike variation in the electron density that has the same frequency and wavenumbers as the original gravity wave provided there are no non-uniformities or boundaries in the ambient plasma. This wave may be regarded as the forced response of the ionosphere to the gravity wave's excitation. At the same time the gravity wave excites a free response in the ionization. Normally this free response is a transient that dies out but under certain conditions it can become unstable (Beer and Moorcroft, 1972a). Both the forced and the unstable free response are believed to instigate disturbances in the E region known as sporadic E.

Sporadic E

In addition to the regular E region echo on ionosondes, it is sometimes possible to observe an extra return from the E region. This return emanates from an ionization irregularity whose occurrence is a random event so that this extra return is labelled as sporadic E, which is generally abbreviated to *Es*. The *Es* observed at high latitudes, mid-latitudes and equatorial latitudes are all believed to be of different origin. High latitude *Es* is liable to be related to charged particle precipitation whereas low latitude *Es* is due to a plasma instability. On the other hand mid-latitude *Es* appears to result from shears in the E region neutral wind. It is very likely that these shears are a result of the gravity waves.

When a wind blows perpendicular to the magnetic field line then it moves the charged particles initially in the same direction. These are then subject to $\mathbf{V} \times \mathbf{B}$ drifts which make it possible for the ionization to converge at one height provided that there are wind shears, as there always will be when there are gravity waves present (Fig. 6.4). This theory is known as the wind shear theory of sporadic E. It is unable to explain the occurrence of equatorial *Es* because the electrons are always constrained to move along the magnetic field lines. At mid-latitudes the electrons can follow the ions vertically along the oblique magnetic field lines. At the equator this is impossible. Chimonas and Axford (1968) point out that the shear will tend

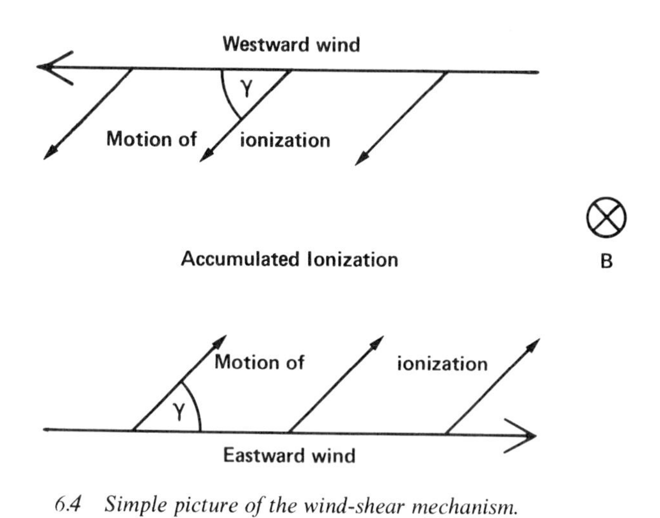

6.4 *Simple picture of the wind-shear mechanism.*
$\tan \gamma \approx \mu_2^+/\mu_1^+$

to converge the ionization to the gravity wave node, which, in turn, moves downwards. This dumping of the ionization from higher to lower heights is known as the corkscrew effect. Simultaneous observations of *Es* and wind shears confirm the importance of the wind shear mechanism, though many of its finer points are still under active discussion.

The wind shear mechanism produces a thin strong layer of ionization. At times one can also observe weaker more diffuse sporadic E that is sometimes called spread E. Many workers have explained this diffuse *Es* as being due to a plasma instability and Beer and Moorcroft (1972b) have pointed out that the unstable free response of the ionization to a gravity wave's passage could generate this type of sporadic E.

Further Effects

In order to make quantitative estimates of the gravity waves, effect on the E region ionization we need to know the resultant amplitude of the ionization fluctuations. In the previous section mention was made that in the D region observations are accumulating which show that N_1 varies as $\exp(z/2H)$. A similar conclusion holds for the E region as well. Fig. 6.5 depicts an experimentally obtained plot of N_1/N_0 as a function of height. If we assume that N_1 varies as $\exp(z/2H)$ then the results of Fig. 6.5 produce a plot of $N_0(z)$ that agrees well with the known afternoon E region profile. This provides further confirmation for the idea of N_1 varying as $\exp(z/2H)$.

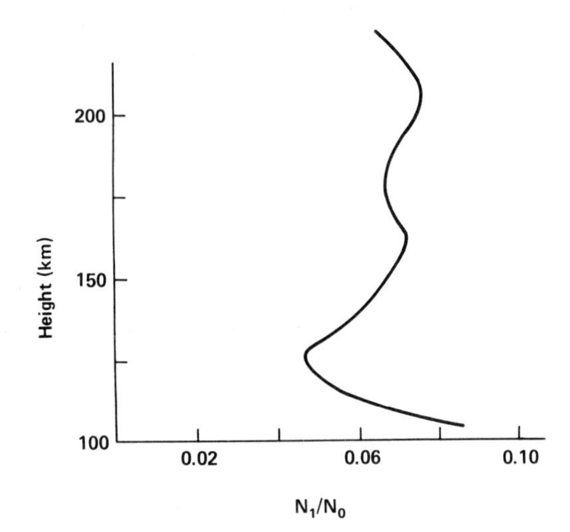

6.5 Average of the peak amplitudes observed at 1605–1650 LT at Arecibo

Gravity waves, through their heating effects, also affect the ionospheric loss processes. In the E region the dominant loss processes are $NO^+ + e \rightarrow N + O$ and $O_2^+ + e \rightarrow O + O$. These will produce changes in the electron density, dN/dt, that are proportional to the density of ions and the density of electrons. Because of charge neutrality these two densities are the same so that the loss processes are proportional to N^2. The proportionality constant, known as the recombination coefficient is usually denoted by α. For the nitrogen reaction

$$\alpha_{NO^+} = 4{\cdot}6 \times 10^{-13} \left(\frac{T^-}{300} \right)^{-1{\cdot}2} \ \mathrm{m^3/s}$$

and for the oxygen reaction

$$\alpha_{O_2^+} = 2{\cdot}2 \times 10^{-13}\left(\frac{T^-}{300}\right)^{-0{\cdot}5} \text{ m}^3/\text{s}$$

where T^- is the electron temperature in degrees Kelvin. The perturbations in T^- due to the gravity wave then produce perturbations in α given by

$$\frac{\alpha_1}{\alpha_0} = \frac{T_1^-\delta}{T_0^-} = \frac{T_1\delta}{T_0}$$

since the electron, ion and neutral particle temperatures are all equal in the E region. For the dominant nitrogen reaction $\delta = -1{\cdot}2$ whereas for the oxygen reaction $\delta = -0{\cdot}5$.

As a wave passes through the ionosphere it perturbs the density of the neutral particles. Now the theory of the formation of ionized layers is based on the solar radiation ionizing a certain percentage of the available neutral particles. If there are fluctuations in the density of neutral particles then there will be fluctuations in the source term of the continuity equation. The theory for this is rather lengthy and has been dealt with by Hooke (1968, 1969a).

Winds in the E *region*

Before passing on to the F region, I wish to help clear up some confusion that has arisen in the literature over the relative strengths of the tidal winds and the irregular winds. Certain authors have claimed that the tidal winds are dominant in the E region. This is incorrect (Whitehead, 1972). Typical results for 30° latitude are a diurnal amplitude of 40 m/s at 105 km, a semidiurnal amplitude of 25 m/s at the same altitude and an irregular wind of about 80 m/s. Thus the direct evidence indicates that the irregular wind amplitude is greater than either of the tidal components and it seems accepted that the irregular winds are generated by gravity waves. The confusion seems to have arisen from Haurwitz's analysis (1964, p. 2) comparing observed mean winds to deduced tidal amplitudes. The mean meridional wind is smaller than the tidal amplitudes in an overwhelming number of cases, but the mean zonal wind is smaller than the tidal amplitudes in only 50 per cent of the cases. Haurwitz, quite correctly, offers the cautious interpretation that these results indicate the importance of upper atmospheric tides. We should further note that they show that the atmospheric response to gravity waves has a strong zonal preference. This suggests that zonal propagating gravity waves may be dominant.

Haurwitz fails to mention whether his mean wind is a spatial or temporal average. In either case the effects of small wavelength or short period waves will be underestimated. Furthermore if gravity wave generation is not a continuous occurrence, then a temporal average will also tend to underestimate their importance.

In the F region the situation is more complex. Firstly the dissipative effects act primarily on smaller modes so that in all likelihood the tidal modes really will be dominant there. Secondly the effect of the magnetic field makes the response of the F region ionization highly anisotropic. Even if there is an isotropic spectrum of gravity waves, which seems unlikely from the results in the previous paragraph, then the ionization will show a marked anisotropy, with the meridional components dominating (Hooke, 1970). Therefore observations of the motion of F region ionization cannot be used to determine the directional spectrum of the initiating waves.

6.4 Waves in the F Region

There are two common types of F region ionization disturbance that appear on ionograms. The first is Spread *F*, which is the name given to the complete blurring of the F region trace. A tentative explanation of equatorial Spread *F* as a wave induced phenomena will be given in § 6.7. The second disturbance is wavelike in nature. It is observed as distortions on ionogram traces in the form of kinks near the *F*2 cusps and these kinks gradually move down towards lower frequencies. These disturbances also show up as scintillations in radio stars. They are believed to be wavefronts of disturbances, many hundreds of kilometres in extent. The disturbances are sometimes called transitoria but more usually travelling ionospheric disturbances (TID's). They have been observed to travel for horizontal distances up to 3000 km. The apparent downward motion seen on ionograms can be accounted for if the wavefronts are inclined to the vertical.

F region TID's can be categorized into three separate classes:

(1) Very large types which usually follow magnetic storms, (2)

medium scale TID's which are common during the day, and (3) a type due to incoherent superposed gravity waves propagating to ionospheric heights from below.

Very Large TID's

TID observations by Thome (1964, 1966), Chan and Villard (1962) and many others show disturbances with horizontal wavelengths of the order of 1000 km moving southwards in the northern hemisphere with speeds in the 200 to 400 m/s range. The wavefronts are found to be nearly horizontal with the phase propagating vertically downwards.

Experimental verification of the fact that large scale TID's occur during geomagnetically disturbed periods has been available for over ten years (see the Bibliography in Georges, 1967). However, the exact link between these disturbances and the mechanism that supplies their energy is not yet available. Infrasonic waves have been related to polar substorms and this has led Davis (1971) to postulate polar substorms as the source of large-scale travelling ionospheric disturbances. Related ideas have been advanced by Hines (1965b) and Testud (1970), who postulate auroral heating through the intense energy input into the auroral thermosphere during a magnetic substorm. Thome (1966) suggests that the auroral electrojet may provide the energy.

Georges (1968a) points out that ducting processes may not be necessary to explain the large horizontal distances the disturbances travel, as energy propagation is inherently horizontal for long-period waves; enhanced interaction with ionization occurs as these waves travel equatorwards since the magnetic dip is decreasing and the air motion in the wave is nearly horizontal. A fully ducted mode would have a high enough speed but should have a nearly vertical wavefront, which is contrary to most observations. This suggests that the disturbances are free internal gravity waves with properties determined by the thermospheric parameters. In this case the limiting angle (θ) of the wavefront with the vertical is given by (Tolstoy, 1963)

$$\cos\theta = \frac{\omega}{\omega_B}. \tag{6.4.1}$$

For a wave period of 20 minutes, the tilt would be 45° and for 2 hours it would be 83°, roughly in agreement with George's (1968a) observations.

Medium Scale TID's

Friedman (1966), besides considering completely ducted waves, also examined the possibility of internal gravity wave guiding by leaky ducts. This is accomplished by allowing K_x to have an imaginary part leading to attenuation in the x direction. The attenuation is supposed to be a manifestation of upwards leakage of energy out of horizontal ducts and into the F region of the ionosphere. This was shown in Fig. 3.11. Gossard (1962) has also treated the imperfect ducting of internal gravity waves in the troposphere and he calculated the upward energy flux of gravity waves leaking out of the troposphere.

Medium scale TID's fit Friedman's theory rather well. They tend to have horizontal wavelengths in the hundreds of kilometres

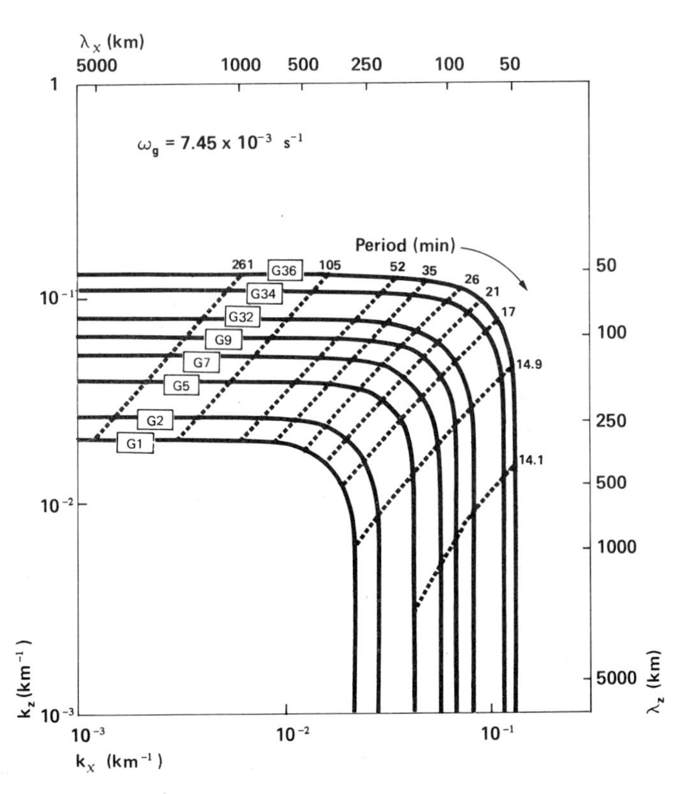

6.6 *Curves of k_z versus k_x for several modes as labelled in boxes. The dashed contours connect modal points with the same period (after Friedman, 1966)*

and speeds in the range 100–200 m/s. Wavefront tilts are in the range 30°–60° and the medium scale TID's show a tendency to come from the general direction of the winter polar region with, in addition, an eastward velocity component. These TID's tend to lose their shape within 200 km.

The apparent paradox that these waves could transport energy over global distances yet not be coherent over horizontal distances greater than about 200 km can be explained by the dispersion properties of gravity waves in this period range. While the horizontal phase and group velocities of long period gravity waves are nearly equal, those of short period waves are not, hence if energy is transported in wave packets, the phase fronts would pass rapidly through the packets, disappearing at the front. A given wave cycle would thus not travel long distances, but the energy might do so, if ducting is effective.

It appears that these TID's are generated in the polar region and travel long distances over the earth. The eastward velocity component that is observed would then be a consequence of the effects of the prevailing winds. The occurrence of mesospheric heating in the winter in the polar regions (Maeda and Young, 1966) lends support to this idea, since gravity wave energy penetrates upwards much more easily in the absence of the mesospheric temperature decline (Friedman, 1966). Wickersham (1968) shows that the thermal perturbation of the principal duct (between 100 and 165 km) by the long winter solstice sunrise in the polar regions is a source of ducted gravity waves consistent with the seasonal variations in direction of TID's observed in Australia.

Random Fluctuations

In order to obtain routine monitoring of ionospheric motions it is possible to use continuous wave doppler soundings. This technique makes use of records of the frequencies of stable high frequency radio transmissions reflected from the E or F layer. Ionization motions show up as doppler shifts in the received frequency. With a spectrum analysis technique the spectral content of the frequency fluctuations is displayed on a record of fluctuation period versus time. These spectrograms, or fluctuation spectra, are interpreted as representing the spectra of ionospheric motions in the vicinity of the radio reflection height.

Georges (1968a) used this technique and claims to have observed incoherent, superposed gravity waves propagating to ionospheric heights from below. This interpretation is supported by the gap in the nightime fluctuation spectrum between 5 and 15 minutes. Gravity waves with periods shorter than the longest Brunt period (about 14 minutes in the thermosphere) and acoustic waves with periods longer than the shortest acoustic cutoff period (about 5 minutes in the troposphere) would not be expected to propagate upward from below since the atmosphere effectively acts as a filter against the upward propagation of these waves.

The correlation between TID's and atmospheric waves is generally accepted nowadays and often observations of TID's are used to make statements about atmospheric wave generation and propagation. The two reports by Georges (1967, 1968b) contain extensive bibliographies and detailed treatments of the interaction between atmospheric waves and the ionization—especially in the F region—though they deal with other heights as well. The May, 1968 issue of the *Journal of Atmospheric and Terrestrial Physics* also contains many papers on these topics.

Ion Drag

In the F region the collision frequency of ions and of electrons is very much smaller than their gyrofrequencies, so that the ionization is usually constrained to move only longitudinally to the magnetic field lines. The neutral particles which constitute the atmospheric wave motion will then collide with the almost stationary ionization and the atmospheric wave will lose energy, predominantly to the heavier ions. This process is known as ion drag and it affects gravity and tidal waves with periods longer than 30 minutes. Ion drag is most significant in the F region but it will play some role at all heights above about 100 km.

Ion drag acts in a manner very similar to viscosity. It dissipates the wave energy, it can produce partial reflections and it can alter the dispersion relation for the waves (Hines and Hooke, 1970). It alters the equation of motion by adding a term $d\mathbf{U}/dt = -\nu^n_+ \mathbf{U}$ to the right-hand side. The coefficient ν^n_+ has already been met as the effective ion–neutral collision frequency. It could also be called the ion–drag coefficient* and it seems to have a value of

$$\nu_+^n = 2{\cdot}6 \times 10^{-15} M^{-\frac{1}{2}} \left(\frac{T}{700}\right)^{0\cdot4} N\,\mathrm{s}^{-1} \qquad (6.4.2)$$

where T is the neutral temperature in degrees Kelvin, M the molecular weight in kg/kmole and N the ion density in particles per cubic metre. In general N is a function of both position and time, varying greatly in the course of a day.

Despite the apparent severity of the ion drag process, recent results by Lindzen (1970) suggest that at low latitudes the dissipative effects of viscosity and thermal conduction dominate well before the effect of ion drag does. In other words, those waves liable to severe changes in their dispersion relation due to ion drag are liable to have been dissipated at lower heights. This is certainly true in the equatorial night-time F region since recent work (Rishbeth, 1971) suggests that at night the ions there are no longer bound to the magnetic field. The combination of Lindzen's and Rishbeth's results argue strongly against the dominance of ion drag at the equator. Even though Lindzen's work is strictly applicable only to the equator the effects of ion drag and viscosity are to a large extent determined by the ratio of dissipative time scale to wave period, so that it seems reasonable to assume that viscosity overrides ion-drag at non-equatorial latitudes for all waves except zonally propagating waves.

The dissipative time scale associated with ion drag is of the order of $1/\nu_+^n$ which is close to 30 minutes in the F region and to four hours in the E region. Thus at high latitudes, zonally propagating waves in the F region whose periods are longer than thirty minutes are strongly influenced by ion drag (Klostermeyer, 1969).

In the F region the dominant ionization loss process is a charge exchange reaction followed by a recombination process. The rate of the reactions

$$O^+ + O_2 \rightarrow O_2^+ + O \quad \text{and} \quad O^+ + N_2 \rightarrow NO^+ + N$$

dominates and the loss term is given by βN where β itself depends on the neutral concentration and is markedly height dependent. It is called the attachment-like coefficient and geophysical and laboratory evidence suggests that β decreases with increasing

* Often written ν_{ni}.

temperature. The attachment-like coefficient at an altitude z km may be written in terms of the scale height H, measured in km, as

$$\beta = 10^{-4}\left(\frac{50}{H}\right)\exp\left[\frac{300-z}{H}\right]\ \mathrm{s}^{-1} \tag{6.4.3}$$

so that during the passage of a gravity wave

$$\frac{\beta_1}{\beta_0} = \frac{\rho_1}{\rho_0} - \frac{T_1}{T_0} \tag{6.4.4}$$

since the exponential term in (6.4.3) arises from the exponential height decay of the density of neutral particles.

Observations of Gravity Waves

Virtually all the observations of F region waves come from travelling ionospheric disturbances. Other observations have used density gauges on satellites (Newton *et al.* 1969) or F region temperature changes (Harris *et al.* 1969) to measure gravity wave parameters. These seem to have detected waves whose properties would coincide with the medium-scale TID's. The observed wave parameters do not agree with the wave parameters that one would expect to observe if the gravity waves had propagated upwards from below the ionosphere, suffering reflection and dissipation en route. The evidence then suggests that gravity waves are generated in the polar ionosphere, though we shall defer further discussion of this to the next chapter. Newton *et al.* also provided some measure of confirmation of the idea of TID's being ducted waves by their observation that the gravity waves had integrally related wavelengths: a fundamental and the second, third and fourth harmonics were observed.

6.5 Drift Motion of Irregularities

In the next section we shall treat some simple examples of the interaction between the ionization and a gravity wave. So far we have de-emphasized the role of the perturbation electric field that is set up when irregularities form. In the presence of a background wind or electric field the ions and electrons will move at different speeds because of their different collision frequencies. This will set up a perturbation electric field that will significantly modify the subsequent motion of the ionization. For example, it is a perturbation field of this sort that confines the dynamo currents to a horizontal layer.

Ionization irregularities once they are formed will also be affected by the background electric fields and winds but they will drift in a different manner to the background ionization. Their drift motion will depend on their alignment with respect to the magnetic field. The following simple rule has been formulated which, except for a few exceptions listed below, gives the motion of all ionization irregularities once they have somehow been created.

*The MCKT Rule**

Field aligned irregularities move with the electron drift velocity. Non-aligned irregularities move with the ion drift velocity.

The demarcation between aligned and non-aligned irregularities occurs when

$$\frac{k_{\|}}{k_{\perp}} = \left(\frac{\omega_H^+ \nu^-}{\omega_H^- \nu^+}\right)^{\frac{1}{2}}$$

where $k_{\|}$ is the wavenumber component parallel to the magnetic field **B** and $k_{\perp}$ is the wave number perpendicular to **B** of the fourier component of the irregularity being considered. In the ionosphere the demarcation occurs when $k_{\|}/k_{\perp} \approx 10^{-2}$.

The dependence on the orientation of the irregularity with the magnetic field line arises because electrons and ions both have different components of mobility. If an ionization irregularity is field aligned so that it has no variation in the direction of the magnetic field then the irregularity will move as if it were completely composed of electrons. This arises because the electrons, which have greater mobility along the field lines, have to travel a much greater distance to neutralize the space charge that is set up than do the ions, which have a greater mobility perpendicular to the field lines. The space

* MCKT is an acronym of the names Martyn, Clemmow, Cole, Kato and Tsedilina and was suggested by Kato *et al.* (1970).

charge electric fields that are set up, which may be regarded as perturbation electric fields, move the ions rather than the electrons and the irregularity is moved by the background electric field as if it were composed of electrons. A similar argument applies to non-aligned irregularities, in which case the electrons are the ones to move in order to neutralize the space charges.

The MCKT rule enables us to understand the intuitively puzzling statement in the discussion on sporadic E when it was pointed out that the wind shear mechanism will not work at the magnetic equator. The electrons can be visualised as beads strung onto the magnetic field lines so that it will be the ions that will move under the influence of the $\mathbf{V} \times \mathbf{B}$ forces. The electrons are then normally able to follow the motions of the ions by sliding up or down the magnetic field lines. At the magnetic equator, where the field lines are completely horizontal, this becomes impossible so that the space charge electric field that is set up prevents the ions from moving as well.

The MCKT rule strictly only applies to small-scale irregularities. Furthermore it is only valid provided that $\nu^+\nu^- \ll \omega_H^+\omega_H^-$. This inequality will not be satisfied in the D region where there are a large number of collisions. Below 100 km the Pedersen mobility of the electrons μ_1^- is greater than that of the ions so that it will be the electrons rather than the ions that will move to neutralize field aligned irregularities. This indicates that below 100 km all irregularities move with the ion drift velocity. Below 90 km the ion drift velocity will equal the velocity of the neutral air, $\mathbf{U}$, since the collisions are so numerous that the effects of the magnetic field are negligible. However between 90 km and 100 km it seems that all irregularities move with a velocity approaching the ion drift velocity which is different from the neutral air motion.

High in the F region of the ionosphere, Pedersen effects are negligible and the Hall coefficients for both ions and electrons become equal. Thus the ion and electron drift velocities are both equal to

$$\frac{\mathbf{E} \times \mathbf{B}}{B^2} + \frac{(\mathbf{U} \cdot \mathbf{B})\mathbf{B}}{B^2}$$

and all irregularities have the same natural drift. In the E region it is crucial to know whether the irregularities are field aligned or not.

6.6 Ionization—Gravity Wave Interaction

The interaction between the ionization and the gravity wave is an extremely complex problem that is still largely unsolved. There is a two way interaction. The gravity wave produces ionization irregularities, but the ionization itself can also modify the gravity wave.

Let us restrict ourselves to the study of the effects of a zonally propagating atmospheric wave on the ionization at the magnetic equator. This simple case provides a suitable illustration of the method of attack. We are going to try to neglect the effects of prevailing winds by transforming into a coordinate system that is moving with the neutral wind, so that the wind velocity $\mathbf{v}$ in the reference frame is

$$\mathbf{v} = \mathbf{V} - \mathbf{U}. \tag{6.6.1}$$

The normal way to start is to use the continuity equation for the ionization which in this case is

$$\frac{\partial N}{\partial t} + \nabla . (N\mathbf{v}) = q - \alpha N^2$$

in the E region, or

$$\frac{\partial N}{\partial t} + \nabla . (N\mathbf{v}) = q - \beta N \tag{6.6.2}$$

in the F region, and to assume a known solution for the equation of motion of the ionization which then gives an expression for $\mathbf{v}$.

For waves of small scale size we must include the effects of diffusion. This arises from the gradients in the plasma pressure $N k T^{\pm}$ set up by the density fluctuations. The ionization will, if left to itself, diffuse into an equilibrium state. In this case equation (5.6.7) which describes the charged particle velocities becomes the tensor equation

$$\mathbf{v}^{\pm} = -\frac{1}{N}\underline{\underline{\mathbf{D}}}^{\pm} . \nabla N \pm \underline{\underline{\boldsymbol{\mu}}}^{\pm} . (\mathbf{E} + \mathbf{U} \times \mathbf{B}) \tag{6.6.3}$$

where the diffusion tensor is much like the mobility tensor

$$\underline{\underline{\mathbf{D}}} = \begin{pmatrix} D_1^{\pm} & 0 & \mp D_2^{\pm} \\ 0 & D_0^{\pm} & 0 \\ \pm D_2^{\pm} & 0 & D_1^{\pm} \end{pmatrix} \tag{6.6.4}$$

where

$$D_0^{\pm} = \frac{kT^{\pm}}{m^{\pm}v_n^{\pm}} \tag{6.6.5}$$

and D_1 and D_2 are defined by substituting D for μ in equations (5.6.8) and (5.6.9). The diffusion coefficient is equivalent to the kinematic viscosity for the plasma, so that we may deduce that the viscosity of a plasma is not isotropic.

In the subsequent pedagogic analyses we will assume that gradients in N are so small that diffusion can be ignored. This can be divided into two parts. Firstly that ∇N_0 is small and secondly that ∇N_1 is small. The background gradients ∇N_0 will be small in the horizontal direction except at sunrise and sunset. There may be large vertical gradients which could also induce plasma instabilities but by focussing attention onto the ionization peaks we will ignore them. The neglect of atmospheric wave induced perturbations ∇N_1 seems reasonable since the wavelengths of the relevant waves are generally likely to be quite large. For waves with smaller wavelengths it appears that diffusion introduces a small time-dependent phase shift. Even in many cases where ∇N_0 is not small the diffusive effects are negligible in certain directions owing to the anisotropy of the tensor (6.6.4). For example, in the F region charged particles are very tightly bound to the magnetic field lines so that zonal diffusion is impossible. Furthermore this means that at the equatorial F region vertical diffusion is also going to be inhibited.

Let us take coordinate axes (me, mn, z) in the magnetic eastward, northward and vertical directions respectively. Then we assume that

$$N = N_0 + N_1 \tag{6.6.6}$$

and attempt to solve the first order perturbed continuity equation

$$\begin{aligned} &\frac{\partial N_1}{\partial t} + \nabla \cdot (N_0 \mathbf{v}_1)^{\pm} + \nabla \cdot (N_1 \mathbf{v}_0)^{\pm} \\ &= q_1 - \alpha_1 N_0^2 - 2\alpha_0 N_1 \quad \text{in the E region} \\ &= q_1 - \beta_1 N_0 - \beta_0 N_1 \quad \text{in the F region} \end{aligned} \tag{6.6.7}$$

Equation (6.6.7) is really four equations: one for each of the electrons and ions in both the E and F regions.

We have assumed field aligned irregularities by taking a zonally propagating gravity wave so that $\partial/\partial\text{mn} = 0$. To further simplify matters let us ignore q_1. If hydrostatic equilibrium operates, then $U_{0z} = 0$, and we will assume that the plasma velocity stays constant with height and $\partial \mathbf{v}_0/\partial z = 0$.

F *region*

The problem is simplest in the F region where the wavelengths of the gravity waves are so large that diffusion can be ignored, and where collisions are so infrequent that Hall effects dominate. Thus we take

$$\begin{aligned} v_{\text{me}} &= -\mu_2(E_z + U_{\text{me}}B) \\ v_z &= \mu_2(E_{\text{me}} - U_z B) \end{aligned} \tag{6.6.8}$$

since at these heights $\mu_2^+ = \mu_2^- = \mu_2 = 1/B$, and we assume that $v_{\text{mn}} = 0$.

The perturbed continuity equation, ignoring horizontal variations in background quantities is now

$$\begin{aligned} &\frac{\partial N_1}{\partial t} + N_0 \frac{\partial v_{1z}}{\partial z} + v_{1z}\frac{\partial N_0}{\partial z} + N_0 \frac{\partial v_{1\text{me}}}{\partial \text{me}} \\ &+ v_{0\text{me}}\frac{\partial N_1}{\partial \text{me}} + v_{0z}\frac{\partial N_1}{\partial z} + \beta_1 N_0 + \beta_0 N_1 = 0. \end{aligned} \tag{6.6.9}$$

Since the gravity wavelength is rather large, and since the ions and electrons are drifting together we will ignore the perturbation electric fields E_1. Then,

$$v_{0\text{me}} = -\frac{E_{0z}}{B} - U_{0\text{me}}; \qquad v_{0z} = \frac{E_{0\text{me}}}{B}$$

$$v_{1\text{me}} = -U_{1\text{me}}; \qquad v_{1z} = -U_{1z} = \frac{-\mathrm{Z}U_{1\text{me}}}{\mathrm{X}}$$

in terms of the polarization terms Z and X. It is unfortunate that even when the charged particle velocities are taken relative to the neutral wind, then **U** (measured relative to the ground) remains in the equations because of the movement of the **B** field with respect to the coordinate axes.

We can see from (6.6.9) that the response of the ionization will depend on the vertical gradients of N_1, N_0 and v_{1z}. $\partial N_0/\partial z$ is the background electron density gradient which may be regarded as a known quantity. The requisite variations in N_1 will depend on what scale size is being examined. For the large scale variations we are dealing with, there appears to be some slight justification for assuming that ionization perturbations follow the gravity wave so that

$$\frac{N_1}{\mathcal{N}} = \frac{U_{1\mathrm{me}}}{\mathrm{X}} = A \exp i[\omega t - k_{\mathrm{me}}\mathrm{me} - K_z z] \tag{6.6.10}$$

where $\mathcal{N}$ is the polarization relation for the ionization.

The perturbations of the attachment-like coefficient in terms of the polarization relations are, from (6.4.4) and (6.1.1)

$$\frac{\beta_1}{\beta_0} = \frac{2\mathrm{R}-\mathrm{P}}{\mathrm{X}} U_{1\mathrm{me}} \tag{6.6.11}$$

so that substituting into (6.6.9)

$$\begin{aligned} \mathcal{N}\left[i\omega + ik_{\mathrm{me}}\left\{\frac{E_{0\mathrm{me}}}{B} + U_{0\mathrm{me}}\right\} - \frac{iK_z E_{0\mathrm{me}}}{B} + \beta_0\right] \\ = \mathrm{Z}\left[\frac{dN_0}{dz} - iK_z N_0\right] - iN_0 k_{\mathrm{me}}\mathrm{X} + \beta_0 N_0 \mathrm{P} - 2\beta_0 N_0 \mathrm{R} \end{aligned} \tag{6.6.12}$$

which may be solved for $\mathcal{N}$ by using the polarization relations.

Approximate Polarization Relations

Several approximate relations can be found for internal waves. Provided that the medium does not vary significantly over a wavelength, or more strictly* if $k_z^2 \gg 1/(4H^2)$, then the real part of the isothermal dispersion relation (2.3.13) becomes

$$\omega^2 k_z^2 = (\omega_g^2 - \omega^2)k_x^2 \tag{6.6.13}$$

The F region is predominantly isothermal so that for low-frequency waves when

$$\omega \ll \frac{g}{c}$$

$$\omega^2 k_z^2 = \omega_g^2 k_x^2 \tag{6.6.14}$$

---

* This is not going to be true for the whole spectrum of gravity waves in the F region.

whence one can obtain

$$\frac{\mathrm{P}}{\mathrm{X}} \approx \frac{\gamma\omega}{c^2 k_x} = \frac{\gamma\omega_g}{c^2 k_z} \tag{6.6.15}$$

$$\frac{\mathrm{R}}{\mathrm{X}} \approx \frac{i\sqrt{\gamma - 1}}{c} \tag{6.6.16}$$

$$\frac{\mathrm{Z}}{\mathrm{X}} \approx \frac{-k_x}{k_z} = \frac{-\omega}{\omega_g}. \tag{6.6.17}$$

We can eventually get

$$\frac{\mathcal{N}}{\mathrm{X}} \doteqdot \frac{\left\{\left(\frac{k_{\mathrm{me}}}{k_z}\right)\left[iK_z N_0 - \frac{dN_0}{dz}\right] - ik_{\mathrm{me}} N_0 + \frac{\beta_0 N_0 \gamma \omega_g}{(c^2 k_z)} - \frac{2i\beta_0 N_0 \sqrt{\gamma - 1}}{c}\right\}}{\left\{i\omega + ik_{\mathrm{me}}\left[\frac{E_{0z}}{B} + U_{0\mathrm{me}}\right] - \frac{iK_z E_{0\mathrm{me}}}{B} + \beta_0\right\}}$$

so that

$$\frac{N_1}{N_0} \doteqdot \frac{U_{1\mathrm{me}}\left\{\frac{\beta_0 \gamma \omega_g}{c^2 k_z} - \frac{\gamma g}{2c^2}\frac{k_{\mathrm{me}}}{k_z} - \frac{1}{N_0}\frac{dN_0}{dz}\frac{k_{\mathrm{me}}}{k_z} - \frac{2i\beta_0\sqrt{\gamma - 1}}{c}\right\}}{\left\{\beta_0 + \frac{\gamma g}{2c^2}\frac{E_{0\mathrm{me}}}{B} + i\left[\frac{\omega_g k_{\mathrm{me}}}{k_z} - \frac{k_z E_{0\mathrm{me}}}{B} + \frac{k_{\mathrm{me}} E_{0z}}{B} + k_{\mathrm{me}} U_{0\mathrm{me}}\right]\right\}}$$

At 300 km altitude we can take as typical values

$$\beta_0 = 10^{-4}\,\mathrm{s}^{-1}$$

$$\frac{c^2}{\gamma g} = 50\,\mathrm{km} = 5 \times 10^4\,\mathrm{m}$$

$$\omega_g = 7{\cdot}55 \times 10^{-3}\,\mathrm{s}^{-1}$$

$$B = 0{\cdot}3 \times 10^{-4}\,\text{tesla}.$$

Furthermore we have pointed out that the E region electric fields propagate directly up the magnetic field lines so that $E_{0\mathrm{me}}$ and E_{0z} will each have values of the order of 1 millivolt/m. Let us assume that $E_{0\mathrm{me}} = E_{0z} = 10^{-3}$ volt/m; then for typical wavelengths of about 10 km vertically and 200 km horizontally $|k_{\mathrm{me}}| = 3 \times 10^{-5}\,\mathrm{m}^{-1}$ and $|k_z| = 6 \times 10^{-4}\,\mathrm{m}^{-1}$. The sign of the wave numbers depends on the direction of phase propagation. If we assume an upward flow of energy then $k_z < 0$ since the vertical phase velocity for a gravity wave is in the opposite direction to the vertical group velocity. In the absence of any compelling evidence on the sign of k_{me} I shall make the *a priori* assumption that the phase propagation is westward so that $k_{me} < 0$ as well. In this case we can take the square root of equation (6.6.14) and still get the correct, positive, sign for ω. For further simplicity let us assume that the F region peak is at 300 km so that $dN_0/dz = 0$ there. Finally, U_{0me} varies in value during the course of the day reaching magnitudes as high as 250 m/sec. Most of the time though $|k_{me}U_{0me}| \ll |k_z E_{0me}/B|$ for the gravity wave we are dealing with.

We can then see that the dominant terms are

$$\frac{N_1}{N_0} = \frac{-U_{1me}\left\{\dfrac{2i\beta_0\sqrt{\gamma-1}}{c}+\dfrac{\gamma g k_{me}}{2c^2 k_z}\right\}}{\left\{\beta_0+\dfrac{\gamma g E_{0me}}{2c^2 B}-\dfrac{ik_z E_{0me}}{B}\right\}}$$

$$= \frac{-1{\cdot}5\times 10^{-7}i-5\times 10^{-7}}{4\times 10^{-4}+2\times 10^{-2}i}U_{1me}. \qquad (6.6.18)$$

The quantity $N_1/N_0 U_{1me}$ is a complex number. Since any complex number $A+iB$ say, can be written in exponential form $C\exp(i\phi)$ where $C = \sqrt{A^2+B^2}$ and $\tan\phi = B/A$, the complex N_1/N_0 gives the phase relation between N_1 and U_{1me}. Rewriting (6.6.18) as

$$\frac{N_1}{N_0} = (2{\cdot}5i-0{\cdot}75)10^{-5}U_{1me}$$

shows N_1 and U_{1me} to be approximately 75° out of phase.

The parameter that provides information about the gravity waves' effects upon the ionization is the modulus $|N_1/N_0| \approx 2{\cdot}5\times 10^{-5}U_{1me}$. Then for 100 m/s gravity wave induced winds there will be ionization perturbations that are 1/4 per cent of the background ionization, and these ionization irregularities are large enough to be detectable.

Even though one may argue that a wind of 100 m/s is no longer of perturbation magnitude, the resulting $\frac{1}{4}$ per cent ionization fluctuations certainly are of perturbation magnitude. Not all the gravity waves are going to be as well behaved as this. Gravity waves for which $k_{me}U_{0me} = k_z E_{0me}/B$ are going to produce ionization perturbations 100 times larger. At heights other than the F region peak these fluctuations may be even larger still. If the variations in the ionization density are too large, and especially if $|N_1| > |N_0|$ then the perturbation technique that was used becomes invalid and a gravity wave which may itself be of perturbation magnitude may produce ionization variations that are non-linear, as Clark *et al.* (1971) also pointed out.

Obviously the derivation in this section is liable to contain a large number of approximations and is likely to be of use only in certain very special areas. The method employed is quite general but the approximations to be made will vary.

Hooke (1969a) and Beer and Moorcroft (1972a) considered the ionization perturbations due to the passage of a gravity wave through the E region. In this case both Pedersen and Hall forces are important, and both these forces are different for each species of charged particle. Hooke allowed for variations in q and α and showed that these play an important role. Beer and Moorcroft limited themselves to the night-time situation when there is no ionization production* and allowed for the possible occurrence of plasma instabilities by using a complex angular frequency for the ionization perturbations. They found that the gravity wave produces a steady state ionization perturbation that may be described by the method outlined above. There is also a free response, much like the transient response of a mechanical system except that under certain conditions the free response may become unstable. The instability depends on the magnitudes and directions of the E region electric fields and on the background ionization gradient dN_0/dz. The requisite fields for instability can only occur at night.

In the F region the whole problem is greatly complicated by the effects of viscosity and heat conduction. In order to work out ionization perturbations one should note that.

(i) Dissipative processes reduce U_{1me} which affects N_1. They may also affect the dispersion relation governing the wave propagation, rendering the solution for N_1 even more complex. At the same time there will be partially reflected atmospheric waves due to viscous and thermal conductive losses. Below 500 km altitude these partially reflected waves

* This may not be completely correct. It has been suggested that there is a weak night-time production of about 3×10^6 ions $m^{-3} s^{-1}$ in the E region by ultra-violet radiations from the night sky and by the scattered radiation from the sunlit side of the earth.

are decoupled from the original upward propagating wave and, furthermore, are evanescent (Volland, 1969a, c).

(ii) When the perturbations in electron and ion density are quite large, this will increase the importance of ion drag forces which could in turn affect the original neutral wave parameters. Furthermore the expression for N_1/N_0 contains a contribution from the ionization gradient. This will in turn be modified by the gravity wave passage, especially if the ionization perturbations are severe.

The solution to all of these problems is as yet not forthcoming. Contributions to the problem have been made by a large number of authors, though the confusion and conflicting importance placed on the role of ion drag renders the conclusions of many of these authors suspect.

Exact Solutions of the Continuity Equation

It was pointed out that atmospheric wave parameters of perturbation magnitude could produce ionization fluctuations that were larger than perturbation magnitude. The only way to solve this dilemma is to seek exact solutions of the continuity equation for the ionization. This is not always possible and in general an exact solution must be solved numerically.

One method of attack is to use a coordinate system that moves with the phase velocity of the wave. Once again consider the simple case of a zonally propagating wave with an eastward and upward phase velocity then

$$\frac{\partial \mathrm{me}}{\partial t} = V_{\mathrm{pme}} \qquad \frac{\partial z}{\partial t} = V_{\mathrm{pz}}$$

so that the continuity equation can be reduced solely to a function of time. The F region ionization continuity equation is

$$\frac{\partial N}{\partial t} + N\frac{\partial V_{\mathrm{me}}}{\partial \mathrm{me}} + N\frac{\partial V_z}{\partial z} + V_{\mathrm{me}}\frac{\partial N}{\partial \mathrm{me}} + V_z\frac{\partial N}{\partial z} = q - \beta N$$

which becomes

$$\frac{\partial N}{\partial t}\left[1 + \frac{V_{\mathrm{me}}}{V_{\mathrm{pme}}} + \frac{V_z}{V_{\mathrm{pz}}}\right] + N\left[\frac{1}{V_{\mathrm{pme}}}\frac{\partial V_{\mathrm{me}}}{\partial t} + \frac{1}{V_{\mathrm{pz}}}\frac{\partial V_z}{\partial t} + \beta\right] = 0$$

if we pretend it is night and take $q = 0$.

Now we must decide which motion we are interested in: large or small scale. We shall only consider one case, namely large-scale variations.

Large-scale Diurnal Variations

We have already mentioned that at night at the magnetic equator, F region ions drift almost with the neutral wind. The dominant neutral wind will be the diurnal wind set up by the large diurnal temperature variations in the thermosphere. This wind, which may be thought of as a tidal mode, will be evanescent so that V_{pz} is infinite. If it seems surprising that an absence of vertical movement gives an infinite rather than a zero phase velocity one should recall that V_{pz} is the trace velocity, and does not represent a component of the true phase velocity. Since $V_{pz} = \omega/k_z$ and $k_z = 0$ for an evanescent wave, $V_{pz} = \infty$. Furthermore as the dominant diurnal tide retains the same position in relation to the sun, it appears to an observer on the earth as if it is moving westward with the velocity of the terminator (the sunrise–sunset line) whose speed we shall denote by V_s. Then $V_{pme} = -V_s$ since the terminator moves westward.

Then

$$\frac{\partial N}{\partial t}\left[1-\frac{V_{me}}{V_s}\right]-N\left[+\frac{1}{V_s}\frac{\partial V_{me}}{\partial t}-\beta\right]=0 \qquad (6.6.20)$$

at night.

However if the ionization moves with the diurnal wind then

$$V_{me} = +V_0 \sin(\omega t)$$

where $\omega = \Omega_E = 2\pi/24\ \text{hours}^{-1}$, and $t = 0$ occurs when the westward wind becomes eastward.

If we denote V_0/V_s by a, where it should be recalled that $a < 0$, then the solution of the continuity equation is

$$N = \frac{N_0 C \exp\left\{\dfrac{-2\beta}{\omega\sqrt{1-a^2}}\tan^{-1}\left[\dfrac{\tan(\omega t/2)-a}{\sqrt{1-a^2}}\right]\right\}}{1-a\sin(\omega t)} \qquad (6.6.21)$$

where C is chosen so that $N = N_0$ when $t = 0$. Equation (6.6.21) is plotted in Fig. 6.7. The maxima and minima of this equation may

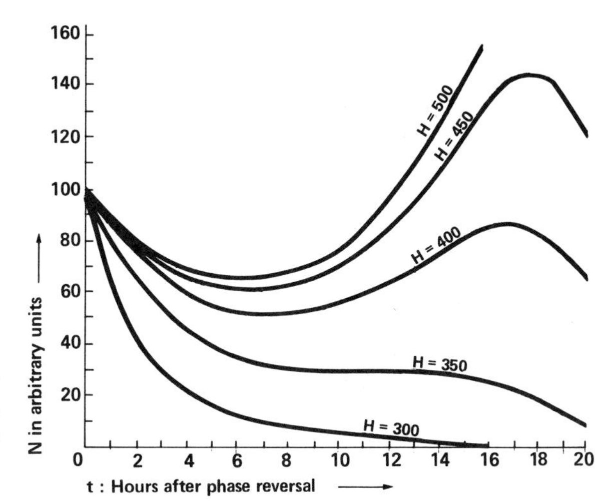

6.7 A plot of the solution of equation (6.6.21) for various values of the height of the F *layer*

easily be found by using (6.6.20) when $\partial N/\partial t = 0$. These occur when

$$\omega t = \pm\cos^{-1}\left|\frac{\beta}{a\omega}\right|$$

so that there are no extrema when

$$\beta > |a\omega|.$$

The results obtained at Legon, Ghana, which is close to the magnetic equator disclose, during the months September to December, a night-time ionization minimum followed by a small maximum that subsequently decays. These extrema are absent in other months when presumably the F layer is lower and β is higher. Attempts to fit equation (6.6.21) to the Legon data indicate (Koster and Beer, 1973):

(a) The attachment-like coefficient β is a function of time. This arises because the F region has vertical drift motions that alter its height and the consequent appropriate value of β. These variations of β are controlled by the E region electric fields which propagate up the highly conducting magnetic field lines and which then drift the F layer vertically by means of $\mathbf{E}\times\mathbf{B}$ forces. Since the electric fields are themselves set up by dynamo action with the tidal winds in the E region, the variations in β will also be periodic with a diurnal periodicity. Their phase will differ from the phase of V_{me} however.

(b) Meridional motions should be taken into account in order to provide a true picture of the nocturnal equatorial F region.

A similar method can be used to solve the more complicated case when internal wave motions are considered, though in this case rather more difficult integrals need to be evaluated. Nelson (1968) and Broche (1971) have also analyzed the exact solution of the continuity equation when sources and losses are in balance. Nelson points out that

$$\frac{\partial N}{\partial t}+\frac{\partial}{\partial z}(NV)=0$$

can be rewritten as

$$1\,.\frac{\partial N}{\partial t}+V\frac{\partial N}{\partial z}=-\frac{\partial V}{\partial z}\,.\,N \tag{6.6.22}$$

so that there are three ordinary differential equations corresponding to (6.6.22), given by

$$\frac{dt}{1}=\frac{dz}{V}=\frac{dN}{-(\partial V/\partial z)N}. \tag{6.6.23}$$

Any two of these can be solved to yield relations of the form

$$f_1(z,t,n)=C_1; \qquad f_2(z,t,n)=C_2$$

where C_1 and C_2 are arbitrary constants. The general solution of (6.6.23) is then

$$f_1(z,t,n)=F[f_2(z,t,n)]$$

where F is an arbitrary function that is determined from the prescribed values of $N(z, t)$ along some curve. In general the problem may be solved numerically though Nelson gives some simple forms amenable to exact solutions.

6.7 Spatial Resonance

When an internal wave propagates through the ionosphere then it will produce wave-like fluctuations in the electron and ion densities. These irregularities arise both from the internal wave-induced changes in the rate at which the loss processes operate and from the effect of collisions between the charged and neutral particles. These collisions will sweep the ionization into a harmonic form.

If the subsequent drift velocity of the ionization irregularity matches the phase velocity of the internal wave then the irregularity will continue to grow by receiving more ionization from the wave. This process is a self-induced spatial resonance of the ionization. Forced spatial resonance occurs when an ionization irregularity formed by some other means (e.g. wind shears) has a drift velocity that coincides with the phase velocity of an internal wave, and is also positioned so that the layer coincides with the peaks of the ionization irregularities produced by the atmospheric wave (Whitehead, 1971).

Self-induced spatial resonance can be examined by solving the perturbed continuity equation. At low and mid-latitudes the longitudinal electric field is negligible so that by treating the electric fields as background, zero-order quantities, and gravity wave winds as first order perturbations, then the perturbed continuity equation may be written in terms of the gravity wave amplitude A and the electron density polarization term $\mathcal{N}$ as

$$(i\omega + S^{\pm})\mathcal{N} + C^{\pm}A = 0 \tag{6.7.1}$$

where ω is the internal wave frequency, $C^{\pm}$ is a constant and $S^{\pm}$ is given by

$$\begin{aligned} S^{\pm} = {} & \beta_0 + D_1^{\pm}(K_{\text{me}}^2 + K_z^2) + D_0^{\pm} K_{\text{me}}^2 \\ & + \frac{\partial}{\partial z}(\pm \mu_1^{\pm} E_{0z} + \mu_2^{\pm} E_{0\text{me}}) \\ & - iK_{\text{me}}(\pm \mu_1^{\pm} E_{0\text{me}} - \mu_2^{\pm} E_{0z}) \\ & - iK_z(\pm \mu_1^{\pm} E_{0z} + \mu_2^{\pm} E_{0\text{me}}) \end{aligned}$$

in the F region. In the E region $2\alpha_0 n_0$ replaces β_0 but in the E region space charge effects which set up perturbation electric fields also become important, so that (6.7.1) needs to be modified.

The choice of sign for $S^{\pm}$ is determined by the MCKT rule. If the irregularity is field aligned then the negative sign is chosen because the irregularities drift as if they were composed of electrons. For non-aligned irregularities the positive sign is used. In deriving equation (6.7.1) it was assumed that the amplification factor of the ionization irregularities is the same as the amplification factor for the gravity waves that set up the irregularities.

From (6.7.1) the electron density amplitude is

$$\mathcal{N} = \frac{-C^{\pm}A}{(i\omega + S^{\pm})}. \tag{6.7.2}$$

The self-induced spatial resonance criteria are then that

$$i\omega + \mathrm{Im}\,(S^{\pm}) = 0 \tag{6.7.3i}$$

and

$$\mathrm{Real}\,(S^{\pm}) = 0. \tag{6.7.3ii}$$

The first condition is equivalent to $\omega = \mathbf{k}\,.\,\mathbf{V}_0^{\pm}$ where $\mathbf{V}_0$ is the charged particle velocity. This may be recognized as the requirement that the component of the drift velocity in the gravity wave direction equals the gravity wave phase velocity. The second condition (6.7.3ii) is satisfied when the effects of diffusion, recombination and the upward gradient of ionization are balanced (Beer, 1973a).

These conditions are quite stringent for tropospherically generated waves, which are liable to severe reflections and dissipation before reaching the ionosphere. Spatial resonance for these waves is only likely to occur at the base of the night-time F region and in the night-time equatorial E region valley. Thermospherically launched gravity waves seem able to induce spatial resonance throughout the whole F region.

The Pousse-Café Effect

A pousse-café is, according to Webster's 7th New Collegiate Dictionary, 'a cocktail consisting of several liqueurs of different colours and specific gravities poured so as to remain in separate layers'. When the spatial resonance mechanism acts on gravity and acoustic waves launched from the equatorial electrojet, then discrete layers of enhanced ionization are formed which show up on ionograms as secondary traces. This is the pousse-café effect. The spatial resonance mechanism will only produce the pousse-café effect near the magnetic equator at night. During the day the horizontal velocity of the ionization is not greát enough to match the horizontal phase velocity of the gravity wave. At night the equatorial ionization seems to move with the rapid diurnal winds and the spatial resonance condition can be satisfied. The pousse-café effect is limited to regions near the dip equator because the gravity waves launched by the electrojet travelling to mid-latitudes will be inclined to the vertical sufficiently to increase their vertical trace phase velocity beyond the maximum capable of generating spatial resonance.

Eventually the electron and ion density in these discrete layers of ionization becomes so great that the layers become unstable. The layers then get mixed together in the same way that a pousse-café does when one tries to pick it up and the resulting turbulence in the equatorial ionization appears on ground based ionograms near the equator as a greatly blurred fuzzy set of traces which is known as spread *F*.

A full mathematical description of the pousse-café effect is most easily done by solving the continuity equation in a reference frame moving with the phase velocity of the gravity wave. This reveals that it is the variations in the loss coefficient, induced by the heating effects during the passage of a wave, that are responsible for the equatorial spatial resonance. If the wave produces an ionization perturbation that is a fraction, a, of the background ionization in each period $2\pi/\omega$ of the wave then the ionization grows until the extra loss rate $\beta_0 N_1$ balances the extra ionization growth rate so that

$$\beta_0 N_1 = \frac{N_0 a\omega}{2\pi}$$

or

$$\frac{N_1}{N_0} = \frac{a\omega}{2\pi\beta_0}.$$

Instability then occurs when the condition

$$\frac{1}{N}\frac{dN}{dz} > \frac{\beta_0 v_n^+}{g}$$

is satisfied (Liu and Yeh, 1966), or in other words when

$$\frac{k_z a\omega}{\beta_0 \pi^2} > \frac{\beta_0 v_n^+}{g}.$$

6 Background Reading

Beer, T. (1972) Atmospheric waves and the ionosphere, *Contemporary Physics*, **13**, 247.
A discussion on D and E region winds over Europe, *Philosophical Transactions of the Royal Society of London*, No. 1217, **A 271**, 455 (1972).
Hines, C. O. (1963) The upper atmosphere in motion, *Quarterly Journal of the Royal Meterological Society*, **89**, 1.
St. Gallen Symposium proceedings, *Journal of Atmospheric and Terrestrial Physics*, **30**, No. 5 (1968).

Two important papers on the generation of TID's have recently appeared:

Francis, S. H. (1973) Lower atmospheric gravity modes and their relation to medium scale travelling ionospheric disturbances, *Journal of Geophysical Research*, **78**, 8289.
Swift, D. W. (1973) The generation of infrasonic waves by auroral electrojets, *Journal of Geophysical Research*, **78**, 8205.

7

Non-linear Effects

7.1 Sources

There are three areas in atmospheric wave studies that are still largely unexplored. These are the sources that supply the energy for the waves, the waves' ability to propagate upwards and the non-linear interactions that atmospheric waves are capable of. To some extent all these points have been briefly dwelt upon in earlier sections. However as these topics offer the greatest opportunity for further research I would like to discuss them a bit further.

The identification of energy sources for the smaller scale waves is done by trying to correlate wave-induced effects with the postulated source mechanism. This procedure is going to be greatly hampered by ducting, reflections and coupling. To make matters even more complicated, it often appears that a postulated tropospheric source yields ionospheric effects but does not show up on ground based microbarographs. To give an example of this, let me point out that there are a number of observations of wavelike disturbances in the ionosphere that were found to correlate with severe local storms. Pierce and Coroniti (1966) have even proposed that the violent convection currents associated with the formation of large cumulus clouds produces the waves. However one set of extensive microbarographic observations over a period of three years failed to confirm any correlation between severe weather and sinusoidal pressure oscillations. There could be many possible reasons for this, one for example being that waves are reflected upwards between the clouds and the ground; but it seems most likely that the correlation between the two processes was fortuitous and that the observed ionospheric perturbations were due to plasma waves propagating down from the magnetosphere.

Planetary Waves

We have already shown how isolated changes in topography can produce planetary waves when an eastward wind is blowing. In this respect the effects of the continents are the most important ones and long wavelength planetary waves are produced by the airflow over the continents and by the differential heating over continents and oceans. It should be noted that the differential heating produces a change in the height of the upper boundary of an assumed incompressible atmosphere by changing the scale height. The topographic effects are equivalent to changes in the lower boundary conditions. Both are capable of generating planetary waves. Up to altitudes of 3 km, orographic features like mountain ranges can generate these waves, up to 20 km the effects of the oceans and continents dominate whereas above about 30 km (the equivalent depth of an atmosphere in adiabatic equilibrium) ground features do not generate planetary waves. Planetary waves generated above this height are due to baroclinic instability. This mechanism generates much shorter wavelengths and is most probably responsible for the shorter planetary waves at lower altitudes as well.

Planetary waves carry a substantial part of the dynamic energy of the troposphere. A representative energy density of 10^2 joules/m^3 and a vertical component of group velocity of the order of 10^{-2} m/s typifies the long (forced) planetary waves (Charney and Drazin, 1961) so that they presumably have an upward flux of energy of 1 watt/m^2. This would lead to an extreme heating of the upper atmosphere which is not observed in practice, so that most of this power must be reflected.

In the stratosphere one of the effects of the thermal gradients, which are quite large at high latitudes, is the production of a polar vortex (Boville, 1967). This vortex, probably because of baroclinic instability, is subject to large oscillations in its movements, particularly in winter. This can manifest itself as a stratospheric warming in which the temperature at around the 30 km level rises by tens of degrees above its normal seasonal value, the increase lasting for some days. If we regard this as a planetary wave-induced phenomenon then we could expect to find effects occurring at higher altitudes. Ionospheric effects do occur but they are not at all like the ones we would expect, thus raising doubts as to whether the planetary wave extends into the ionosphere. A correlation is found between the stratospheric warming, an increase in radio wave absorption and an increase in

D region electron density. This could be interpreted by saying that the D region dropped by about 10 km. However if an external wave is propagating concomitant with a temperature increase, one would expect the D region to rise. The mechanism of this stratospheric–ionospheric coupling is still not understood.

Atmospheric Tides

For tidal motions the choice of a source mechanism narrows down to either gravitational attractions or to thermal excitation. For the atmospheric tides solar insolation appears to be the dominant mechanism. This occurs predominantly in four regions of the atmosphere:

(1) The molecular oxygen heating region at altitudes greater than 90 km.
(2) The ozone heating region centred at 45 km altitude.
(3) The low altitude heat absorbers such as water vapour and carbon dioxide.
(4) The ground.

There also remains the possibility that some minor constituents could also play a role in the absorption of the incoming and reradiated spectrum, but the success of present tidal theory in accounting for most of the major observed features indicates that this role, if it exists, will be a small one.

Acoustic and Gravity Waves

In the ionosphere there seems to be an almost continuous occurrence of gravity waves whose source remains as yet unidentified. There appears little doubt that spectacular events such as the eruption of volcanoes and the detonation of nuclear blasts generate acoustic and gravity waves that appear in both the lower and upper atmospheres. There is even evidence that the energy from the ground launched by the Alaskan earthquake of 1964 managed to reach ionospheric heights and produce wavelike ionization fluctuations (Davies and Baker, 1965). Early barometric observations showed that atmospheric waves were launched by the great Siberian meteorite of 1908. It is likely however that the meteors that fall on the upper atmosphere are too small to generate appreciable atmospheric waves.

Though the oscillation amplitude of these waves is large in the upper atmosphere, the energy flux is small by tropospheric standards, being only of the order of 10^{-3} Watts/m^2. Observations of Gossard (1962) show that a flux of 10^{-1} Watts/m^2 commonly leaves the troposphere so that waves in the upper air could have propagated up from the troposphere even if there were substantial partial reflections en route. The presence or absence of critical layers which absorb all the wave energy of a particular component will determine the parameters of the dominant wave at ionospheric heights. Furthermore the most important critical layers would be those occasioned by jet streams. Since jet streams are strong eastward flows this will introduce an anisotropy into any thermospheric sources. It is possible that jet streams themselves are also capable of generating gravity waves which are not subject to critical layer absorption to such a degree. Two further possibilities: mountain waves and ocean tides. The former definitely produce gravity waves but it is only going to be extensive chains of mountains that are likely to produce waves of sufficient strength to propagate to the ionosphere. Ocean tides act like a large piston at the lower boundary of the atmosphere. At certain locations, such as in the Bay of Fundy, the occurrence of a funnel shaped shore line can produce a tidal range as great as 20 metres. This certainly tends to generate low period waves.

Some of the ionospheric wavelike disturbances are the result of thermospherically generated gravity waves. Large and medium scale TID's appear to be generated in the polar regions by either a supersonic displacement of an auroral arc or a sudden surge of the auroral electrojet current following a polar substorm. The equatorial electrojet is also likely to generate waves.

It has been suggested that the cooling action resulting from a solar eclipse can generate bow waves as the shadow of the moon passes across the surface of the earth at a supersonic speed. Fluctuations in the ionosphere following a solar eclipse have been interpreted as being due to the atmospheric wave generated by it. The atmospheric waves generated by the eclipse will emanate from any or all of the four regions mentioned in connection with tidal sources. If they emanate from the lower two then oscillations after an eclipse should show up on ground based recorders, whereas oscillations emanating from the higher regions of insolation may not be detected at the ground.

If the supersonic shadow of a solar eclipse can generate atmospheric waves then so can the terminator (the line separating the dark side of the earth from the sunlit side) (Beer, 1973b; Raitt and Clark, 1973). The terminator is supersonic between $\pm 45°$ latitude at all altitudes below the mesopause. At the mesopause the latitude belt increases to about $\pm 55°$, but above 100 km it is no longer supersonic because the high temperatures in the thermosphere increase the sound speed there to a value greater than the earth's equatorial rotational speed. The waves generated by the supersonic terminator will appear from the regions of solar insolation as well. Since gravity waves in the troposphere are rather rare events, it thus seems likely that the ozone heating region and the bottom of the molecular oxygen heating region will generate atmospheric waves every twelve hours and also when there is a solar eclipse. This mechanism seems capable of generating the observed continuous atmospheric wave spectrum in the upper atmosphere.

A large number of authors have dealt with the generation of waves by various excitation mechanisms. Row (1967) produced a theory capable of explaining ionospheric observations following a nuclear detonation and an earthquake. Dickinson (1969b) has comprehensively studied the transient excitation of these waves by point impulses and switched-on sources. The non-linear effects are extremely important near a source and so mode analysis becomes inapplicable. Liu and Yeh (1971) suggest that the non-linear terms in the hydrodynamic equations can be taken to the right-hand side of the equations and treated as sources. The far-field response to these sources could then be calculated from known formulae given by them. The effect of the ground will be to introduce a reflected wave and a surface wave. The surface wave is equivalent to the Lamb wave, or to an edge wave in a non-isothermal atmosphere.

7.2 Resonant Triads

Non-linearities in a wave system may show up in myriad ways. They may be due to the presence of a source or to the effects of a strong dissipative process. They can also arise from the wave amplitude becoming too large or through the interaction between two waves.

Resonant Interaction

Resonant interaction is concerned with wave–wave interactions that satisfy a certain resonant condition. If the waves interact as a resonant trio with frequencies ω_1, ω_2 and ω_3 and with their corresponding complex wave vectors $\mathbf{K}_1$, $\mathbf{K}_2$ and $\mathbf{K}_3$, the resonance conditions are the simultaneous relations

$$\omega_1+\omega_2+\omega_3 = 0 \tag{7.2.1}$$

$$\mathbf{K}_1+\mathbf{K}_2+\mathbf{K}_3 = 0. \tag{7.2.2}$$

Since the values of ω and $\mathbf{K}$ can be positive or negative, the third member of the trio can have two values if ω_1, ω_2, $\mathbf{K}_1$ and $\mathbf{K}_2$ are known, namely

$$\omega_3 = \omega_1 \pm \omega_2$$

$$\mathbf{K}_3 = \mathbf{K}_1 \pm \mathbf{K}_2$$

though the choice of values will be further restricted by the dispersion relation between ω and $\mathbf{K}$. The ordinary perturbation analysis of a resonant trio leads to an amplitude that grows linearly with time.

If two evanescent waves whose wave number magnitudes are almost equal interact, then the difference frequency $\omega_1-\omega_2$ can be arbitrarily small whilst the difference wave number is comparable in magnitude with $\mathbf{K}_1$ or $\mathbf{K}_2$. Though the magnitudes of $\mathbf{K}_1$ and $\mathbf{K}_2$ are nearly equal their directions need not be, so that the resonance condition (7.2.2) may be represented on a vector diagram as a closed triangle (Fig. 7.1). By varying the magnitude of $\mathbf{K}_2$ about $\mathbf{K}_1$ it would be possible to obtain an internal mode for $(\mathbf{K}_3, \omega_3)$. This provides a further source mechanism for internal waves since the $(\mathbf{K}_3, \omega_3)$ wave

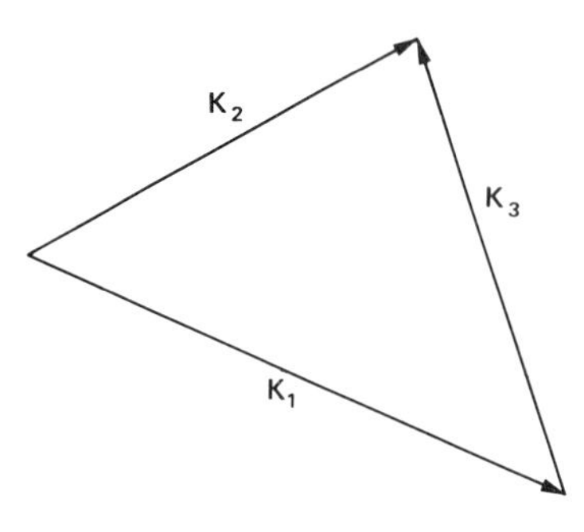

7.1 Horizontal wave number vectors for a resonant triad (after Yeh and Liu, 1970)

slowly extracts energy from the evanescent waves to which it is coupled. This mechanism is important in the ocean and it is possible that internal waves can be extracted from the principal evanescent atmospheric waves, namely evanescent tidal modes, the Lamb wave and the characteristic surface wave. Since it seems probable that there are many other more important source mechanisms for internal waves, the resonant interaction is more likely to occur between two existing internal modes to produce an evanescent one.

Yeh and Liu (1970) studied the resonant interaction between two acoustic gravity waves with frequencies ω_1 and ω_2, one of which is taken to have $\omega_1 = 0$. This hypothetical zero frequency gravity wave can then be chosen so as to have any real vertical wave number that satisfies the resonance criterion. It therefore represents a sinusoidal background wind. In this case there is produced a new internal wave whose frequency ω_3 is the same as ω_2. The wave number will differ. The results obtained by treating this problem are the same as would result from alternative methods of analysis. The original gravity wave becomes trapped if the zero frequency gravity wave is sufficiently strong due to the critical layer absorptions.

Spizzichino (1970) launched a thorough investigation into non-linear resonant interactions and decided that most of the observed gravity waves at meteor heights arise from an interaction between the diurnal tide and gravity waves trapped below the mesopause. Other conclusions were that the prevailing wind (a zero frequency wave) produces an additional meridional circulation through non-linear interactions with the atmospheric oscillations and that the interaction between the S_1 and S_2 tides produces a downward propagating diurnal tide. The generation of weak higher order resonant waves can also lead to turbulence.

Resonant interactions between planetary waves will also take place (Longuet-Higgins and Gill, 1967), leading to energy exchanges between the various forced and free modes.

7.3 Instability

We will recall that if the perturbation quantities become of comparable magnitude to the background quantities so that $\rho_1/\rho_0 \sim 1$ or $p_1/p_0 \sim 1$ then the wave will start to lose its sinusoidal character. If the perturbation velocities were to attain very high speeds comparable to the speed of sound, so that $U_{1x}/c \sim 1$ then a shock wave will result. This seems unlikely to occur in practice because of the dissipative effects and because there is likely to be a self induced critical layer set up by the wave.

Preliminary results by Johnston (1967) indicated that in regions of the atmosphere where $\omega_B > \omega_a$ gravity waves may become unstable. As we recall from equation (3.7.8) the variable q can be thought of as a vertical wavenumber where

$$q^2 = \left(\frac{\omega_B^2}{\Omega^2} - 1\right)k_x^2 + \frac{(\Omega^2 - \omega_a^2)}{c^2} \tag{7.3.1}$$

If $\omega_B > \omega_a$ then complex values of the wave frequency Ω can satisfy this equation. This implies that convective instabilities may exist.

The (q, k_x) plot is depicted in Fig. 7.2, for various values of Ω/ω_B given in rectangular boxes. The acoustic modes, whose vertical group and phase velocities are in the same direction, are represented by

7.2 Propagation surfaces when $\omega_a = 0.6\omega_B$. Values of Ω/ω_B for each mode are given inside the rectangular boxes (after Johnston, 1967)

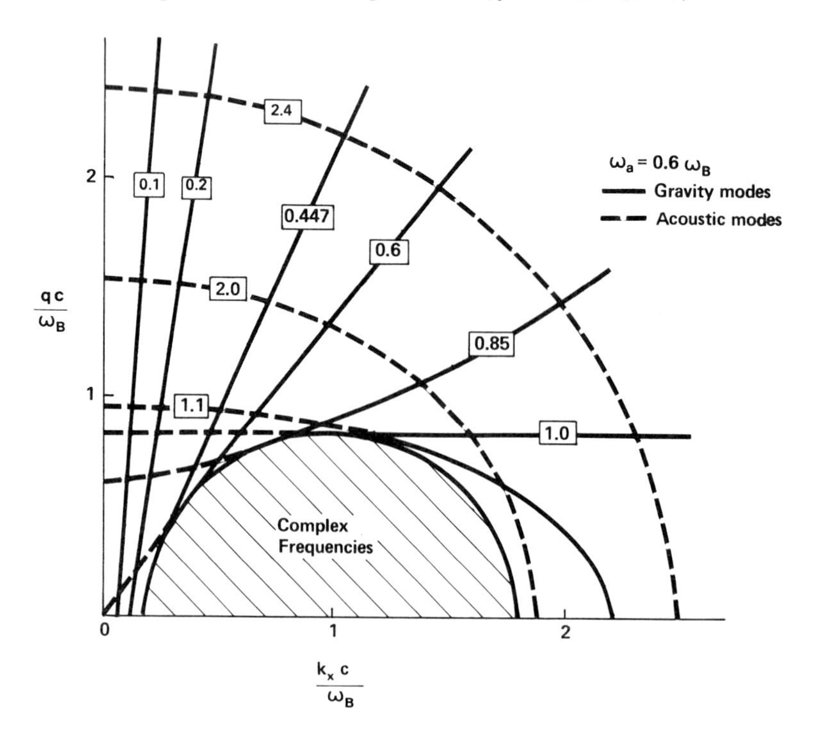

dashed lines. The gravity modes for which the vertical phase and group velocities are oppositely directed are shown by solid lines. The transition occurs for values at which $\omega^2 = ck_x\omega_B$. At these transition points the group velocity becomes infinite. The locus of the transition points bounds a closed area within which the frequency can be complex and hence instabilities may occur. This occurs for values of the frequency that lie between

$$\sqrt{\omega_B^2 - \omega_B(\omega_B^2 - \omega_a^2)^{\frac{1}{2}}} < \Omega < \sqrt{\omega_B^2 + \omega_B(\omega_B^2 - \omega_a^2)^{\frac{1}{2}}}. \quad (7.3.2)$$

When $\omega_a = 0{\cdot}6\omega_B$, this corresponds to

$$0{\cdot}447 < \frac{\Omega}{\omega_B} < 1{\cdot}3416.$$

Fig. 7.1 should be compared to Fig. 3.4 which denotes the propagation surfaces in a non-anomalous atmosphere in which $\omega_a > \omega_B$. Johnston indicates that this instability may not really occur. The reason is that the WKB approximation on which the analysis is based requires q to be greater than $(1/H)\,dH/dz$ where H is the scale height. However instability is indicated only for those values of q which are small enough to violate the WKB condition. Nevertheless, the results of illegitimate WKB calculations can sometimes be surprisingly reliable, so that the indications point to the existence of some sort of instability. Intuitively we would guess that $\omega_B > \omega_{an}$ is likely to be a more realistic requirement for the instability condition and should replace ω_a in (7.3.1) and (7.3.2). Tolstoy and Lau (1970) claim that this instability enhances the amplitude of certain oscillations which have passed through the regions of the upper atmosphere that are anomalous (i.e. $\omega_B > \omega_{an}$), so that they produce stronger microbarograph traces than one would deduce in the absence of the instability.

The best known gravity wave instability is the one that leads to clear air turbulence. It is believed that this follows from a manifestation of an instability mechanism known as the Kelvin–Helmoltz instability (KHI). The KHI occurs in sheared flows in the presence of gravity. It can be induced by gravity waves in two different ways. One way occurs when the path of an upward propagating gravity wave intersects with a jet stream or other strong wind which acts as a critical layer. The gravity wave energy is absorbed which increases the jet stream speed and so leads to instability. The other way occurs when the gravity acts directly in the fast moving flow, producing unstable oscillations in it directly.

7 Background Reading

Cho, H. R. (1972) A study of nonlinear atmospheric waves, *Department of Electrical Engineering Report No. 48*, University of Illinois, Urbana.

Lindzen, R. S. (1973) Wave-mean flow interactions in the upper atmosphere, *Boundary-Layer Meteorology*, **4**, 327.

Micrograph observations of eclipses are now available. These are summarized in:

Chimonas, G. (1973) Lamb waves generated by the 1970 solar eclipse, *Planetary and Space Science*, **21**, 1843.

Records of the 1973 Eclipse, *Nature*, **246**, 376 (1973).

References

Alaka, M. A. (editor) (1960) The airflow over mountains, *WMO Technical note*, **34**, Geneva.

Balachandran, N. K. (1968) Acoustic gravity wave propagation in a temperature and wind stratified atmosphere, *J. Atmos. Sci.*, **25**, 818.

Barrett, E. W. (1958) Eccentric circumpolar vortices in a barotropic atmosphere, *Tellus*, **10**, 395.

Batchelor G. (1953) *Theory of homogeneous turbulence*, Cambridge University Press.

Beer, T. (1971) *Atmospheric wave induced instability in the E region*, Ph.D. thesis, University of Western Ontario, London, Canada.

Beer, T (1972a) *D* Region parameters from one component of partial reflections, *Ann. Geophys.*, **28**, 341.

Beer, T. (1972b) Atmospheric waves and the ionosphere, *Contemp. Phys.*, **13**, 247.

Beer, T. (1972c) Lamb waves and the conservation of energy, *Am. J. Phys.*, **40**, 774.

Beer, T. (1973a) Spatial resonance in the ionosphere, *Planet. Sp. Sci.*, **21**, 297.

Beer, T. (1973b) Supersonic generation of atmospheric waves, *Nature*, **242**, 34.

Beer, T. & Moorcroft, D. R. (1972a) Atmospheric wave induced instability in the nighttime E region, *J. Atmos. Terr. Phys.*, **34**, 2025.

Beer, T. & Moorcroft, D. R. (1972b) Nighttime sporadic *E*, *J. Atmos. Terr. Phys.*, **34**, 2045.

Berkshire, F. H. & Warren, F. W. G. (1970) Some aspects of linear lee wave theory for the stratosphere, *Q. J. Roy. Met. Soc.*, **96**, 50.

Booker, J. R. & Bretherton, F. P. (1967) The critical layer for internal gravity waves in a shear flow, *J. Fluid Mech.*, **27**, 513.

Boville, B. W. (1967) Planetary waves in the stratosphere and their upward propagation, in *Space Research VII*, North Holland Publ. Co., Amsterdam.

Brekhovskikh, L. M. (1960) *Waves in layered media*, Academic Press, New York.

Bretherton, F. P. (1966) The propagation of groups of internal gravity waves in shear flow, *Q. J. Roy. Met. Soc.*, **92**, 466.

Bretherton, F. P. (1969a) Waves and turbulence in stably stratified fluids, *Radio Sci.*, **4**, 1279.

Bretherton, F. P. (1969b) Lamb waves in a nearly isothermal atmosphere, *Q. J. Roy. Met. Soc.*, **95**, 754.

Bretherton, F. P. & Garrett, C. J. R. (1971) Wave trains in homogeneous moving media, *Proc. Roy. Soc.*, **A302**, 75.

Brinkmann, W. A. R. (1971) What is a Foehn?, *Weather*, **26**, 230.

Broche, P. (1971) Variation de densité électronique au passage d'une onde de gravité dans l'ionosphere, *Ann. Geophys.*, **27**, 75.

Browning, K. A., Harrold, T. W. & Starr, J. R. (1970) Richardson number limited shear zones in the free atmosphere, *Q. J. Roy. Met. Soc.*, **96**, 40.

Budden, K. G. (1961) *The wave-guide mode theory of wave propagation*, Logos Press, London.

Burpee, R. W. (1972) The origin and structure of easterly waves in the lower troposphere over North Africa, *J. Atmos. Sci.*, **29**, 77.

Butler, S. T. & Small, K. A. (1963) The excitation of atmospheric oscillations, *Proc. Roy. Soc.*, **A274**, 91.

Chan, K. L. & Villard, O. G. (1962) Observations of large scale travelling ionospheric disturbances by spaced path high frequency instantaneous frequency measurements, *J. Geophys. Res.*, **67**, 973.

Chandrasekhar, S. (1961) *Hydrodynamic and hydromagnetic instability*, Oxford University Press.

Chapman, S. (1932) On the theory of lunar tidal variation of atmospheric temperature, *Mem. Roy. Met. Soc.*, **4**, 35

Chapman, S & Bartels, J. (1940) *Geomagnetism*, Oxford University Press.

Chapman, S. & Cowling, T. G. (1952) *The mathematical theory of non uniform gases*, Cambridge University Press.

Chapman, S. & Kendall, P. C. (1970) Tidal generation of electric currents and magnetic fields, *Planet. Sp. Sci.*, **18**, 1597.

Chapman, S. & Lindzen, R. S. (1970) *Atmospheric tides*, D. Reidel Publ. Co., Dordrecht.

Charney, J. G. (1947) The dynamics of long waves in a baroclinic westerly current, *J. Met.*, **4**, 135.

Charney, J. (1971) Geostrophic turbulence, *J. Atmos. Sci.*, **28**, 1087.

Charney, J. G. & Drazin, P. G. (1961) Propagation of planetary-scale disturbances from the lower into the upper atmosphere, *J. Geophys. Res.*, **66**, 83.

Chimonas, G. (1968) *The launching of low frequency travelling disturbances by auroral currents*, in T. M. Georges (1968b), p. 101.

Chimonas, G. (1970a) The extension of the Miles–Howard theorem to compressible fluids, *J. Fluid Mech.*, **43**, 833.

Chimonas, G. (1970b) Internal gravity wave motions induced in the earth's atmosphere by a solar eclipse, *J. Geophys. Res.*, **75**, 5545.

Chimonas, G. & Axford, W. I. (1968) Vertical movements of temperate zone sporadic *E* layers, *J. Geophys. Res.*, **73**, 111.

Chimonas, G. & Hines, C. O. (1971) Atmospheric gravity waves induced by a solar eclipse, *J. Geophys. Res.*, **76**, 7003.

CIRA (1961) & CIRA (1965) *Cospar international reference atmosphere*, North Holland Publ. Co., Amsterdam.

CIRA (1972) *Cospar international reference atmosphere*, Akademie-Verlag, Berlin.

Clark, R. M., Yeh, K. C. & Liu, C. H. (1971) Interaction of internal gravity waves with the ionospheric F_2-layer, *J. Atmos. Terr. Phys.*, **33**, 1567.

Clemmow, P. C. & Dougherty, J. P. (1969) *Electrodynamics of particles and plasmas*, Addison-Wesley, Reading, Mass.

Covez, L. (1971) On the theory of atmospheric tides, *J. Atmos. Terr. Phys.*, **33**, 1273.

Dalgarno, A. & Smith, F. J. (1962) The thermal conductivity and viscosity of atomic oxygen, *Planet. Sp. Sci.*, **9**, 1.

Davies, K. & Baker, D. M. (1965) Ionospheric effects observed around the time of the Alaskan earthquake of March 28, 1964., *J. Geophys. Res.*, **70**, 2251.

Davies, K., Baker, D. M. & Chang, N. J. F. (1969) Comparison between formulas for ionospheric radio propagation and atmospheric wave propagation, *Radio Sci.*, **4**, 231.

Davis, M. J. (1971) On polar substorms as the source of large scale travelling ionospheric disturbances, *J. Geophys. Res.*, **76**, 4525.

Davis, M. J. & Da Rosa, A. V. (1970) Possible detection of atmospheric gravity waves generated by the solar eclipse, *Nature*, **226**, 1123.

Deland, R. J. (1965) Some observations of the behavior of spherical harmonic waves, *Mon. Weath. Rev.*, **93**, 307.

Deland, R. J. (1970) The vertical structure of planetary-scale Rossby waves, *Q. J. Roy. Met. Soc.*, **96**, 756.

Deland, R. J. & Friedman, R. M. (1972) Correlation of fluctuations of ionospheric absorption and atmospheric planetary scale waves, *J. Atmos. Terr. Phys.*, **34**, 295.
Dickinson, R. E. (1969a) Propagation of planetary Rossby waves through an atmosphere with Newtonian cooling, *J. Geophys. Res.*, **74**, 929.
Dickinson, R. E. (1969b) Propagation of atmospheric motions, *Rev. Geophys.*, **7**, 483.
Dickinson, R. E. (1970) Development of a Rossby wave critical level, *J. Atmos. Sci.*, **27**, 627.
Donn, W. L. & Shaw, D. M. (1967) Exploring the atmosphere with nuclear explosions, *Rev. Geophys.*, **5**, 53.
Doodson, A. T. (1922) The harmonic development of the tide-generating potential, *Proc. Roy. Soc.*, **A100**, 305.
Drazin, P. G. (1970) Non-linear baroclinic instability of a continuous zonal flow, *Q. J. Roy. Met. Soc.*, **96**, 667.
Dutton, J. A. (1971) Clear air turbulence, aviation and atmospheric science, *Rev. Geophys. Sp. Sci.*, **9**, 613.
Eady, E. T. (1949) Long waves and cyclone waves, *Tellus*, **1**, 35.
Eckart, C. (1960) *Hydrodynamics of oceans and atmospheres*, Pergamon Press, New York.
Eckart, C. & Ferris, H. G. (1956) Equations of motion of the ocean and atmosphere, *Rev. Mod. Phys.*, **28**, 48.
Ekman, V. W. (1905) On the influence of the earth's rotation on ocean-currents, *Ark. Met. Astr. Fys.*, **2**, 1.
Eliasen, E. & Machenhauer, B. (1965) A study of the fluctuations of the atmospheric planetary flow patterns represented by spherical harmonics, *Tellus*, **17**, 220.
Eliassen, A. & Kleinschmidt, E. (1957) Dynamic Meteorology, in *Handbuch der Physik XLVIII*, Springer Verlag, Berlin.
Fejer, J. A. (1964) Atmospheric tides and associated magnetic effects, *Rev. Geophys.*, **2**, 275.
Fjortoft, R. (1951) Stability properties of large scale atmospheric disturbances, in *Compendium of meteorology*, American Meteorological Society, Boston, Mass.
Fogle, B. & Haurwitz, B. (1966) Noctilucent clouds, *Space Sci. Rev.*, **6**, 279.
Francis, S. H. (1973) Acoustic-gravity modes and large scale travelling ionospheric disturbances of a realistic, dissipative atmosphere, *J. Geophys. Res.*, **78**, 2278.
Frank, N. L. (1970) Tropical easterly waves due to easterly airflow over Ethiopian mountains, *Mon. Weather Rev.*, **98**, 307.
Friedman, J. P. (1966) Propagation of internal gravity waves in a thermally stratified atmosphere, *J. Geophys. Res.*, **71**, 1033.
Frith, R. (1968) The Earth's higher atmosphere, *Weather*, **23**, 142.
Fultz, D. & Long, R. R. (1951) Two dimensional flow around a circular barrier in a rotating spherical shell, *Tellus*, **3**, 61.
Garrett, C. J. R. (1969) Atmospheric edge waves, *Q. J. Roy. Met. Soc.*, **95**, 731.
Georges, T. M. (1967) *Ionospheric effects of atmospheric waves, ESSA Tech. Report IER 57-ITSA 54*, U.S. Govt. Printing Office, Washington, D.C.
Georges, T. M. (1968a) HF Doppler studies of travelling atmospheric disturbances, *J. Atmos. Terr. Phys.*, **30**, 735.
Georges, T. M. (1968b) *Acoustic gravity waves in the ionosphere*, (*Symposium Proceedings*), U.S. Govt. Printing Office, Washington, D.C.
Giwa, F. B. A. (1968) The response curve in atmospheric oscillations, *Q. J. Roy. Met. Soc.*, **94**, 192.
Giwa, F. B. A. (1969) The influence of vertical variation of zonal winds on the response curve in atmospheric oscillations, *Q. J. Roy. Met. Soc.*, **95**, 771.
Goldstein, H. (1950) *Classical mechanics*, Addison-Wesley, Mass.
Golitsyn, G. S. (1965) Damping of small oscillations in the atmosphere due to viscosity and thermal conductivity, *Izvestiya, Atmos. & Ocean Phys.*, **1**, 82 (English translation).

Gossard, E. E. (1962) Vertical flux of energy into the lower ionosphere from internal gravity waves generated in the troposphere, *J. Geophys. Res.*, **67**, 745.

Gossard, E. E. & Munk, W. H. (1954) On gravity waves in the atmosphere, *J. Met.*, **11**, 259.

Greenhow, J. S. & Neufeld, E. L. (1956) The height variation of upper atmospheric winds, *Phil. Mag.*, **1**, 1157.

Greenhow, J. S. & Neufeld, E. L. (1959) Measurements of turbulence in the 80 to 100 km region from the radio echo observations of meteors, *J. Geophys. Res.*, **64**, 2129.

Greenhow, J. S. & Neufeld, E. L. (1960) Large scale irregularities in high altitude winds, *Proc. Phys. Soc.*, **75**, 228.

Groves, G. V. (1969a) Tidal oscillations and gravity waves, *Annals of the IQSY*, **5**, 263.

Groves, G. V. (1969b) Wind models from 60 to 130 km altitude for different months and latitudes, *J. Brit. Interplanet. Soc.*, **22**, 285.

Harkrider, D. G. (1964) Theoretical and observed acoustic gravity waves from explosive sources in the atmosphere, *J. Geophys. Res.*, **69**, 5295.

Harkrider, D. G. & Wells, F. J. (1968) *The excitation and dispersion of the atmospheric surface wave*, in Georges (1968b).

Harris, I. & Priester, W. (1965) On the dynamical variation of the upper atmosphere, *J. Atmos.* Sci., **22**, 3.

Harris, K. K., Sharp, G. W. & Knudsen, W. C. (1969) Gravity waves observed by ionospheric temperature measurements in the F region, *J. Geophys. Res.*, **74**, 197.

Haurwitz, B. (1940) The motion of atmospheric disturbances on the spherical earth, *J. Marine Res.*, **3**, 254.

Haurwitz, B. (1951) The perturbation equations in meteorology, in *Compendium of Meteorology*, American Meteorological Society, Boston, Mass.

Haurwitz, B. (1956) The geographical distribution of the solar semidiurnal pressure oscillations, *Met. Papers*, **2**, No. 5, New York University.

Haurwitz, B. (1964) Tidal phenomena in the upper atmosphere, *WMO Tech. Note 58*, Geneva.

Haurwitz, B. (1965) The diurnal surface pressure oscillation, *Arch. Met. Geophys. Bioklimat.*, **A14**, 361.

Helliwell, R. A. (1965) *Whistlers and related ionospheric phenomena*, Stanford University Press, Palo Alto, Calif.

Herbert, F. (1971) Statische und quasistatische Bewegungen in der Atmosphare, *Beitrage z. Phys. der Atmos.*, **44**, 17.

Herron, T. S. & Montes, H (1970) Correlation of atmospheric pressure waves with ionospheric doppler signals, *J. Atmos. Sci.*, **27**, 51.

Herzfeld, K. F. & Litovitz, T. A. (1959) *Absorption of ultrasonic waves*, Academic Press, New York.

Hide, R. (1970) Laboratory experiments on free thermal convection and their relation to the global atmospheric circulation, in *Proceedings of the Conference on the Global circulation of the atmosphere*, Royal Meteorological Society.

Hines, C. O. (1960) Internal atmospheric gravity waves at ionospheric heights, *Can. J. Phys.*, **38**, 1441.

Hines, C. O. (1963) The upper atmosphere in motion, *Q. J. Roy. Met. Soc.*, **89**, 1.

Hines, C. O. (1964a) Correction to internal atmospheric gravity waves at ionospheric heights, *Can. J. Phys.*, **42**, 1424.

Hines, C. O. (1964b) Comments on a paper by A. F. Wickersham Jr., *J. Geophys. Res.*, **69**, 2395.

Hines, C. O. (1964c) Minimum vertical scale sizes in the wind structure above 100 kilometres, *J. Geophys. Res.*, **69**, 2847.

Hines, C. O. (1965a) Atmospheric gravity waves: A new toy for the wave theorist, *Radio Sci.*, **69D**, 375.

Hines, C. O. (1965b) Dynamical heating of the upper atmosphere, *J. Geophys. Res.*, **70**, 177.
Hines, C. O. (1966) Diurnal tide in the upper atmosphere, *J. Geophys. Res.*, **71**, 1453.
Hines, C. O. (1967) Tidal oscillations, shorter period gravity waves and shear waves, *Space Research VII*, 30, North Holland Publ. Co.
Hines, C. O. (1971) Generalizations of the Richardson criterion for the onset of atmospheric turbulence, *Q. J. Roy. Met. Soc.*, **97**, 429.
Hines, C. O. (1972) Motions in the ionospheric D and E regions, *Phil. Trans. Roy. Soc.*, **A271**, 457.
Hines, C. O. (1973) A critique of multilayer analyses in application to the propagation of acoustic-gravity waves, *J. Geophys. Res.*, **78**, 265.
Hines, C. O. & Hooke, W. H. (1970) Discussion of ionization effects on the propagation of acoustic gravity waves in the ionosphere, *J. Geophys. Res.*, **75**, 2563.
Hines, C. O. & Reddy, C. A. (1967) On the propagation of atmospheric gravity waves through regions of wind shear, *J. Geophys. Res.*, **72**, 1015.
Hodges, R. R. (1967) Generation of turbulence in the upper atmosphere by internal gravity waves, *J. Geophys. Res.*, **72**, 3455.
Hodges, R. R. (1969) Eddy diffusion coefficients due to instability in internal gravity waves, *J. Geophys. Res.*, **74**, 4087.
Holton, J. R. (1970) A note on forced equatorial waves, *Mon. Weather. Rev.*, **98**, 614.
Holton, J. R. & Wallace, J. M. (1971) On the nature of large-scale eddies in the tropical stratosphere, in *Observational aspects of the tropical general circulation*, edited by R. E. Newell, M. I. T., Camb. Mass.
Hooke, W. H. (1968) Ionospheric irregularities produced by internal atmospheric gravity waves, *J. Atmos. Terr. Phys.*, **30**, 795.
Hooke, W. H. (1969a) E region irregularities produced by internal atmospheric gravity waves, *Planet. Sp. Sci.*, **17**, 749.
Hooke, W. H. (1969b) Radar thomson scatter observations of E region temperature interpreted as reversible heating by atmospheric tides, *J. Geophys. Res.*, **74**, 1870.
Hooke, W. H. (1970) Ionospheric response to an isotropic spectrum of internal gravity waves, *Planet. Sp. Sci.*, **18**, 1793.
Hoskins, B. J. (1973) Stability of the Rossby–Haurwitz wave, *Q. J. Roy. Met. Soc.*, **99**, 723.
Hoult, D. P. (1967) Dispersive waves in the upper atmosphere, *Space Research VII*, 1059, North Holland Publ. Co., Amsterdam.
Howarth, L. (1953) *Modern developments in fluid dynamics*, Oxford University Press.
Hunt, J. N., Palmer, R., & Penney, W. (1960) Atmospheric waves caused by large explosions, *Phil. Trans. Roy. Soc.*, **A252**, 273.
Jackson, J. S. (1971) On the diurnal variation of the magnetic field, *J. Geophys. Res.*, **76**, 6896; **76**, 6909.
Johannessen, A. *et al.* (1972) Water cluster ions in the high-latitude summer mesophere, *Nature* **235**, 215.
Johnston, T. W. (1967) Atmospheric gravity wave instability, *J. Geophys. Res.*, **72**, 2972.
Jones, D. & Maude, A. D. (1965) Evidence for wave motions in the E region in the ionosphere, *Nature*, **206**, 177.
Jones, D. & Maude, A. D. (1972) Dispersive motions in the ionosphere, *J. Atmos. Terr. Phys.*, **34**, 1241.
Jones, M. N. (1970, 1971, 1972) Atmospheric oscillations, *Planet. Sp. Sci.*, **18**, 1393; **19**, 609; **19**, 1359; **20**, 1627.
Jones, W. L. (1970) A theory for quasi-periodic oscillations observed in the ionosphere, *J. Atmos. Terr. Phys.*, **32**, 1555.
Jones, W. L. (1971) Energy momentum tensor for linearized waves in material media, *Rev. Geophys. Space Sci.*, **9**, 917.

Jones, W. L. & Houghton, D. D. (1972) The self destructing internal gravity wave, *J. Atmos. Sci.*, **29**, 844.
Kato, S. (1966) Diurnal atmospheric oscillations, *J. Geophys. Res.*, **71**, 3201 : **71**, 3211.
Kato, S., Reddy, C. A., & Matsushita, S. (1970) Possible hydromagnetic coupling between the perturbations of the neutral and ionized atmosphere, *J. Geophys. Res.*, **75**, 2540.
Kellog, W. W. (1961) Chemical heating above the polar mesopause in winter, *J. Met.*, **18**, 373.
Kirchhoff, G. (1868) Uber den Einfluss der Warmeleitung in einem Gase auf die Schallbewegung, *Ann. Phys. Chem.* (*Poggendorf Ann.*), **134**, 177.
Klostermeyer, J. (1969) Gravity waves in the F region, *J. Atmos. Terr. Phys.*, **31**, 25.
Koster, J. R. & Beer, T. (1973) *An interpretation of ionospheric Faraday rotation observations at the equator*, Sci. Report, Physics Dept., Univ. of Ghana.
Kuo, H. L. (1949) Dynamic instability of a two-dimensional non divergent flow in a barotropic atmosphere, *J. Met.*, **6**, 105.
Lamb, H. (1910) On atmospheric oscillations, *Proc. Roy. Soc.*, **A84**, 551.
Lamb, H. (1945) *Hydrodynamics*, 6th. ed., Dover, New York.
Larsen, J. C. (1968) Electric and magnetic fields induced by deep sea tides, *Geophys. J. Roy. Astron. Soc.*, **16**, 47.
Liller, W. & Whipple, F. L. (1954) High altitude winds by meteor train photography, *J. Atmos. Terr. Phys.*, (*spec. supp.*) **1**, 112.
Lindzen, R. S. (1966) On the theory of the diurnal tide, *Mon. Wea. Rev.*, **94**, 295.
Lindzen, R. S. (1967a) Thermally driven diurnal tide in the atmosphere, *Q. J. Roy. Met. Soc.*, **93**, 18.
Lindzen, R. S. (1967b) Planetary waves on a beta-plane, *Mon. Wea. Rev.*, **95**, 441.
Lindzen, R. S. (1968a) Application of classical atmospheric tidal theory, *Proc. Roy. Soc.*, **A303**, 299.
Lindzen, R. S. (1968b) Vertically propagating waves in an atmosphere with Newtonian cooling, *Can. J. Phys.*, **46**, 1835.
Lindzen, R. S. (1968c) Rossby waves with negative equivalent depths, *Q. J. Roy. Met. Soc.*, **94**, 402.
Lindzen, R. S. (1970, 1971a) Internal gravity waves in an atmosphere with realistic dissipation and temperature, *Geophys. Fluid Dyn.*, **1**, 303; (with Blake, D. (1971)) **2**, 31; **2**, 89.
Lindzen, R. S. (1971b) Atmospheric Tides, in *Mathematical Problems in the Geophysical Sciences and Lectures in Applied Mathematics*, American Mathematical Society, **14**, 293.
Lindzen, R. S. & Blake, D. (1971) Mean heating of the thermosphere by tides, *J. Geophys. Res.*, **75**, 6868.
Lindzen, R. S. & Blake, D. (1972) Lamb waves in the presence of realistic distributions of temperature and dissipation, *J. Geophys. Res.*, **77**, 2166.
Liu, C. H. (1970) Ducting of acoustic gravity waves in the atmosphere with spatially periodic wind shears, *J. Geophys. Res.*, **75**, 1339.
Liu, C. H. and Yeh, K. C. (1966) Gradient instabilities as a possible cause of irregularities in the ionosphere, *Radio Sci.*, **1**, 1283.
Liu, C. H. & Yeh, K. C. (1969) Effects of ion drag on the propagation of acoustic gravity waves in the atmospheric F region, *J. Geophys. Res.*, **74**, 2248.
Liu, C. H. & Yeh, K. C. (1971) Excitation of the acoustic gravity waves in an isothermal atmosphere, *Tellus*, **23**, 150.
Longuet-Higgins, M. S. (1964) Planetary waves on a rotating sphere, *Proc. Roy. Soc.*, **A279**, 446.
Longuet-Higgins, M. S. (1968) The eigenfunctions of Laplace's tidal equation over a sphere, *Phil. Trans. Roy. Soc.*, **A262**, 511.
Longuet-Higgins, M. S. & Gill, A. E. (1967) Resonant interactions between planetary waves, *Proc. Roy. Soc.*, **A229**, 120.

Lorenz, E. N. (1967) *The Nature and theory of the general circulation of the atmosphere*, World Meteorological Organization, Geneva.
Lorenz, E. N. (1972) Barotropic instability of Rossby wave motion, *J. Atmos. Sci.*, **29**, 258.
Ludlam, F. H. (1967) Characteristics of billow clouds and their relation to clear air turbulence, *Q. J. Roy. Met. Soc.*, **93**, 419.
Lyra, G. (1940) Uber den Einfluss von Bodenerhebungen auf die Stromung einer stabil geschichteten Atmosphare, *Beitr-z. Phys. Atmos.*, **26**, 197.
Lyra, G. (1943) Theorie der stationaren Leewellenstromung in freier Atmosphare, *Z. angew. Math. Mech.*, **23**, 1.
Maeda, K. & Young, J. M. (1966) Propagation of the pressure wave produced by auroras, *J. Geomag. Geoelec.*, **18**, 275.
Martyn, D. F. (1950) Cellular atmospheric waves in the ionosphere and troposphere, *Proc. Roy. Soc.*, **A201**, 216.
Matsuno, T. (1966) Quasi-geostrophic motions in the equatorial area, *J. Met. Soc. Japan.*, **44**, 25.
Matsushita, S. (1967) Solar quiet and lunar daily variation fields, in *Physics of Geomagnetic Phenomena* (eds. Matsushita, S. & Campbell, W. H.), p. 301, Academic Press, New York.
Matsushita, S. (1969) Dynamo currents, winds and electric fields, *Radio Sci.*, **4**, 771.
Mayr, H. G. & Volland, H. (1973) A two component model of the diurnal variation in the thermospheric composition, *J. Atmos. Terr. Phys.*, **35**, 669.
McIntyre, M. E. (1965) A separable non-geostrophic baroclinic stability problem, *J. Atmos. Sci.*, **22**, 730.
McIntyre, M. E. (1970) On the non-separable baroclinic parallel flow instability problem, *J. Fluid. Mech.*, **40**, 273.
Menkes, J. (1961) On the stability of a heterogeneous shear layer subject to a body force, *J. Fluid. Mech.*, **11**, 284.
Midgley, J. E. & Liemohn, H. B. (1966) Gravity waves in a realistic atmosphere, *J. Geophys. Res.*, **71**, 3729.
Moffett, R. (1973) On the diurnal variations of total mass density, number density and temperature in the upper thermosphere, *Planet. Sp. Sci.*, **21**, 1457.
Murgatroyd, R. J. (1957) Winds and temperatures between 20 km and 100 km, *Q. J. Roy. Met. Soc.*, **83**, 417.
Muller, H. G. (1972) Long period meteor wind oscillations, *Phil. Trans. Roy. Soc.*, **A271**, 585.
Naito, K. (1966) Internal gravity shear waves in the troposphere, *Can. J. Phys.*, **44**, 2259.
Nelson, R. A. (1968) Response of the ionosphere to the passage of atmospheric waves, *J. Atmos. Terr. Phys.*, **30**, 825.
Newell, R. E., Mahoney, J. & Lenhard, D. R. (1966) A pilot study of small scale wind variations in the stratosphere and mesosphere, *Q. J. Roy. Met. Soc.*, **92**, 41.
Newton, G. P., Pelz, D. T. & Volland, H. (1969) Direct *in situ* measurements of wave propagation in the neutral thermosphere, *J. Geophys. Res.*, **74**, 183.
Palmer, C. E. (1952) Tropical meteorology, *Q. J. Roy. Met. Soc.*, **78**, 126.
Paulin, G. (1970) A study of the energetics of January 1959, *Mon. Weather Rev.*, **98**, 795.
Pekeris, C. L. (1948a) The propagation of a pulse in the atmosphere, *Phys. Rev., 2nd Ser.*, **73**, 145.
Pekeris, C. L. (1948b) Theory of propagation of explosive sound in shallow water, *Geolog. Soc. Am., Memoir*, 27.
Pettersen, S. (1940) *Weather analysis and forecasting*, McGraw-Hill, New York.
Pettersen, S. (1969) *Introduction to meteorology, 3rd edn.*, McGraw-Hill, New York.

Pfeffer, R. L. & Zarichny, J. (1963) Acoustic gravity wave propagation in an atmosphere with two sound channels, *Geofisica Pura e Applicata*, **55**, 175.

Phillips, N. A. (1964) An overlooked aspect of the baroclinic instability problem, *Tellus*, **16**, 268.

Phillips, N. A. (1966) The equations of motion for a shallow rotating atmosphere and the traditional approximation, *J. Atmos. Sci.*, **23**, 626.

Phillips, N. A. (1968) Reply, *J. Atmos. Sci.*, **25**, 1155.

Pierce, A. D. & Coroniti, S. C. (1966) A mechanism for the generation of acoustic-gravity waves during thunderstorm formation, *Nature*, **210**, 1209.

Pitteway, M. L. V. & Hines, C. O. (1963) The viscous damping of atmospheric gravity waves, *Can. J. Phys.*, **41**, 1935.

Pitteway, M. L. V. & Hines, C. O. (1965) The reflection and ducting of atmospheric acoustic gravity waves, *Can. J. Phys.*, **43**, 2222.

Platzman, G. (1968) The Rossby wave, *Q. J. Roy. Met. Soc.*, **94**, 225.

Press, F. & Harkrider, D. (1962) Propagation of acoustic gravity waves in the atmosphere, *J. Geophys. Res.*, **67**, 3889.

Price, A. T. (1967) Electromagnetic induction within the earth, in *Physics of geomagnetic phenomena, vol. I*, p. 235, Academic Press, New York.

Rai, D. B. & Fejer, B. G. (1971) Evidence of internal gravity waves in the lower ionosphere, *Planet. Sp. Sci.*, **19**, 561.

Raitt, W. J. & Clark, D. H. (1973) Wave-like disturbances in the ionosphere, *Nature*, **243**, 508.

Ratcliffe, J. A. (1959) *The magnetoionic theory and its applications to the ionosphere*, Cambridge University Press.

Reed, R. J. & Recker, E. (1971) Structure and propagation of synoptic scale wave disturbances in the equatorial western Pacific, *J. Atmos. Sci.*, **28**, 1117.

Revah, I. (1970) Partial reflections of gravity waves observed by a meteor radar, *J. Atmos. Terr. Phys.*, **32**, 1313.

Rhines, P. B. (1970a) Wave propagation in periodic media, *Rev. Geophys. Sp. Sci.*, **8**, 303.

Rhines, P. B. (1970b) Edge and Rossby waves in a rotating stratified fluid, *Geophys. Fluid Dynamics*, **1**, 273.

Richardson, L. F. (1920) The supply of energy from and to atmospheric eddies, *Proc. Roy. Soc.*, **A67**, 354.

Richardson, L. F. (1922) *Weather prediction by numerical process*, Cambridge University Press.

Richmond, A. D. (1971) Tidal winds at ionospheric heights, *Radio Sci.*, **6**, 175.

Riehl, H. (1954) *Tropical meteorology*, McGraw-Hill, London. Chapters 9 & 10.

Rishbeth, H. (1971) Polarization fields produced by winds in the equatorial F region, *Planet. Sp. Sci.*, **19**, 357.

Rogers, R. H. (1959) The structure of the jet-stream in a rotating fluid with a horizontal temperature gradient, *J. Fluid. Mech.*, **5**, 41.

Rossby, C. G. (1940) Planetary flow pattern in the atmosphere, *Q. J. Roy. Met. Soc. Supplement*, **66**, 68.

Rossby, C. G. (1945) On the propagation of frequencies and energy in certain types of oceanic and atmospheric waves, *J. Met.*, **2**, 187.

Rossby, C. G., *et al.* (1939) Relations between the variations in zonal circulation of the atmosphere and the displacements of the semi-permanent centres of action, *J. Marine. Res.*, **2**, 38.

Row, R. V. (1967) Acoustic gravity waves in the upper atmosphere due to a nuclear detonation and an earthquake, *J. Geophys. Res.*, **72**, 1599.

Sawada, R. (1956) The atmospheric lunar tides and the temperature profile in the upper atmosphere, *Geophys. Mag.* **27**, 213.

Schoedel, J. P., Klostermeyer, J. & Roettger, J. (1973) Atmospheric gravity waves observed after the solar eclipse of June 30, 1973, *Nature*, **245**, 87.
Scorer, R. S. (1949) Theory of waves in the lee of mountains, *Q. J. Roy. Met. Soc.*, **75**, 41.
Scorer, R. S. (1950) The dispersion of a pressure pulse in the atmosphere, *Proc. Roy. Soc.*, **A201**, 137.
Scorer, R. S. (1951) Billow clouds, *Q. J. Roy. Met. Soc.*, **77**, 235.
Scorer, R. S. (1953) Gravity waves in the atmosphere, *Arch. Met. Geophys. Biokl.*, **A4**, 176.
Shapiro, R. (1970) Smoothing, filtering and boundary effects, *Rev. Geophys. Sp. Sci.*, **8**, 359.
Siebert, M. (1961) Atmospheric tides, *Advances in Geophys.*, **7**, 105, Academic Press, New York.
Sivian, L. J. (1947) High frequency absorption in air and other gases, *J. Acoustic. Soc. Am.*, **19**, 914.
Spizzichino, A. (1969, 1970) Étude des interactions entre les différentes composantes du vent dans la haute atmosphere, *Ann. Geophys.*, **25**, 693; **25**, 755; **25**, 773; **26**, 9; **26**, 25.
Stokes, C. G. (1845) On the theories of the internal friction of fluids in motions and of the equilibrium motion of elastic bodies, *Trans. Phil. Soc. Camb.*, **81**, 287.
Synge, J. L. (1933) Stability of Heterogeneous Liquids, *Trans. Roy. Soc. Can.*, **27**, 9.
Taffe, W. J. (1969) Hydromagnetic effects on atmospheric tides, *J. Geophys. Res.*, **74**, 5575.
Tarpley, J. D. (1970) The ionospheric wind dynamo, *Planet. Sp. Sci.*, **18**, 1075; **18**, 1091.
Tchen, C. M. (1970) *Turbulence and gravity waves in the upper atmosphere*, Research paper, P-595, Institute for Defence Analyses, Arlington.
Testud, J. (1970) Gravity waves generated during magnetic substorms, *J. Atmos. Terr. Phys.*, **32**, 1793.
Thome, G. D. (1964) Incoherent scatter observations of travelling ionospheric disturbances, *J. Geophys. Res.*, **69**, 4047.
Thome, G. D. (1966) *A study of large scale travelling disturbances in the ionosphere using the Arecibo UHF radar*, Report CRSR 236, Cornell University, Ithaca.
Thompson, P. D. (1961) *Numerical weather analysis and prediction*, Macmillan, New York.
Tolstoy, I. (1963) The theory of waves in stratified fluids including the effects of gravity and rotation, *Rev. Mod. Phys.*, **35**, 207.
Tolstoy, I. & Lau, J. (1970) Ground level pressure fluctuations connected with ionospheric disturbances, *J. Atmos. Sci.*, **27**, 494.
Vergeiner, I. (1971) An operational lee wave model for arbitrary basic flow and two-dimensional topography, *Q. J. Roy. Met. Soc.*, **97**, 30.
Vergeiner, I. and Ogura, Y. (1972) A numerical shallow fluid model including orography with a variable grid, *J. Atmos. Sci.*, **29**, 270.
Vincent, R. A. (1969) A criterion for the use of the multilayer approximation in the study of acoustic gravity wave propagation, *J. Geophys. Res.*, **74**, 2996.
Volland, H. (1969a) Full wave calculations of gravity wave propagation through the thermosphere, *J. Geophys. Res.*, **74**, 1786.
Volland, H. (1969b) A theory of thermospheric dynamics, *Planet. Sp. Sci.*, **17**, 1581.
Volland, H. (1969c) The upper atmosphere as a multiple refractive index medium for neutral air motions, *J. Atmos. Terr. Phys.*, **31**, 491.
Volland, H. & Mayr, H. G. (1972) The problem of the boundary condition in thermospheric dynamics, *J. Atmos. Sci.*, **29**, 1143.
Wallace, J. M. (1969) Some recent developments in the study of tropical wave disturbances, *Bull. Amer. Met. Soc.*, **50**, 792.
Wallace, J. M. (1971) Spectral study of tropospheric wave disturbances in the tropical western Pacific, *Rev. Geophys. Sp. Sci.*, **9**, 557.
Wallace, J. M. & Hartranft, F. R. (1969) Diurnal wind variations surface to 30 km, *Mon. Weather Rev.*, **96**, 446.

Wand, R. H. & Perkins, F. W. (1968) Radar thompson scatter observations of temperature and ion-neutral collision frequency in the E region, *J. Geophys. Res.*, **73**, 6370.
Wand, R. H. & Perkins, F. W. (1970) Temperature and composition of the ionosphere: diurnal variations and waves, *J. Atmos. Terr. Phys.*, **32**, 1921.
Warren, F. W. G. (1968) A note on Howard's proof of Mile's theorem, *J. Mech. Appl. Math.*, **21**, 433.
Weston, V. H. (1961) The pressure pulse produced by a large explosion in the atmosphere, *Can. J. Phys.*, **39**, 993; **40**, 431.
Whitehead, J. D. (1971) Ionization disturbances caused by gravity waves in the presence of an electrostatic field and background wind, *J. Geophys. Res.*, **76**, 238.
Whitehead, J. D. (1972) Winds in the E region, *Radio Sci.*, **7**, 403.
Whittaker, E. T. & Watson, G. N. (1962) *A course of modern analysis*, 4th edn., Cambridge University Press.
Wickersham, A. F. Jr. (1966) Acoustic-gravity wave modes from ionospheric range-time observations, *J. Geophys. Res.*, **71**, 4551.
Wickersham, A. F. Jr. (1968) The origin and propagation of acoustic gravity waves ducted in the thermosphere, *Aust. J. Phys.*, **21**, 671.
Wilkes, M. V. (1949) *Oscillations of the earth's atmosphere*, Cambridge University Press.
Willmore, A. P. (1970) Electron and ion temperatures in the ionosphere, *Sp. Sci. Rev.*, **11**, 607.
Witt, G. (1962) Height, structure and displacement of noctilucent clouds, *Tellus*, **14**, 1.
Yamamoto, R. (1957) Microbarographic oscillation produced by the Soviet explosion, *J. Met. Soc. Japan*, **35**, 288.
Yanai, M. (1971) A review of recent studies of tropical meteorology relevant to the planning of GATE, in *GATE Experimental Design proposals*, **2**, 1 (WMO, Geneva).
Yanowitch, M. (1967) Effect of viscosity on gravity waves and the upper boundary condition, *J. Fluid. Mech.*, **29**, 209.
Yeh, K. C. & Liu, C. H. (1970) On resonant interactions of acoustic gravity waves, *Radio Sci.*, **5**, 39.
Young, J. R. C. (1971) Breaking gravity waves on a jet stream over the North Pacific, *Weather*, **26**, 306.
Zimmerman, S. P. (1964) Small scale wind structure above 100 kilometres, *J. Geophys. Res.*, **69**, 784.

Subject Index

Author Index

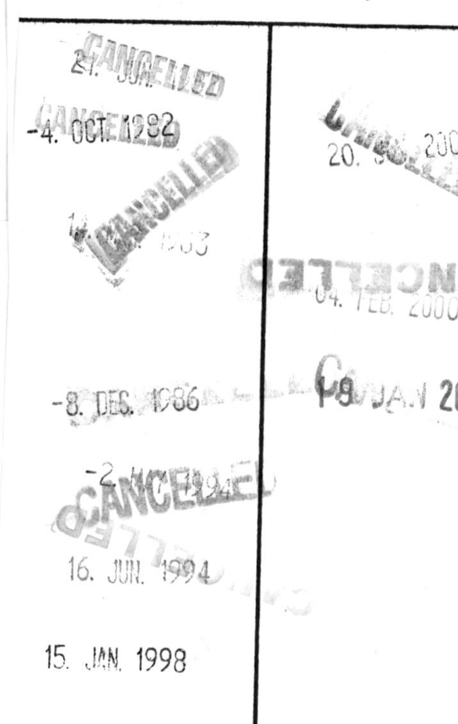

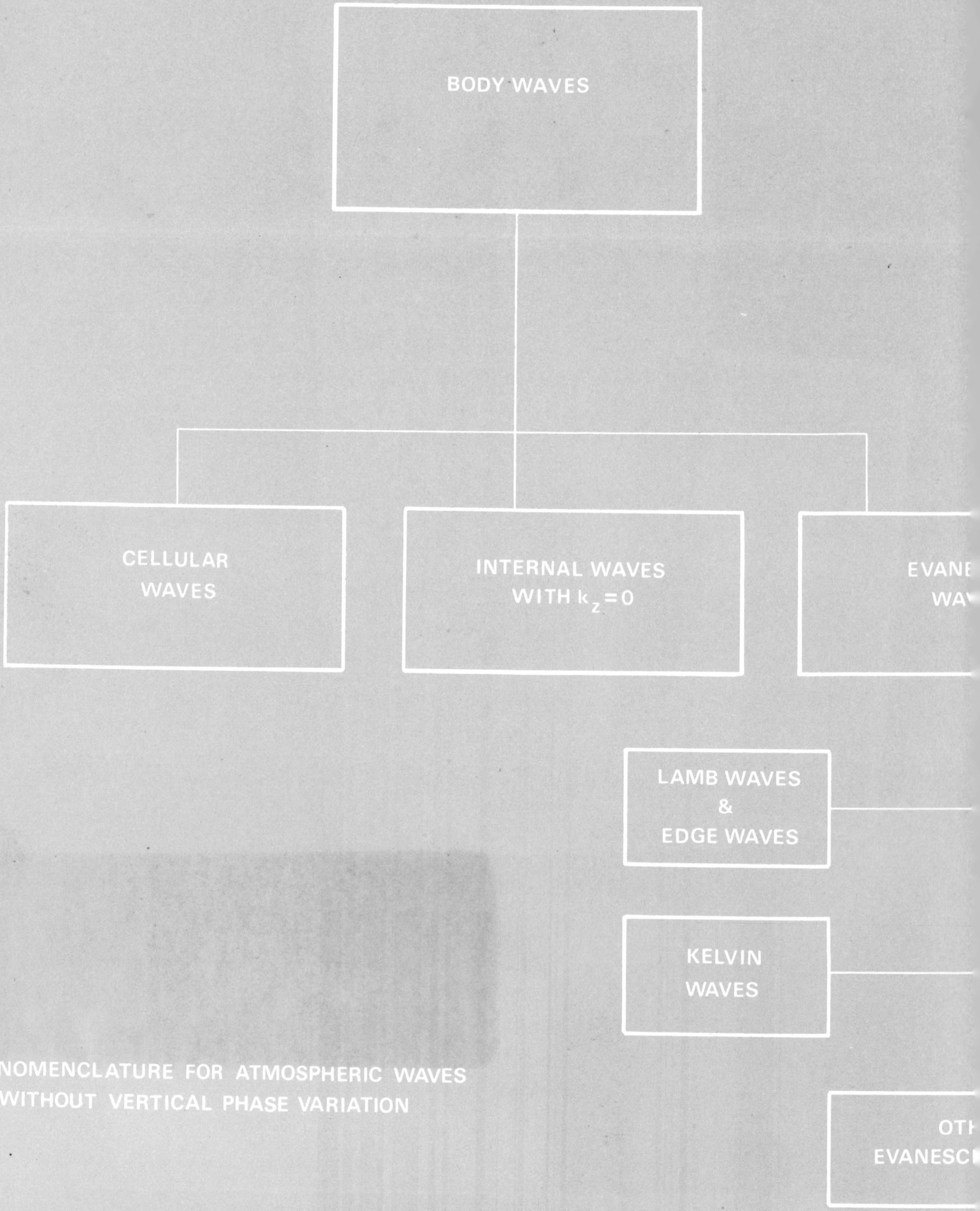

BODY WAVES

CELLULAR WAVES

INTERNAL WAVES WITH $k_z = 0$

EVANE
WA

LAMB WAVES & EDGE WAVES

KELVIN WAVES

OTH
EVANESC

NOMENCLATURE FOR ATMOSPHERIC WAVES WITHOUT VERTICAL PHASE VARIATION